Rare Isotope Beams

Rare Isotope Beams
Concepts and Techniques

Alok Chakrabarti, Vaishali Naik,
and Siddhartha Dechoudhury

CRC Press is an imprint of the
Taylor & Francis Group, an **informa** business

First edition published 2022
by CRC Press
6000 Broken Sound Parkway NW, Suite 300, Boca Raton, FL 33487-2742

and by CRC Press
2 Park Square, Milton Park, Abingdon, Oxon, OX14 4RN

© 2022 Taylor & Francis Group, LLC

CRC Press is an imprint of Taylor & Francis Group, LLC

Reasonable efforts have been made to publish reliable data and information, but the author and publisher cannot assume responsibility for the validity of all materials or the consequences of their use. The authors and publishers have attempted to trace the copyright holders of all material reproduced in this publication and apologize to copyright holders if permission to publish in this form has not been obtained. If any copyright material has not been acknowledged please write and let us know so we may rectify in any future reprint.

Except as permitted under U.S. Copyright Law, no part of this book may be reprinted, reproduced, transmitted, or utilized in any form by any electronic, mechanical, or other means, now known or hereafter invented, including photocopying, microfilming, and recording, or in any information storage or retrieval system, without written permission from the publishers.

For permission to photocopy or use material electronically from this work, access www.copyright.com or contact the Copyright Clearance Center, Inc. (CCC), 222 Rosewood Drive, Danvers, MA 01923, 978-750-8400. For works that are not available on CCC please contact mpkbookspermissions@tandf.co.uk

Trademark notice: Product or corporate names may be trademarks or registered trademarks and are used only for identification and explanation without intent to infringe.

ISBN: [978-1-4987-8878-6] (hbk)
ISBN: [978-1-032-03709-7] (pbk)
ISBN: [978-0-429-18588-5] (ebk)

Typeset in Palatino
by Deanta Global Publishing Services, Chennai, India

Dedicated to our colleagues and collaborators

Contents

Preface xi
Acknowledgments xiii
Authors xv

1. **Rare Isotope Beams—The Scientific Motivation** 1
   1.1 Introduction 1
   1.2 RIBs and Nuclear Physics 3
       1.2.1 The Limits of Nuclear Stability 3
       1.2.2 Nuclear Halo in Drip Line and Near Drip Line Nuclei 9
       1.2.3 Evolution of Shell Structure away from Stability 14
   1.3 Nuclear Astrophysics: The Origin of Elements, the Stellar Evolution
       and the Role of RIBs 18
       1.3.1 Primordial or Big Bang Nucleo-Synthesis 19
       1.3.2 Nucleo-Synthesis in Stars up to Iron 21
       1.3.3 Synthesis of Elements Heavier Than Iron: The S, R and P Processes 25
           1.3.3.1 The S Process 25
           1.3.3.2 The R Process 28
           1.3.3.3 The P Process Nucleo-Synthesis 32
   1.4 RIBs and the Test of Fundamental Symmetries of Nature 36
       1.4.1 The Electric Di-Pole Moment in Atomic Systems and the
             CP Violation 36
       1.4.2 Atomic Parity Violation 38
       1.4.3 The CVC Hypothesis, Nuclear Beta-Decay and the Unitarity of
             CKM Quark Mixing Matrix 40
           1.4.3.1 The CVC and the Nuclear Beta Decay 40
           1.4.3.2 Unitarity of CKM Matrix 41
   1.5 RIBs and Condensed Matter Physics 41
       1.5.1 Mossbauer Spectroscopy 42
       1.5.2 Perturbed Angular Correlation 43
       1.5.3 β--NMR 44
   1.6 RIBs: Medical Physics and Applications 45

2. **Production of Rare Isotope Beams: The Two Approaches** 49
   2.1 Introduction 49
   2.2 The ISOL Post-Accelerator Approach 52
   2.3 The PFS Approach 53
   2.4 Comparison between the ISOL and PFS Approaches 54
   2.5 The Combined Approaches 58

3. **Nuclear Reactions for Production of Rare Isotope Beams** 61
   3.1 Production of RIBs in High-Energy Proton-Induced Reactions
       (Spallation/Target Fragmentation) 61
       3.1.1 Introduction 61
       3.1.2 The Spallation Reaction Process 62

vii

		3.1.3	Production of Neutron-Deficient Exotic Nuclei Using Spallation–Evaporation Reaction 64
		3.1.4	Production of n-Rich Exotic Nuclei in Spallation–Fission Reaction 66
		3.1.5	Highly Asymmetric Fission vs Multi-Fragmentation 67
		3.1.6	Measured Yields of Exotic Species Using Spallation Reaction at ISOLDE 68
		3.1.7	Reaction Codes for Spallation Reaction 69
	3.2	Production of RIBs Using High and Intermediate Energy Heavy Ion Induced Projectile Fragmentation and In-Flight Fission Reactions 70
		3.2.1	Introduction 70
		3.2.2	The PF Reaction Process 70
		3.2.3	Limiting Fragmentation and Factorization 72
		3.2.4	Momentum/Energy Width of the Projectile Fragments 72
		3.2.5	Production of Exotic Species in PF Reaction 74
			3.2.5.1 Production of Neutron-Deficient Nuclei 74
			3.2.5.2 Production of n-Rich Nuclei 75
		3.2.6	Production of n-Rich Nuclei in In-Flight Fission of ^{238}U 77
		3.2.7	Choice of Target Thickness, Target and Projectile Energy 78
		3.2.8	Reaching Closer to the Neutron Drip Line Using Fragmentation of Secondary RIBs 81
		3.2.9	Theoretical Estimation of Production Cross-Sections in PF Reaction 83
	3.3	Fission Induced by Low-Energy Neutrons, Protons and Gamma Rays 85
		3.3.1	The Fission Process 85
		3.3.2	Production of n-Rich Isotopes in Fission Induced by Thermal Neutrons 86
		3.3.3	Production of n-Rich Isotopes in Fission Induced by Energetic Protons/Light Ions 89
		3.3.4	Fission Induced by Energetic Neutrons 91
		3.3.5	Fission Induced by Gamma Rays 93
	3.4	Production of RIBs Using Low-Energy Heavy Ions above the Coulomb Barrier 97
		3.4.1	Fusion–Evaporation Reactions for the Production of Neutron-Deficient Nuclei 97
		3.4.2	Deep Inelastic Transfer Reactions 104

4. Targets for RIB Production 107
4.1 Introduction 107
4.2 High-Power Targets for ISOL Facilities 108
4.3 Types of Target Material 110
4.4 R&D for Future ISOL Targets 113
4.5 Target Station in ISOL Method 114
4.6 Targets for PFS Facilities 114
4.7 High-Power Beam Dumps 117

5. Ion Sources for RIB Production in ISOL-Type Facilities 119
5.1 Introduction 119
5.2 Ion Sources for 1$^+$ Charge State Production 121
    5.2.1 Surface Ion Source 121
    5.2.2 The Resonant Ionization Laser Ion Source for Metallic Ions 122
    5.2.3 Forced Electron Beam Arc Discharge (FEBIAD) Ion Source 124

Contents

    5.3   Electron Cyclotron Resonance (ECR) Ion Source 125
            5.3.1   ECIRS for 1⁺ Charge State 126
            5.3.2   ECIRS for High Charge State Production 127
            5.3.3   ECRIS as Charge Breeder 128
    5.4   The EBIS: For High Charge State Production and as Charge Breeder 130
    5.5   Positioning the First Ion Source away from The Target
           (the HeJRT Technique) 132

6. **Accelerators for RIB Production and Post-Acceleration** 135
    6.1   Introduction 135
           Driver and the Post-Accelerator 135
    6.2   DC Accelerators for RIB Production 138
    6.3   Cyclic Accelerators for RIB Production 138
            6.3.1   Cyclotrons 138
            6.3.2   Synchrotrons 144
    6.4   Linear Accelerators for RIB Production 147
            6.4.1   Radio Frequency Quadrupole (RFQ) Linac 152
            6.4.2   Acceleration to High Energies: Room Temperature Linacs 155
            6.4.3   Acceleration to High Energies: Superconducting Linacs 158
    6.5   Beam Acceleration and Charge Stripper 160
    6.6   Post-Accelerators for Acceleration of RIBs in ISOL Facilities 164

7. **Experimental Techniques** 167
    7.1   Introduction 167
    7.2   Separation of Isotopes in ISOL- and PFS-Type RIB Facilities 167
    7.3   Isotope Separation in ISOL-Type RIB Facilities 169
            7.3.1   Radio Frequency Quadrupole (RFQ) Cooler 173
            7.3.2   High-Resolution Separator—A Typical Example 177
            7.3.3   Identification of Isotopes in ISOL-Type Facilities 177
    7.4   Separation in In-Flight Separators at Intermediate and Relativistic
            Energies (~50 to 1500 MeV/u) 178
            7.4.1   Identification of New Isotopes in the PFS Method 185
    7.5   Measurement of Mass 189
            7.5.1   Indirect Methods for Mass Measurement of Exotic Nuclei 190
                    7.5.1.1 Q_β and Q_α Measurements 191
                    7.5.1.2   Missing Mass Method 192
                    7.5.1.3   Invariant Mass Spectroscopy 194
            7.5.2   Direct Methods of Mass Measurement of Exotic Nuclei 196
            7.5.3   Mass Separation and Measurement in Paul and Penning Traps 197
                    7.5.3.1   Paul Trap 198
                    7.5.3.2   Penning Trap 200
                    7.5.3.3   MR-ToF and Measurement of Mass 208
    7.6   Mass Measurements in Storage Ring 210
            7.6.1   Schottky Mass Spectrometry (SMS) 212
            7.6.2   Isochronous Mass Spectrometry (IMS) 216
    7.7   Measurement of Ground State Properties of Nuclei Using Laser
            Spectroscopic Techniques 217
            7.7.1   The Collinear Laser Spectroscopy (CLS) Technique 220
            7.7.2   The Collinear Resonant Ionization Spectroscopy (CRIS) Technique 223
            7.7.3   Optical Pumping Using Collinear Laser and β–NMR 225

- 7.8 Matter Radii of Drip Line Isotopes through Measurements of Interaction Cross-Sections 227
- 7.9 Measurement of Half-Life of Exotic Nuclei....................229
- 7.10 Coulomb Excitation and Study of Exotic Nuclei 232
  - 7.10.1 Coulomb Break-Up 236
- 7.11 Measurement of Cross-Sections for Nuclear Astrophysics 239
  - 7.11.1 Measurement of Proton Capture Cross-Section, Direct Methods 240
    - 7.11.1.1 Study of Charged Particle Capture Reactions Using Recoil Mass Separators 242
    - 7.11.1.2 Study of Charged Particle Capture Reactions Using Low-Energy Ion Storage Rings 243
    - 7.11.1.3 Direct Measurement of (n, γ) Cross-Sections Using Storage Rings 244
  - 7.11.2 Coulomb Dissociation Technique for Measuring (p, γ) and (n, γ) Reaction Rates 245
- 7.12 EDM Experiments 246

8. Overview of Major RIB Facilities Worldwide 249
- 8.1 Introduction 249
- 8.2 Major ISOL-Type RIB Facilities 249
- 8.3 Major Projectile Fragment Separator (PFS) Type RIB Facilities 257
- 8.4 Specialized Facilities 263

References 271

Index 285

Preface

In the last few decades, Rare Isotope Beams (RIBs) have clearly emerged as the key area of research in the field of low and medium energy nuclear physics. A number of facilities in North America, Europe and Japan have begun to produce new and exciting results, while many new facilities are being planned or built worldwide. Progress in the technology of particle accelerators and detectors, coupled with advances in computer technology, both in hardware and software, have played a major role in the tremendous growth in this field. Essentially, research in nuclear physics would mean, in the coming decades, research using RIBs. This is because RIBs hold the key to understanding the properties of so-called exotic nuclei—nuclei with unusual proton to neutron ratios compared with their stable counterparts lying in the valley of beta stability, and to understand synthesis of elements in various stellar environments including explosive scenarios such as supernovae. RIBs also have rather unique potential to throw light on fundamental interactions and symmetries of nature. Of no less importance are the potential opportunities that RIBs provide in pursuing frontline research in material science, atomic physics, biology, radiation therapy and in the production of new isotopes for medical diagnostics and therapy.

This book attempts to cover the current status of major thrust areas in the field. The task is difficult because we endeavor to include a wide range of topics from particle accelerator–based fundamental sciences to the technology of particle accelerators, while at the same time restricting the number of pages to a moderate limit. In doing this it is difficult to do justice to all the excellent contributions that have been made in the field, and inevitably we have ended up prioritizing some topics over others.

A similar endeavor was made in 1989 in the book *Treatise on Heavy-Ion Science, Vol. 8: Nuclei Far From Stability* (edited by D. Allan Bromley, 1989, Plenum Press, New York, NY). Bromley's book offers an excellent review of the field, and the various topics were written by renowned scientists, mostly pioneers in their respective areas of research. Bromley's book is as relevant today as it was in 1989. However, the field of RIBs has witnessed an almost exponential growth in the last three decades. These advancements have created the need and motivation for writing the present book. One must acknowledge that a good number of excellent review articles on various individual topics have been written since 1989. Our attempt is to put together an updated account of research potential, techniques and applications of rare ion beams in a single compilation.

This book covers the experimental part of RIB science. The wide range of topics covered does not typically fit into the format of a textbook. We have therefore not included a list of exercises to be solved at the end of each chapter. Instead, this book is meant to be a reference book for a wide range of readers—such as practitioners in the field of nuclear physics for a quick glance and brushing up of some facts, beginners aspiring to pursue research in this field for an overall understanding that may help them to identify key research areas, the facility planners and policy-makers for a quick review that might aid in decision making and finally for the inquisitive minds interested to know how this field might contribute to our overall understanding of the cosmos and in pushing the technological frontiers further.

Throughout this book the emphasis is on concepts, and specific details have been kept limited. This is necessary firstly to contain the size of this book, with the second reason being that usually the details evolve faster than concepts and therefore get outdated faster.

This book contains a substantial number of references. These are meant to help the readers quickly master deeper knowledge on the topics of their interest. We have included original references of works of the pioneers in the field along with references on the latest contributions. However, an effort such as this can hardly be complete, and it is possible that some important references have been missed. The authors apologize in advance for any such inadvertent mistakes and shortcomings.

Acknowledgments

This book is an outcome of the authors' own research related to RIB science and technology over the last two decades. The construction of a low energy RIB facility at the Variable Energy Cyclotron Centre (VECC) by the authors and their research group has been a thrilling journey of learning and new insights, which would not have been possible without the support of the leadership at VECC and policy-makers at the Department of Atomic Energy (DAE). Active encouragement of senior colleagues and participation of members from the RIB group have contributed to the progress of the RIB project and the authors gratefully acknowledge the same. The authors have also benefitted greatly from their various collaborations with national and international laboratories, especially with RIKEN and TRIUMF.

The comments and reviews by colleagues during the preparation of the manuscript were valuable and the authors express their gratitude, especially to Prabir Banerjee for reviewing Chapter 1, Osamu Kamigaito for reviewing Chapter 6 and Takahide Nakagawa for reviewing Chapter 5. A special thanks to our colleague Sayed Masum for preparing a good number of figures for this book. This book also contains a number of figures generously sent by colleagues from different institutes around the world. The authors are indebted to them for their kind help—many high-resolution figures needed reworking, and some were freshly prepared by them from experimental data. The authors would especially like to thank Klauss Blaum of GSI & Max Plank Institute, Thomas J. Baumann and Erin O'Donnell of NSCL-MSU, Alahari Navin and Jerome Giovinazzo of SPIRAL-GANIL, Osamu Kamigaito of RIKEN, Mark Loiselet of UC Louvain, Ann Y. W. Fong of TRIUMF and Karl Johnston from ISOLDE-CERN.

Alok Chakrabarti would like to thank DAE for the award of the Raja Ramanna Fellowship and would also like to acknowledge the generous support of Sumit Som, Director, VECC. This has facilitated continued close interaction with colleagues at VECC and enabled him to pursue active research which has greatly helped in writing this book.

Last but not least, this book would not have been completed but for the patience and continuous support of Ms. Aastha Sharma and Ms. Shikha Garg of Taylor & Francis Group. The authors greatly appreciate their contributions and would like to thank them both.

Authors

Alok Chakrabarti, former Director at the Variable Energy Cyclotron Center (VECC), Kolkata, India, and former Raja Ramanna Fellow at VECC, is an accomplished Physicist and a leader who has pioneered the fields of Exotic Nuclei and Rare Isotope Beams (RIB) in India.

Dr. Chakrabarti began his career in 1979 at the Bhabha Atomic Research Center (BARC), Mumbai, India, and earned his Ph.D. from the University of Calcutta, Kolkata, India. A recipient of the Indian Physical Society's Young Physicist award, DAE's Homi Bhabha Science and Technology Award and "Brojendranath Seal Smarak Samman," he has published more than 100 papers in reputed peer-reviewed journals. Dr. Chakrabarti is regularly invited to deliver lectures at institutes of world repute and prestigious conferences.

For his research, Dr. Chakrabarti built India's first on-line isotope separator at VECC and developed many key techniques for experimental study of short-lived nuclei. His major contributions include the discovery of a new isomer, a new technique for beta-delayed proton spectroscopy and an innovative experimental study which proves that asymmetric splitting survives in fission, rather surprisingly, up to quite high excitations and that there is a signature of bi-modal fission.

As a Project Director, Dr. Chakrabarti initiated and led the RIB project at VECC for which he successfully developed a number of advanced particle accelerators. This includes India's first heavy ion Radio Frequency Quadrupole (RFQ) linac, a state-of-the-art machine that efficiently accelerates low-energy ion beams. Dr. Chakrabarti, along with Dr. Dechoudhury, suggested a novel concept on "Alternate Phase focusing in independently phased superconducting resonators," which makes acceleration of particles much more efficient and allows acceleration of ions having different charge states.

Dr. Chakrabarti has supervised more than a dozen Ph.D. students. He also serves as Referee for journals such as *European Physics Journal A*, *Review of Scientific Instruments*, *Pramana*, *Nuclear Instruments* and *Methods and Ceramics International* and serves on the Scientific Committees of major international conferences in the field of particle accelerators.

Vaishali Naik is presently serving as Head of the RIB Facilities Group at VECC, Kolkata, India. A post-graduate in the Department of Physics, University of Pune, India, Dr. Naik graduated from the 33rd batch of training school at BARC, Mumbai, India. She joined VECC in 1991 and began her career as an experimental nuclear physicist. She later started working on particle accelerator development for the RIB program, where she played a leading role in the design and development of ECR ion sources and RFQ linacs and developed a gas-jet ECR technique that led to the first successful production of RIBs using the facility at VECC. A recipient of the DAE Scientific and Technical Excellence Award and Group Achievement Awards, Dr. Naik has published more than 30 papers in reputed peer-reviewed journals. She is presently supervising several Ph.D. students as a Professor at the Homi Bhabha National Institute (HBNI) and is also the Project Leader for the superconducting electron linac at VECC that is being built in collaboration with TRIUMF, Canada.

Siddhartha Dechoudhury joined the RIB Facilities Group at VECC, Kolkata, India, in 2001 as a Scientific Officer after completing his M.Sc. in Physics, followed by a one-year orientation course in the 44th batch of the Training School at BARC, Mumbai, India. Since then, he has been actively involved in the design and development of particle accelerators for the RIB project at VECC. His doctoral thesis, under supervision of Dr. Alok Chakrabarti, was on the development of a heavy ion rod type RFQ, which is the first of its kind in India. Presently he is involved in the development of a superconducting electron linac photo-fission driver, which is a joint collaboration between VECC and TRIUMF, Canada. Dr. Dechoudhury was conferred with a DAE Young Scientist Award in 2007 and a DAE Science and Technology Award in 2010. He has contributed to 30 publications in reputed peer-reviewed journals and about 40 papers in national and international particle accelerator conferences.

1
Rare Isotope Beams—The Scientific Motivation

1.1 Introduction

Rare Isotope Beams (RIBs), also called Rare Ion Beams or Radioactive Ion Beams, are ion beams of β-unstable nuclei. The energy of the ion beams of unstable isotopes (RIB) may vary over a wide range, from a few keV/u to a few GeV/u, depending upon the kind of study being conducted, either on the beam particles themselves or on reaction products produced by the beam through interaction with a target. Although it was shown way back in 1951 by Hansen and Nielsen that short-lived isotopes can be produced on-line, RIB physics has gained its importance mainly due to pioneering studies in the mid-1960s and in the 1970s carried out at Lawrence Berkeley Laboratory (LBL), Berkeley, using three accelerators: the Bevatron, the 88-inch AVF Cyclotron and the BEVALAC; at CERN, ISOLDE using the proton synchrotron; and at Dubna using low energy heavy ions. These studies led to the production and identification of a number of exotic nuclei situated far from the β-stability valley and showed that these nuclei could have different properties and decay modes compared to nuclei that are β-stable or lie very close to β-stability valley. Apart from high energy proton-induced spallation reactions first used at the LBL Bevatron and then at ISOLDE for production of a number of exotic nuclei, two important reaction routes for exotic nuclei production were discovered in the early 1970s: the deep inelastic transfer reaction at Dubna using low energy heavy ions and the projectile fragmentation reaction at Berkeley using high energy light heavy ions from BEVALAC. Both the projectile fragmentation and the deep inelastic transfer reactions involving heavy ion beams as projectiles enabled on-line production of a large number of light n-rich nuclei and created a lot of interest in this topic. The discovery of the neutron halo, one of the most striking properties of light n-rich nuclei, was soon to follow, triggered by a pioneering experiment in 1985 at BEVALAC by Tanihata and his collaborators. This established further the discovery potential of exotic nuclei in terms of their unexpected properties and their importance in the understanding of the atomic nucleus in its iso-spin dimension. By the early 1990s RIBs had emerged as a separate branch of study in nuclear physics and very soon became a priority research area in all the leading accelerator laboratories in the world (for example at ISOLDE, Louvain-la-Neuve, GANIL, GSI, RIKEN, ANL, MSU, TRIUMF, ORNL, to name a few) studying nuclear phenomena.

RIB science, from a nuclear physics point of view, is basically an exploration of the properties of the atomic nucleus in its iso-spin dimension and to ascertain the limits of stability, that is how many neutrons can be added or taken out from a β-stable nucleus of a given element so that the nucleus still remains bound, or in other words what are the different possible proton–neutron combinations that the strong force can bind and what are the loci where this binding ends for too few or too many neutrons as a function of proton number. It is estimated that there are at least 7000 nuclei, which include about 300 β-stable isotopes of different elements from hydrogen to uranium that are available on Earth. A large number of these nuclei are expected to have different and unexpected properties that cannot be predicted by the existing theories, primarily because the nuclear structure theories were initially developed to explain the properties of β-stable isotopes, which were experimentally known. As new data on exotic nuclei started flowing in, the inadequacies of the theoretical descriptions became evident. It was also realized that short-lived exotic nuclei play a crucial role in the nucleo-synthesis of elements beyond iron (also for light nuclei) and that a large number of heavy elements and isotopes that we find on earth are produced in violent stellar events, such as the type 1 or type 2 supernovae, by rapid neutron capture and proton capture processes. In these processes, the short-lived β-unstable nuclei far from the valley of β stability play a key role, and in as much as we are made of many of these heavy elements the story of the synthesis of elements is the story of our origin. To understand nucleo-synthesis in explosive stellar environments, accurate nuclear physics inputs about the mass, half-life and decay modes of exotic nuclei are absolutely essential. The task, therefore, from both nuclear physics and astrophysics points of view, is to study experimentally the properties of as many of exotic nuclei as possible and their neutron/proton capture cross-sections so that suitable theories can be developed which would explain the observed experimental data well and provide the platform to accurately predict the properties of other even more exotic nuclei which might not be accessible for study in the near future.

Since the late 1980s, the field has witnessed development at a tremendous rate, and it has now reached a level which was rather unimaginable a few decades back. The new developments in accelerator technology, experimental techniques and detector technology, including electronic read-out systems, have helped to overcome many of the challenges related to the production and study of these short-lived nuclei that are produced in minuscule quantities in nuclear reactions (involving beams of stable isotopes and stable targets), thereby turning the challenges into opportunities to explore new phenomena. Theoretically, apart from new ideas, the other factor that contributed to our understanding of the atomic nucleus is the tremendous progress in high-performance computing in recent times that has allowed many calculations that were previously impossible. Experimentally, it has now been possible to produce more than 3000 isotopes in the laboratory and study their properties to various extents. This capability of producing new isotopes of varied iso-spins in sufficient numbers has also opened the possibility of carrying out precise measurements on electro-weak interactions using nuclear and atomic techniques (like the β decay of a ^{60}Co nucleus allowed us to discover parity violation in weak interaction in 1956) and plan for more precise experiments. These measurements would enable the standard model to be tested, such as the conserved vector current (CVC) hypothesis and the unitarity of the Cabibbo–Kobayashi–Maskawa (CKM) matrix, and the fundamental symmetries (like charge-parity; CP) of nature or their violation with improved precision.

The availability of various radioactive probes has made RIBs a very important tool for condensed matter physics/material science, especially for studying the surfaces and the interfaces, semiconductors and magnetic materials, which hold great potential for future devices. RIBs also have great potential application in medical physics in terms of producing new and more effective isotopes for diagnostic and therapeutic purposes. The development in accelerator technology for RIBs has helped in the design and construction of high energy heavy ion accelerators for particle therapy. It has been established now that hadron therapy using heavy ions like carbon is very useful in killing the cancer cells in otherwise resistant tumors. Treatment with high energy unstable ion beams (RIBs, such as ^{11}C) is also a new area that apparently has a number of potential advantages over the stable ions (say, ^{12}C) in which RIB development techniques will play a direct role. Apart from ion therapy, RIB developmental efforts are very important for future clean energy options. The technology of accelerating high current protons to about 1 GeV and of the thick targets for RIB production can be applied directly to produce nuclear energy in an inherently safe manner using an accelerator-driven sub-critical reactor (ADS), where the additional neutrons (to make the reactor critical which is needed for energy production) are supplied by spallation neutrons produced in a reaction of high energy protons with a thick heavy (say, Pb-Bi) target. This technique is inherently safe since the proton beam can be switched off within a fraction of a second leaving a sub-critical reactor, and this method would also produce only comparatively much shorter-lived nuclear waste by destroying the longer-lived actinides. The ADS route, being absolutely safe and having (potentially) no major issue with the disposal of waste, is considered one of the most promising routes for green energy production and efforts to develop such an accelerator-driven reactor system on a pilot scale are being pursued at present in Belgium (MYRRHA 18).

The field of RIB science will thus allow us to understand the atomic nucleus and the limits of its existence (through study of nuclear properties and nuclear mass), our origin (in terms of understanding the nucleo-synthesis), the symmetries of nature (in terms of stringent tests on the fundamental symmetries and physics beyond the standard model) and at the same time will create a wealth of new opportunities in material science and medical physics. It will also likely to provide one of the most viable alternatives to future energy solutions. For an aspiring researcher, the best part however is that the field of RIB is in its early stage of development possessing tremendous discovery potential loaded with challenges and opportunities in basic science as well as in technology.

1.2 RIBs and Nuclear Physics

1.2.1 The Limits of Nuclear Stability

It is customary to start any discussion on the role of RIBs in nuclear physics with the chart of nuclides or the N–Z chart as shown in Figure 1.1. In the figure, the proton number of any isotope is plotted along the Y-axis and the neutron number along the X-axis. As mentioned, in nature there are about 300 isotopes that belong to 92 elements from hydrogen to uranium. These isotopes are either stable or have half-lives comparable with the age of the Earth (and solar system). These isotopes are shown in Figure 1.1 by black squares which are distributed around the $N = Z$ line for lighter elements up to $Z = 20$ and then gradually bends towards the x-axis for heavier elements since Coulomb repulsion between protons limits the number of protons in a stable combination of protons and neutrons. Together, these isotopes lie in what is known as the valley of β-stability. For any given isobaric chain (A = N + Z = constant; a diagonal line in the N–Z chart) there is, with few exceptions, only one beta stable nuclide if A is odd, but for even A there are two, sometimes three, stable even (Z)–even (N) isobars varying in natural abundance. The mass of a nucleus with proton number (proton mass) Z (m_p) and neutron number (neutron mass) N (m_n) is given by:

$$M = Zm_p + N\,m_n - E_B\,(\text{binding energy}) \tag{1.1}$$

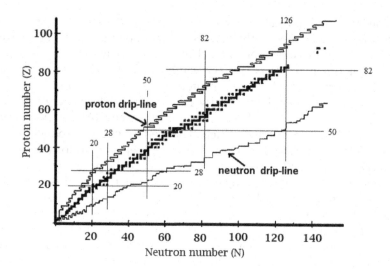

FIGURE 1.1
The N–Z diagram showing the β-stable isotopes (black squares) and the approximate locations of the neutron and proton drip lines. The four isolated black squares beyond N = 126 represent ^{232}Th and ^{234,235,238}U.

The binding energy is more for closed shell nuclei and for even–even nuclei (because of the pairing effect between two protons and two neutrons of opposite spins). Thus, elements having proton numbers corresponding to closed shells (8, 20, 50, 82, etc.) have a good number of β-stable isotopes. For example, tin (Z = 50), with a closed-shell proton number of 50 has 10 β-stable isotopes in the range A = 112 to 124, where the even A isotopes have more natural abundance with respect to their immediate odd A neighbors (for example 14.8% for ^{116}Sn and 7.75% for ^{117}Sn). In general, because of pairing, even Z elements have more than one, often several, β-stable isotopes compared to odd Z elements which have one or at the most two β-stable isotopes, mostly with N even (odd A). The odd Z–odd N stable combinations are rare. There are only a few naturally occurring isotopes of odd–odd type: ^{2}H, ^{6}Li, ^{10}B, ^{14}N, ^{40}K, ^{50}V, ^{138}La, ^{176}Lu, ^{180}Ta, etc., among which only ^{14}N, ^{10}B and ^{6}Li are in high or substantial natural abundance (^{40}K, ^{50}V, ^{138}La, ^{176}Lu, ^{180}Ta, although not stable, have very long decay lifetimes).

Keeping A (= N + Z) constant, if either N or Z is varied, starting from a β-stable isobar, the binding energy decreases and the mass of the isobar increases which makes it unstable with respect to β-decay; β⁻ or β⁺ decay depending on whether the isobar has excess or deficient neutrons with respect to the β-stable isobar. The strong interaction that binds the protons and the neutrons in the nucleus, however, allows, apart from the β-stable ones, many (but not all) combinations of protons and neutrons to remain bound and to form a nucleus. Estimates based on different mass formulae suggest that the number of these stable combinations could be around 7000 or even more (depending upon the mass formula relied upon), which includes the 300 nuclei that lie in the valley of β-stability. The vast majorities of nuclei that are not β stable are therefore bound from the strong interaction perspective but are not found on Earth because of the existence of the weak interaction that gives rise to nuclear β-decay. These nuclei, often termed particle-stable nuclei, if produced in nuclear reactions (in Earth-based laboratories or in stellar interiors), would undergo β-decay (or successive β decays along an isobaric chain until a β-stable isobar is reached) and their β decay half-lives get shorter and shorter as their distance from the valley of stability increases.

Rare Isotope Beams—The Scientific Motivation

The strong interaction (nuclear force), as mentioned already, does not allow any combination of protons and neutrons to be stable. For any given proton number (element) if neutrons continue to be added to the β-stable nucleus, the nuclear force restricts the neutron excess (number of neutrons over and above the number of neutrons present in the stable isotope of that element) to a certain limit defined by the neutron drip line beyond which the nucleus would not be able to bind any additional neutrons. Similarly, if protons continue to be added to a β-stable isotope, the number of protons that can be added is limited by the proton drip line. The neutron and proton drip lines, shown in Figure 1.1 and in somewhat more detail for light nuclei in Figure 1.2, represent the boundaries of the nuclear landscape, that is, the limits of particle stability within which about 7000 particle-stable nuclei lie. The allowed particle-stable combinations of protons and neutrons that lie to the left of the β-stability line are termed as the neutron-deficient or p-rich nuclei. These nuclei decay by β^+ emission or by electron capture (EC). The permitted combinations to the right are termed n-rich nuclei and these nuclei decay by β^- emission. It can be seen (Figure 1.1) that the neutron drip line is much further away from the β-stability valley compared to the proton drip line (because Coulomb repulsion limits the proton excess) and thus the number of n-rich particle-stable nuclei exceeds by far the number of particle-stable neutron-deficient nuclei. It can also be seen that the drip lines or the borders of nuclear existence are not smooth but follow a zigzag path between even and odd A due to pairing interaction and because of structure effects that are especially pronounced for light nuclei (Figures 1.2 and 1.1).

The term drip line implies that if a neutron (proton) is added to a neutron (proton) drip line nucleus, the neutron (proton) will drip out, since it is not bound. However, the definition of drip line is often not straightforward because the stability of a nucleus with respect to strong nuclear force can be defined in a number of ways. For example, the typical strong interaction time is 10^{-22} s and thus any combination of protons and neutrons that can coexist for a time more than 10^{-22} s, can in principle be considered stable from a strong interaction (nuclear force) point of view. So, the drip line can be defined by this time limit. Thus, for example, if the addition of a neutron leads to the new nucleus decaying faster than 10^{-22} s, then the nucleus to which the neutron was added can be defined as the neutron

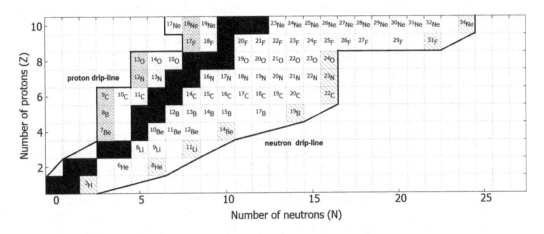

FIGURE 1.2
The neutron and proton drip lines, as established experimentally, for light nuclei up to neon. The drip line is not a smooth curve because of nuclear structure effects. The drip line nuclei or the last particle-stable isotopes are shaded.

drip line nucleus. The angular momentum barrier can also slow down the decay because the neutron would need to tunnel through the centrifugal or angular momentum barrier. But often the angular momentum of the last neutron is not high enough to cause any substantial delay. Experimentally, the typical time limit for detection of a new exotic isotope is about a few tens of nanoseconds (ns). Thus, it is difficult to observe any nuclei decaying by neutron emission faster than a few ns. However, this time limit is short enough with respect to the β-decay half-lives of drip line nuclei that are typically in the range of a few milliseconds (ms) and above. Thus any drip line nucleus that undergoes β-decay can be experimentally observed, at least in principle, and the neutron drip line is decided by the boundary between the experimentally observed and non-observed (which should have been observed given the experimentally sensitivity had it been particle-stable) nuclei. On the neutron-deficient side, however, the experimental determination of proton drip line is more complicated because of the existence of the Coulomb barrier, which often causes delay in the emission of the unbound proton by many orders of magnitude (up to a factor of 10^{20} or so, depending upon Z of the isotope and the energy of the unbound proton since the unbound proton needs to tunnel through the Coulomb barrier), resulting in a half-life of a few hundred ms or even higher. Thus, proton decay would then compete with β+ decay since the β-decay half-life at the proton drip line is typically a few ms or even less. Thus, even a neutron-deficient nucleus beyond the proton drip line can decay predominantly via β+ emission with a half-life ranging from, say, a few μs (microseconds) to a few hundred ms, behaving in the same way as a nucleus lying within the proton drip line. A practical way to define the drip lines is in terms of neutron and proton separation energies (S_n and S_p respectively). The separation energies are given by the difference in the binding energies between two adjacent nuclei:

$$S_n = B.E(Z, N) - B.E(Z, N-1)$$

$$S_p = B.E(Z, N) - B.E(Z-1, N) \qquad (1.2)$$

S_n (S_p) equals zero (or crossing zero) defines the neutron (proton drip line). Because of pairing interaction between identical nucleons which gives extra stability to nuclei with an even number of neutrons or protons, the neutron (proton) drip line in terms of S_{2n} (S_{2p}) equals zero, S_{2n} (S_{2p}), being the separation energy of two neutrons (two protons), also needs to be defined. Helium nuclei represent the most typical case where ^{4,6,8}He are stable while ^{5,7}He are unstable. So, the two-neutron drip line lies further away compared to the one-neutron drip line. ^{11}Li, which is the drip line nucleus for Z = 3, represents another typical case where S_{2n} is needed, since ^{9}Li is stable but ^{10}Li is not. In general, it is customary to show the limits of nuclear existence in a N–Z chart by drawing the loci of both S_n (S_p) and S_{2n} (S_{2p}) equal to zero.

It should be mentioned that the exact locations of the drip lines are not experimentally established, and we can talk about drip lines only in terms of predictions of various mass formulae. This is especially true for the neutron drip line which is much farther away from the β-stability valley and it is difficult to produce the drip line nuclei except for the light nuclei. The proton drip line, being much closer, has been reached experimentally, more or less for all elements up to Z = 82. In contrast, the neutron drip line is reached so far only for light elements up to Z = 10 (Ahn 19) as shown in Figure 1.2. ^{8}He is so far the most n-rich drip line nucleus discovered in terms of the N/Z ratio. As can be seen from Figure 1.1, for heavier nuclei the neutron drip line moves farther away from the stability valley towards more neutron excess. Just to give an example, for the element tin, which has ten stable isotopes in the range $^{112-124}$Sn, some mass formulae predict the drip line to be situated as far as $A = 176$, which has 52 excess neutrons compared to the last β-stable nucleus, ^{124}Sn (Erl 12 and the references therein). So far, experimentally the mass is measured only up to ^{135}Sn (Hak 12). Understandably, reliable theoretical models to predict the drip lines are required.

The first mass formula for calculating nuclear binding energy is the semi-empirical mass formulae of Weizsäcker based on the liquid drop model (Wei 35), which has been reasonably successful in accounting for the masses of β-stable isotopes. The nuclear binding energy in this model has been expressed in terms of A and Z and as a sum of volume, surface, symmetry and the Coulomb energies. The volume energy is the main and the dominant term that contributes to the binding energy; all the other terms having a negative contribution to the binding energy. The main components of the mass formula can be expressed as:

$$E_B = a_{vol}A - a_{surf}A^{\frac{2}{3}} - \frac{a_{coul}Z^2}{A^{\frac{1}{3}}} - \frac{a_{sym}(N-Z)^2}{A} + \delta \tag{1.3}$$

where a_{vol}, a_{surf}, a_{coul} and a_{sym} are the coefficient of the volume, surface, Coulomb and symmetry energy term respectively and δ is the pairing energy term. The volume energy term is proportional to nuclear volume or the mass number A (r, the radius of the spherical liquid drop nucleus is given by $r_0 A^{\frac{1}{3}}$), which is a consequence of the short range and saturation property of the nuclear force. The surface energy term reflects the fact that the saturation would not be complete for nucleons at the surface of the drop and thus would reduce the binding energy based on volume that assumes saturation for all nucleons. The Coulomb energy is the electrostatic energy due to the protons in the nucleus which would reduce the overall binding energy. The symmetry energy term reflects in a way the fact that the strong force favors symmetry between the protons and neutrons since Pauli's exclusion principle does not allow two identical particles (protons or neutrons) to occupy the same orbit. Thus, for light nuclei up to $Z = 20$, where the Coulomb force between protons is still rather weak, equal numbers of protons and neutrons form the most abundant β-stable combinations. However, for $Z > 20$ more neutrons are needed to counterbalance the Coulomb repulsion and thus the β-stability line bends towards $N > Z$. It should be noted that symmetry energy, as well as the Coulomb repulsion between protons, limits the proton excess in nuclei whereas it is only the symmetry energy that limits the neutron excess. This brings the proton drip line much closer to the β-stability line compared to the neutron drip line. The pairing energy term takes care of the fact that the even Z - even N (odd Z - odd N) nuclei are the most (least) stable. The coefficient of different terms in the mass formula can be calculated in terms of the best fit to known experimental masses and then the mass formulae can be used to predict unknown masses that are not determined experimentally.

Beginning with Weizsäcker's semi-empirical mass formula of 1935, many improved versions of this mass formula have since been worked out, based on inputs of new experimental data on exotic nuclei and aiming at accurate predictions on drip line nuclei. These formulae are based on macroscopic + microscopic and purely microscopic approaches. The macroscopic–microscopic finite range mass model (FRDM; Mo 95; Mol 12) and the Weizsäcker–Skyrme energy-dependent functional (Wan 10 or Wei 35) the Infinite Nuclear Matter (INM) based mass models (Sat 83, Nay 99), the microscopic Extended Thomas–Fermi Strutinsky Integral approach (ETFSI; Abo 95), the microscopic Skyrme–Hartree–Fock–Bogoliubov mass formula (Gor 10), the large scale shell model (Ots 10), the density functional-based mass models (Erl 12) and ab initio theories (Hag 12) are but a few typical examples of mass models that have been used for comparison with experimentally determined masses of new exotic nuclei, for the prediction of new magic numbers and for the prediction of drip lines. For masses that have been experimentally measured, these mass models agree with each other to within about 500 keV. But this does not guarantee similar accuracy for the prediction of masses of nuclei that are far away from β-stability. In fact, the mass formulae fail to predict the neutron drip line even for the lighter isotopes. For example, experimentally, the drip line nuclei for oxygen and fluorine are established to be ^{24}O and ^{31}F respectively while the majority of mass models predict ^{26}O and ^{29}F to be the heaviest stable isotopes of oxygen and fluorine. A great deal of effort is being made to understand this sudden change of drip line from ^{24}O and ^{31}F (the addition of a proton allows six more neutrons to be bound!). Further, ^{34}Ne, the heaviest (drip line) bound isotope of Ne determined experimentally and which has now been established as the drip line nucleus for Ne (Ahn 19), is predicted to be unbound in most mass models. In fact, the predictions for masses of heavier exotic nuclei often deviate by several MeVs and the prediction becomes increasingly worse as the distance from the valley of β-stability increases.

The predictive capability of the mass formulae can be compared with an accuracy of better than 100 keV (often better than ~10 keV) necessary to predict whether an exotic nucleus is on or outside the drip line (for weak interaction studies the accuracy required is even more stringent and is about 1 keV). This level of accuracy is needed since the binding energy of the last neutron added to a n-rich nucleus falls off rapidly as the drip line is approached, and it is quite possible that the last (or last two) neutron(s) has (have) a binding energy (or separation energy) in the range of a few 100 keV or less (even as small as few tens of keV; last two neutrons in ^{26}O are unbound by only about 20 keV). Thus, any mass formula attempting to predict the neutron drip line should be able to predict masses with accuracy better than a few tens of keV. To improve the mass models, the nuclear structure effects at the extremes of iso-spin (for nuclei on or close to the drip lines) need to appropriately be taken into account. This requires, in turn, the knowledge of the nuclear potential which must include, as the trend in the theoretical descriptions shows, the tensor and three-body forces, pairing interaction and the coupling between the bound states and the low lying continuum. Some effects might show up quite abruptly as the drip line is approached, as has been experimentally found in the case of a number of light n-rich nuclei such as ^{11}Li, ^{11}Be, ^{14}Be, etc., affecting the nuclear binding. It is of course an open question, due to lack of experimental data, whether the structural changes observed for light n-rich drip line nuclei would have the same trend (universality) as for heavier n-rich nuclei all along the nuclear chart. Obviously, more experimental studies on heavier exotic nuclei are required to address this question. The future development of mass formulae would thus have to depend critically on experimental inputs on structures of heavier exotic nuclei and on new mass measurements.

Establishing the limits of nuclear stability experimentally is one of the major goals of RIB science. To achieve the goal, production of exotic isotopes close to the drip line and a precise determination of masses to determine the separation energies are required. These require a huge advancement in the accelerator technology and production techniques for producing drip line or near drip line nuclei with sufficient intensity (discussed in Chapters 2 and 6). These developments are being addressed at present. However, even assuming these efforts are completely successful, the neutron drip line would probably remain experimentally inaccessible for heavier elements ($Z > 35$ or so) in near future. Nonetheless, these developments would give access to precision mass data of about 1000 new nuclei, most of which are of an exotic brand lying far away from β-stability. At present we have precision mass data on about 2000 isotopes. Thus, the new measurements would lead to a substantial increase in the volume of mass data which would allow fine-tuning of mass formulae for better and more accurate predictions. It should be mentioned that accurate prediction of masses is not only important to determine the limits of nuclear stability but is also important in understanding nucleo-synthesis, the crustal composition of a neutron star and for stringent tests on weak interactions. Since it would not be possible to determine experimentally the nuclear masses for very exotic nuclei in near future, development of theoretical models for predictions of nuclear masses of exotic nuclei assumes great importance. The task ahead is no doubt challenging in terms of the demand on accuracy and universality (applicability over the entire chart and not over a specific limited zone).

1.2.2 Nuclear Halo in Drip Line and Near Drip Line Nuclei

The term "halo" implies a dilute region around a central denser object. As nuclei are systems bound strongly by a short-range nuclear force, halos are in general not expected. However, measurement of matter radii through total interaction cross-sections (total cross-section for nucleon removal) measurement at Bevalac, Berkeley by Tanihata and his collaborators in 1985 (Ta 85, 85a, 88) clearly showed that a number of light n-rich nuclei, including ^{11}Li, whose decay properties were already known experimentally, have a much larger spatial extension than what is expected from the $r = r_0 * A^{1/3}$ law, where r is the half-density radius of matter distribution. The large value of the observed matter radius for ^{11}Li was suggested by the authors to arise from either a large deformation or a long tail in the matter distribution due to low binding energy of the two neutrons (Ta 85, 85a). Subsequently, the measurement of spin and magnetic moment of ^{11}Li at ISOLDE in 1986 (Ar 87), established that ^{11}Li is more or less spherical (the measured spin of 3/2 itself ruled out any strong prolate deformation in which case the spin should have been 1/2) and thus the deformation as the cause for the observed large spatial extension was ruled out. The term "halo" was first used by Hansen and Jonson (Han 87) who showed that by treating ^{11}Li as a binary system of ^{9}Li core and a di-neutron (although a di-neutron is not stable it is the attraction between the two neutrons that results in bound ^{11}Li, since ^{10}Li is unstable, making the assumption of a di-neutron cluster near the nuclear surface not an unreasonable one) that the large interaction radius is a consequence of a long tail in the neutron wave function associated with the small binding energy (~340 keV) of the last two neutrons. The low binding energy of the di-neutron would lead to "a neutronization of the nuclear surface" (Han 87), which would appear as a halo, a region of low neutron density surrounding the ^{9}Li core. The weakly bound neutrons are often termed valance neutrons, which tunnel out from the dense ^{9}Li nuclear core to form the halo.

It should be mentioned that weak binding is a necessary condition for the formation of a halo since nuclei are systems bound by short-range potential, but this is not a sufficient condition. Quantum mechanically, the formation of an inert core comprising of rest of the nucleons (^{9}Li in this case) is additionally needed, from which the valance nucleon(s) (two neutrons in this case) essentially de-couple in order to have low binding energy and large spatial extension beyond the classical limit. The other important condition for halo formation is that the halo neutron(s) should occupy a low angular momentum orbital otherwise the centrifugal barrier would limit the spatial extension through tunneling, inhibiting halo formation. Thus, low binding energy (say, less than 1 MeV) of the valance nucleons, inert core and valance nucleons occupying a low orbital angular momentum state are the three criteria for halo formation. It should be mentioned that a number of stable nuclei have excited states of low orbital angular momentum quite close to the neutron threshold (therefore low binding energy of the neutron in that excited state) and these are, in principle, candidates for halo in excited states. But many of these excited states may not actually exhibit the halo character because of high density of states in that region, which implies that the core is not inert.

The weak binding and the large extension have together allowed investigation of the halo neutrons almost independent of the nuclear core, providing information about the structure of the wave function and correlations. ^{11}Li has only one bound state which is the ground state and thus it should be easily excitable in the nuclear and Coulomb fields of target nuclei to energies beyond the two-neutron emission threshold. In such an experiment, it would be expected that the two extra core valence or halo neutrons representing the low-density neutron matter in ^{11}Li should have small internal momentum corresponding to their large extension or spatial uncertainty (Heisenberg's uncertainty principle). Because of that the core part ^{9}Li should also have narrow momentum distribution, which is often easier to measure experimentally. This was indeed borne out experimentally (Kob 88) through the observation of narrow momentum distribution of ^{9}Li in the breakup reaction of ^{11}Li with a carbon target in inverse kinematics and in the measurement of the angular distribution of neutrons emitted (Ann 93).

Another aspect of halo and low binding energy is the possibility of the existence of electric di-pole excitation (Han 87) mode at low excitation which would result in a large cross-section for Coulomb breakup: ^{11}Li → ^{9}Li + 2n. This E1 mode of excitation, termed the soft di-pole resonance, should be at low excitation energy, roughly estimated to be about 0.9 MeV (Han 87). This conjecture was also verified experimentally (Kob 89) in a reaction of ^{11}Li with a Pb target, that measured a huge Coulomb dissociation cross-section of about 1 barn. The excitation energy of this mode has also been determined experimentally and found to be (for ^{11}Li) about 0.6 MeV (Nak 06). The soft di-pole resonance might be due to a low frequency collective oscillation of the ^{9}Li core against the 2n halo or due to direct break-up from continuum final states above the two-neutron threshold, having sufficient di-pole strength at low excitation due to the large radial extent of the two halo neutrons. The experiments on Coulomb dissociation of ^{11}Li, ^{11}Be, and other halo nuclei (Lek 93, Ann 94, Nak 94), together with theoretical understanding of this phenomenon (Ber 88, Shy 92, Bau 92, Ban 93) have established that the soft di-pole excitation in halo nuclei is predominantly of the non-resonant nature (not a "slow" collective oscillation). It is important to note that this excitation energy is far too small when compared with the excitation energy of the usual Giant di-pole resonance mode (GDR, corresponding to the collective oscillation of all the neutrons against all the protons in a nucleus) given by $E^* \sim 80\ A^{-1/3}$ MeV. Thus, halo nuclei have three characteristic experimental signatures: large interaction cross-section, narrow momentum distribution of the core nucleus and large Coulomb dissociation cross-section.

The halo is just not a unique property of ^{11}Li. The halo structure has been experimentally established in a number of other light n-rich nuclei (Nak 17 and references therein) such as ^{6}He, ^{11,14}Be, ^{17,19}B, ^{15,19,22}C, ^{31}Ne, ^{37}Mg, etc.; ^{11}Be, ^{15,19}C, ^{31}Ne and ^{37}Mg are 1n neutron halo nuclei while the others exhibit 2n halo (Figure 1.3). The 2n halo systems (or the three-body n-n-core systems) are often called Borromean nuclei since Borromean rings are three linked rings where each pair is unlinked, and thus removal of any one ring would result in the system falling apart (for example, for ^{11}Li, ^{10}Li and di-neutron are both unstable). As already mentioned, the halo is formed when the halo neutron or neutrons occupy the low angular momentum orbitals (s wave: $l = 0$ or p wave: $l = 1$) because tunneling is possible

Rare Isotope Beams—The Scientific Motivation 11

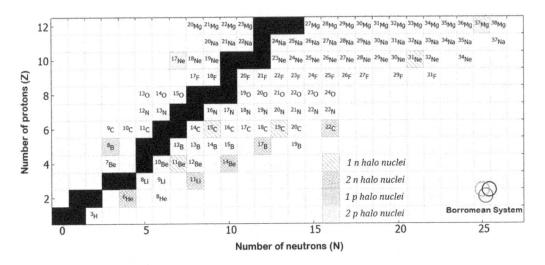

FIGURE 1.3
Experimentally established neutron and proton halo light nuclei. The one-nucleon and the two-nucleon halo nuclei are differently shaded.

only when the angular momentum barrier is zero or low. For halo nuclei from ^{6}He to ^{22}C, the halo is of the *s*-wave type with the halo neutrons occupying the *s* orbital while for ^{31}Ne, ^{37}Mg, the halo is of the *p*-wave type with the last neutron occupying a *p* orbital. Incidentally, ^{37}Mg is the heaviest halo nuclei observed so far. The halo in ^{31}Ne, ^{37}Mg is rather unique in the sense that it is deformation driven. In both cases, $1f_{7/2}$ is the orbital for the valence neutron in a conventional spherical shell model (see Section 1.1.3). The large deformation of the core is considered to be responsible for significant occupancy of the weakly bound neutron in the $p_{3/2}$ orbital which leads to the halo. This kind of halo might therefore occur in some heavier drip line nuclei also, especially in the regions of deformation. Generally speaking, however, the requirement of neutron(s) occupying low angular momentum orbitals may not be satisfied for a majority of the heavier nuclei along the drip line. A halo, therefore, may not be a general feature for all heavier drip line *n*-rich nuclei. The difficulties in producing heavier drip line nuclei did not allow detailed study into halo universality.

Proton halos are expected less since the Coulomb barrier tends to inhibit the formation of halo or wave function of a large spatial extension by quantum mechanical tunneling. However, as shown in Figure 1.3, in the neutron deficient side two-nuclei ^{8}B (*p*-wave; one-proton halo) and ^{17}Ne (*s*-wave; two-proton halo) are found to exhibit the halo character. However, since the Coulomb barrier increases for heavier systems, the proton halo is not expected to continue in heavier systems.

The neutron drip line or near drip line nuclei do not necessarily form a halo. In some drip line nuclei, a thicker neutron skin (neutron skin is expressed in terms of difference between the root mean square radius of neutron ($< r_n^2 >^{1/2}$) and proton ($< r_p^2 >^{1/2}$)) was observed instead of a highly diffuse low-density tail. The neutron skin was first observed in ^{8}He by Tanihata and his collaborators (Ta 92). It was deduced that the skin thickness was unusually large: about 0.8 fm compared to about 0.18 fm for ^{48}Ca, the most *n*-rich β-stable light nuclei, about 0.17 fm for the heavy ^{208}Pb, with 44 excess neutrons, and about 0.23 fm in *n*-rich doubly magic ^{132}Sn (Pie 12). It is clear that just an excess of neutrons over protons does not lead to a thick neutron skin; the neutron excess should be enough to bring about a large difference between the Fermi levels of neutrons and protons. This apparently happens not for stable nuclei with a neutron excess but for exotic n-rich nuclei. The large difference between the neutron and proton Fermi energies reflects the difference of the orbital sizes of the last few neutrons and the protons and that gives rise to a thicker neutron skin. In Figure 1.4 the Fermi levels of protons and neutrons are schematically shown for three cases: (a) for heavy stable nuclei, (b) for n-rich exotic nuclei exhibiting a thick neutron skin, and (c) for drip line n-rich nuclei exhibiting a neutron halo (along with the corresponding density distributions for protons and neutrons). It can be seen that near the drip line, the Fermi levels of protons and neutrons change dramatically and the Fermi level for neutrons lies very close to the continuum. The large difference between Fermi energies of neutrons and protons near the neutron drip line leads to the near decoupling of proton and neutron distributions, which in turn shows up forming either a neutron skin or halo. For heavier stable nuclei, for example, ^{208}Pb which contains 82 protons and 126 neutrons, the difference in Fermi energies between neutrons and protons is not large and this results only in a relatively more extended density distribution for neutrons compared to protons. This difference, or the neutron skin, is also a feature of most stable heavy nuclei. However, in this case, the density distributions of both protons and neutrons have the same overall nature both in the interior of the nucleus and on the surface, in sharp contrast to a low-density long-tail neutron distribution in n-rich halo nuclei. Neutron distribution for the neutron skin comes in between these two. Coming back to helium nuclei, ^{6}He is a halo nucleus with an alpha core plus two valance neutrons and ^{7}He is unstable. ^{8}He is thus a candidate for a possible four-neutron halo outside an inert alpha core. However, the matter distribution of four extra core valance neutrons does not extend far enough to qualify for a typical halo nucleus. It is thus considered as a nucleus with an extremely thick skin suggesting that multi-neutron halo formation has not been favored in this system, possibly due to an increase in mutual attraction between the four valance neutrons.

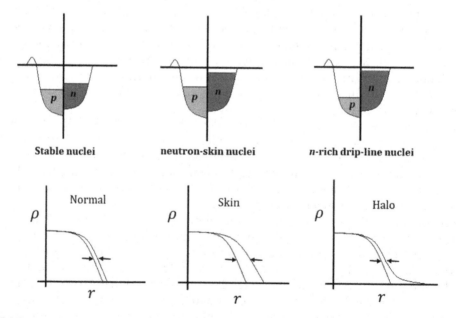

FIGURE 1.4
Fermi levels of protons and neutrons and the corresponding density profiles for heavy stable nuclei and light n-rich nuclei with neutron skin and neutron halo.

Rare Isotope Beams—The Scientific Motivation

The neutron skin nuclei are expected to show pygmy di-pole resonances (PDR) due to collective vibration of the neutron skin against the core. The excitation energy for PDR is expected to be higher compared to that of the soft E1 mode for the halo nuclei, because the neutron fluid in the case of a neutron skin is much more strongly bound to the core than it is with a halo nucleus. Experimentally, the PDR has been observed at excitations of around 8 to 10 MeV (beyond the one-neutron separation threshold), which is, however, still significantly lower than the conventional GDR bump that occurs in between 15 to 20 MeV. The pygmy resonance has been observed both in light and heavier systems, for example in ^{26}Ne (Gib 08) and the doubly magic ^{132}Sn (Adr 05). The di-pole response of halo nuclei, neutron skin nuclei and the conventional GDR in stable nuclei is represented schematically in Figure 1.5.

As mentioned, the halo structure observed in light nuclei might not continue to show up all the way up to heavy nuclei, but a neutron skin of appreciable thickness might turn out to be quite a general property of drip line and near n-rich nuclei. The presence of a thick skin would obviously have its effect on nuclear binding as well as on the single particle states in n-rich nuclei. Further, in neutron skin nuclei, the presence of PDR at much lower excitation than the GDR (but above the neutron separation energy) is likely to accelerate neutron capture rates in r-process nucleo-synthesis occurring in explosive stellar environments. The other importance of the neutron skin comes through the connection between the dynamics of atomic nuclei and the neutron stars through the symmetry pressure, allowing both to be described in a unified framework (Hor 01). In the case of a neutron star the symmetry pressure works against the gravity to prevent a collapse, while for nuclei it works against the surface tension of the liquid drop that favors the excess neutrons to be placed in the core. Thus, the larger the symmetry pressure, the larger the skin thickness in a n-rich nuclei will be and the larger the radius of neutron stars will be (Roc 11), especially the ones with comparatively low mass in which the density in the interior is slightly more than the saturation density (~ 0.15 fm^{-3}). For heavier neutron stars, the density in the interior can reach up to about twice the saturation density and the "neutron skin–radius of neutron star" correlation in these stars would depend upon how the symmetry energy

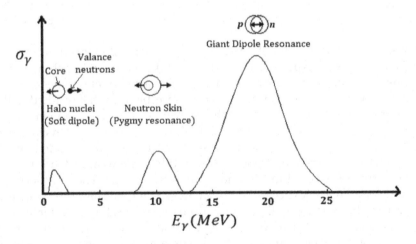

FIGURE 1.5
Schematic representation showing the E1 responses of light halo nuclei, neutron-skin nuclei and β-stable normal nuclei. The stable nuclei only show a GDR bump, the neutron-skin nuclei show bumps due to both pygmy resonance and the GDR and the halo nuclei show the soft E1 and the GDR.

behaves at densities higher than the saturation density. The RIB facilities would provide a unique opportunity for studying this dependence or the equation of state (*EOS*) of asymmetric *n*-rich matter in terrestrial laboratories in high energy collisions involving nuclei with thick neutron skins.

1.2.3 Evolution of Shell Structure away from Stability

The shell structure of nuclei proposed in 1949 has been the most successful approach to understand atomic nuclei. The observed extra stability and high natural abundance of nuclei with proton or neutron numbers equal to 2, 8, 20, 28, 50, 82 and 126 (for neutrons only) could be explained by assuming nucleons to move in a central (spherically symmetric) mean field potential created by all the nucleons in a nucleus which includes a spin-orbit term that is concentrated more on the surface. The magic numbers (2, 8, 20, 28, etc.) are reproduced well if the single particle energy levels are filled up following the Pauli exclusion principle. At magic numbers, large energy gaps appear between the occupied orbitals and the valance orbitals. The nuclei with magic numbers are called closed shell nuclei and these are all spherical in shape as the mean field mimics the distribution of the nucleons in the nucleus. However, the mean field potential that reproduces the magic numbers does not contain the pairing interactions between like nucleons with opposite spin. It was found that it is necessary to add residual interactions like pairing and p–n interactions to the mean field potential in order to explain the ground state and excited state properties, especially those of mid-shell nuclei. The residual interactions result in partial occupancy (particle-hole picture) of mean field single-particle states and lead to description of the nucleus in terms of superposition of many configurations (called configuration mixing) and often it is the deformed but not the spherical shapes that are found to be energetically more stable for nuclei outside closed shells, especially for mid-shell nuclei. Overall, the success of the independent particle model (which includes the residual interaction and is often called the shell model) has been overwhelming in explaining the ground state as well as excited state properties of β-stable nuclei. However, it soon became clear, as the experimental data on exotic nuclei started pouring in, that the shell structure evolves differently in exotic nuclei. It is observed experimentally that some of the erstwhile magic numbers lose their magic properties, while new magic numbers seem to appear.

It should be mentioned that qualitatively, the re-ordering of magic numbers away from β-stability is not totally unexpected. This is because the nature of bunching of the energy levels or non-uniformity in the distribution of single particle states depends on the mean field potential in which the nucleons move. The spin-orbit term that resulted in the correct conventional magic numbers depends on the derivative of the radial potential and would therefore be most prominent at the nuclear surface where the nuclear density is decreasing. Since exotic nuclei far away from β-stability are likely to have an extended and more diffuse nuclear surface, the spin-orbit potential would become weaker. Once the mean field changes, the nature of residual interaction would also change. The re-ordering of magic numbers is therefore not unexpected in nuclei away from the stability level. However, as will be clear from the experimental evidence discussed below, for nuclei far away from stability the disappearance of conventional magic numbers (or the quenching of shell gaps) and appearance of new magic numbers has become more of a local (for example a particular shell gap for neutrons might quench only for a few neutron rich isotopes but not for other isotones) than a global feature. Thus, for exotic nuclei, the magic numbers have apparently lost their global validity.

Experimentally, the appearance or disappearance of magic numbers is indicated by a change in the nuclear binding energy (or single- or two-nucleon separation energies). There are other signatures too. One of these is the excitation energy of the first excited $J^\pi = 2^+$ state in even–even nuclei and the transition probability which is proportional to B (E2; 0^+ to 2^+) or to simplify the notation B (E2). A magic nucleus would be difficult to excite because of a large energy gap. Thus, for a magic nucleus the first 2^+ excited state is expected to be located at a higher excitation energy and therefore the probability of exciting it, expressed in terms of B (E2) value, is expected to be small. For β-stable doubly magic nuclei, E (2^+) typically lies above 4 MeV. For exotic nuclei and for stable nuclei with only proton or neutron magic number, the E (2^+) value may not be as high as 4 MeV. In these cases, to confirm the magicity E (2^+) should be comparatively much higher than that of isotopes of the same element or isotones and also the corresponding value of B (E2) has to be comparatively much lower. Usually for highly deformed nuclei E (2^+) lie below 1 MeV. For less deformed nuclei E (2^+) can exceed 1 MeV. The value of E (2^+) alone is not always sufficient to make conclusions about the deformation. B (E2) also needs to be determined. One typical example is tin isotopes for which E (2^+) remains more or less constant at around 1200 keV for a broad range of isotopes from $A \sim 104$ to 130, although the shape changes from spherical to deformed, which is reflected in the B (E2) values. The E (2^+) for the double magic ^{132}Sn is suddenly very high at 4 MeV, confirming its magicity.

The fact that doubly magic ($Z = 2, N = 8$) ^{10}He is not stable while ^{8}He is, raises some doubt about the status of $N = 8$ as a magic number for light n-rich nuclei. However, the first clear indication of the re-ordering of conventional orbitals in light neutron-rich exotic nuclei leading to a breakdown of magic number $N = 8$ was provided by the $2n$ halo nucleus ^{11}Li. In ^{11}Li, the two-valance neutrons are expected to occupy the $1p_{1/2}$ orbital as per the conventional shell ordering (Figure 1.6). It was found, however, that the two-halo neutrons must also occupy the $2s_{1/2}$ orbital ($l = 0$) instead of only the $1p_{1/2}$ orbital. The observed halo thus suggests a reordering of shell model orbitals where the energy of $2s_{1/2}$ orbital becomes lower in energy than the $1d_{5/2}$ level, thereby intruding into the p shell, resulting in large mixing between the $1p_{1/2}$ and $2s_{1/2}$ orbitals. This mixing is required to explain the experimentally observed large radius and the narrow momentum distribution. This trend of lowering of the $2s_{1/2}$ orbital continues through n-rich beryllium (Be) nuclei. For example, all the experiments on ^{11}Be clearly show that the valance neutron must occupy the $2s_{1/2}$ orbital instead of the $1p_{1/2}$ orbital. For ^{12}Be the excitation of the first 2^+ state was found to be low which signifies the vanishing of the $N = 8$ shell closure for n-rich nuclei. The study of proton halo nuclei reveals that the shell orbitals may not behave exactly in the same way as the proton drip line is approached. It has been found for ^{8}B, there is no reordering of conventional shell levels and the halo proton occupies the $p_{3/2}$ orbital, while for ^{17}Ne, a lowering of the $2s_{1/2}$ orbitals takes place (although not to the extent it comes down for neutron drip line nuclei) and the two protons occupy the s-wave orbital (Kan 03).

The breakdown of the $N = 20$ shell closure was indicated by the anomalous binding energy of n-rich Na nuclei (Thi 75) and later more clearly by the observation of the low excitation energy ~880 keV of the $J^\pi = 2^+$ state in ^{32}Mg populated in the β decay of ^{32}Na (Det 79). Following these, a number of n-rich nuclei of Ne, Na and Mg have been studied with a neutron number of around 20 and very rapid changes in nuclear structure has been observed leading to the vanishing of the $N = 20$ gap for these n-rich nuclei, which constitutes what is called the "island of inversion." The exact size of the island of inversion is currently an open question and needs to be determined by future experiments, but based on experimental evidence the island of inversion so far comprises a number of Ne, Na and Mg isotopes ($^{30,\,32}$Ne, $^{30-33}$Na, $^{31-34,\,36}$Mg). The energy of the first excited states of even–even Ne,

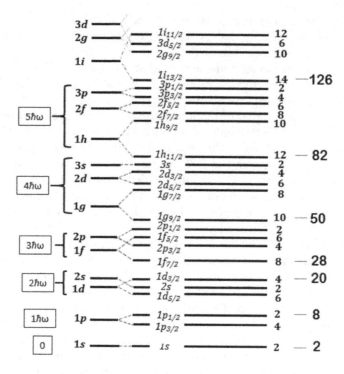

FIGURE 1.6
Energy levels in a three-dimensional isotropic harmonic oscillator potential with spin-orbit coupling representing a conventional shell model ordering of energy levels.

Mg and Si isotopes are shown in Figure 1.7. The low values of $E(2^+)$ for $N = 20$ Ne and Mg isotopes clearly show the vanishing of the major shell gap at $N = 20$ and that the ground states of these isotopes are deformed.

Shell model calculations can explain the behavior of n-rich nuclei in the island of inversion only if the fp shell is taken into account in addition to the sd shell. The calculations show that the vanishing of the $N = 20$ gap is not so much due to the lowering of the $f_{7/2}$ orbital, reducing the energy gap between the $f_{7/2}$ and the $d_{3/2}$ orbitals, but due to the 2p–2h intruder configurations $\nu(sd)^{-2}(fp)^2$ (Figure 1.8) that reduce to very low energy to form deformed ground states in these nuclei. The deformed ground state provides an explanation for the observed low excitation energy (<1 MeV) of the first 2^+ state in the even–even isotopes lying within the island of inversion.

Regarding the next higher ($N = 28$) shell gap, the most n-rich doubly magic stable nucleus, ^{48}Ca, was found to have large gaps consistent with the expectation. This is expected since ^{48}Ca has about 50 times more natural abundance than ^{46}Ca which has two neutrons less. However, as protons are being removed from ^{48}Ca, resulting in n-rich nuclei, experimental evidence suggests substantial weakening of the shell structure in ^{46}Ar (Sch 96) and ^{44}S nuclei (Gla 97). ^{46}Ar shows a signature of collective vibration and ^{44}S has a prolate-spherical shape coexistence. There are conflicting experimental findings (Frid 05; Bas 07) regarding the magic status of ^{42}Si ($Z = 14$; $N = 28$)—whether the nucleus is spherical or deformed. However, the present consensus is in favor of a well deformed oblate shape. The more exotic ^{40}Mg, however, seems to have a prolate ground state. Overall, it is clear that the shell gap is reduced for n-rich nuclei with $N = 28$ and that the magicity tends to collapse for very exotic species. Moving still higher, evidence for shell closures for ^{68}Ni (Sor 02), ^{78}Ni

Rare Isotope Beams—The Scientific Motivation

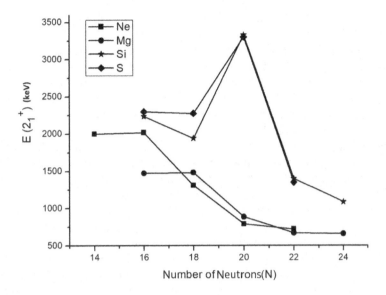

FIGURE 1.7
The energy of the first 2+ excited states E (2₁+) in even–even nuclei of Ne, Mg, Si, S as a function of neutron number (N).

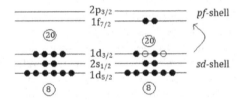

FIGURE 1.8
The neutron occupation of orbitals for N = 20 corresponding to spherical shape (left) and deformed shape due to 2p–2h intruder configuration (right).

and ^{132}Sn is observed. It is interesting to note that ^{68}Ni corresponds to N = 40 harmonic oscillator shell closure without the spin-orbit potential. However, the shell gap in ^{68}Ni as evidenced from rather high E (2+) and small B (E2) values is not supported by the observed 2n separation energies. Also, the gap erodes very fast if two protons are added or taken away since both ^{70}Zn and ^{66}Fe have small E (2+) values: 885 and 573 keV respectively. Both the doubly magic nuclei ^{78}Ni (Z = 28 and N = 50) and ^{132}Sn (Z = 50 and N = 82) are found to be particle-stable with ^{132}Sn showing all the experimental signatures of a doubly magic nucleus (Jon 10) with energy of the first excited state E (2+) exceeding 4 MeV. The magicity of ^{78}Ni has also been claimed in a recent experiment (Taniu 19) where the first excited 2+ state was found to be at around 2.6 MeV along with a second low-lying 2+ state originating from a competing deformed configuration suggesting coexistence of spherical and deformed configurations. More n-rich nuclei, however, are predicted to be deformed.

Along with the disappearance of the conventional magic numbers, there is also enough evidence for new shell closures at N = 16, 32 and 34. The drip line nucleus ^{24}O with N = 16 indeed behaves like a doubly magic nucleus with its first excited 2+ state at 4.65 MeV. The N = 16 can become a new magic number only if all the valance neutrons outside the $d_{5/2}$ shell occupy the $2s_{1/2}$ orbital and there is a large energy gap between $1d_{3/2}$ and the $2s_{1/2}$

orbital. Experimentally this was found to be the case (Kan 09), establishing the new magic number at $N = 16$. The appearance of new magic numbers at $N = 32$ and 34 was inferred/suggested based on studies on ^{52}Ca (Hu 85), ^{54}Ca (Step 13) and ^{52}Ar (Liu 19).

It can be said that experiments on exotic nuclei have clearly established that the shell evolves differently (and in a quite local manner) in the case of exotic nuclei. Theoretically, this evolution has been traced to the effects of central, spin-orbit, tensor and three-body forces, pairing interaction and change in the forces due to coupling with the continuum. For example, it has been shown that understanding of shell evolution away from stability needs the inclusion of two-body tensor forces (Ots 05) that can affect the spin-orbit splitting and this combination of central and tensor forces can explain/predict in quite a general way the shell evolutions for lighter as well as heavier exotic nuclei. It is also shown that in addition to tensor forces, three-body forces also play an important role in exotic nuclei and it is the three-body forces that can provide an explanation for the observation of ^{24}O as the neutron drip line nucleus (Ots 13) for oxygen isotopes, which is unexpectedly close to stability when compared to the drip line isotopes (^{23}N and ^{31}F) of its neighboring elements.

However, a proper understanding of shell evolution far away from β-stability is yet to emerge and a wealth of new experimental inputs is needed, especially for heavier exotic nuclei where the available data is only sparse. This understanding, it should be stressed again, is not only important to understand the atomic nuclei in its iso-spin dimension but is also important to chalk out exactly the nucleo-synthesis routes in explosive stellar environments. One of the major factors on which these new experimental inputs critically depend is the intensity of the primary beam from the primary or driver accelerator which decides whether or not it would be possible to produce these exotic nuclei in sufficient numbers necessary for these studies. It is likely that a good number of these new exotic nuclei would be amenable for study in the third-generation Projectile Fragment Separator (PFS) as well as in advanced Isotope Separator On Line (ISOL) type RIB facilities which promise to deliver primary beams (heavy/light ions) of several orders of magnitude higher intensity than what is available at present. In this context, it should be noted that a majority of the recent studies on exotic nuclei close to the drip line (especially those with $A > 50$), providing information on shell evolution for exotic nuclei beyond the *sd* shell, have been carried out at the RIKEN RI Beam Factory, which is at present the only third-generation PFS-type RIB facility delivering the highest intensity (and continuously pushing the intensity frontier through upgrades) of 345 MeV/u stable heavy ion beams of several isotopes such as ^{48}Ca, ^{78}Kr, ^{238}U, etc. (Kam 19).

1.3 Nuclear Astrophysics: The Origin of Elements, the Stellar Evolution and the Role of RIBs

Nuclear astrophysics, as discussed here, is concerned with tracing back the origin of about 5% of visible matter (the other 95% comprises about 23% dark matter which neither emits nor reflects radiation and about 72% dark energy which is even more mysterious and is thought to be responsible for the accelerated expansion of the universe) in the cosmos: how and where these are produced. It was established long ago that nuclear fusion is the source of energy in stars, including our sun, which makes the stars shine (Bethe 39). The subsequent observations related to the solar abundance of various elements and observation of technetium, which has no stable isotope, in red giants clearly established that most of the elements, especially the heavier ones, have their origin in stars rather than in the Big Bang. Of course, stars need seed materials for production of energy in exothermic nuclear reactions, which is provided by the big bang nucleo-synthesis (BBN). A wide variety of nuclear reactions are continuously taking place in stellar interiors and in explosive stellar scenarios involving, depending upon the stellar site, both β-stable and short-lived exotic nuclei and thus these nuclei play a key role in stellar evolution and the associated nucleo-synthesis (Burb 57, Cam 57; books: Rolf 88, Boyd 08). In fact, the stellar evolution and nucleo-synthesis are inseparably intertwined, and both need to be understood as a whole and not in part. The spectacular advances in experimental sciences in recent times have allowed the study, using ground- as well as space-based observatories, of a wide variety of stellar objects, starting from normal main sequence stars like our sun, to red giants and white dwarfs that are of much higher and lower luminosities respectively when compared to main sequence stars, to most violent stellar objects like core collapse supernovae, novae and x-ray bursters, and type-1a supernovae. In order to understand this huge body of observational facts revealing the dynamics of these stellar objects, as well as the element synthesis in them, a detailed knowledge of properties of a huge number (a few thousand) of nuclei is crucial. The nuclear properties that need to be known are also many and include masses, β-decay half-lives, β-delayed fission and particle emission probabilities, level densities, neutron and proton radiative capture rates, neutron skin thickness and the existence of di-pole resonances at low excitations. Nuclear astrophysics deals with these nuclear inputs. It should be noted that it is the short-lived exotic nuclei that play a key role in the dynamics of violent stellar objects and the associated nucleo-synthesis that takes place in a timescale of 1 to 100 s, and the study of these nuclei is one of the main objectives of any RIB facility.

1.3.1 Primordial or Big Bang Nucleo-Synthesis

It may be worthwhile to outline briefly our present understanding about the origin of elements in our universe. Nucleo-synthesis started a minute after the big bang that happened roughly 13.7 billion years ago creating the universe, which has been expanding since then. After the creation, the extremely hot and dense universe started expanding and consequently, cooling down. The evolution of the universe during its very early period from about 10^{-43} s to about 1 µs is a topic of current research and therefore not exactly known. However, scientists agree, more or less, that during this period the universe cooled down from a temperature of about 10^{19} GeV to about 1 GeV (1 GeV = 10^{13} K); the hot and dense primordial soup contained only fundamental particles, the leptons and the quarks, not distinguishable from one and another initially since the strong and electro-weak interactions were unified into a single interaction. Also, during this period matter and anti-matter particles were continually created and annihilated and were equal in numbers. But as the universe cooled down, first the strong interaction was decoupled and then when the temperature decreased further the electro-weak interaction became separated out into electromagnetic and weak interactions. Also, at some point in time during the cooling down process, it is conjectured that the CP symmetry or the matter–antimatter symmetry was violated to the extent that it resulted in a universe where for every billion particle–antiparticle pairs there was one extra matter particle. CP violation to this extent, although its origin is not known, offers a possible explanation to the mystery of why our universe contains matter and not just radiation and that the experimental values for the baryon to photon ratio in the universe lie in the range 10^{-9} to 10^{-10}.

Hadronic matter existed in the form of de-confined quarks and gluons (Quark Gluon Plasma, QGP) until about 10 μs. When the temperature of the universe dropped below about 100 MeV (10^{12} K), a phase transition from QGP to hadrons took place and quarks were confined, forming the basic building blocks of atomic nuclei—the neutrons and the protons. This was also the beginning of the hadron–lepton era with primordial plasma soup now comprising of hadrons, 10^9 to 10^{10} times as many photons and numerous leptons. As the universe cooled further to a temperature of about 1 MeV, the neutrinos and anti-neutrinos could no longer interact to convert neutrons into protons or vice-versa through inverse β-decay reactions and they ceased to participate in the further evolution of the universe. However, at this temperature the number of high energy photons sitting at the tail of the Maxwell distribution was still high enough to produce electron–positron pairs and protons (neutrons) could still be converted into neutrons (protons) through charged-current weak interaction. As the universe cooled further to about 10^9 K, the pair production became insignificant and neutron to proton ratio froze out with about 13% neutrons and 87% protons.

The difference between the quantities of neutrons and protons is a consequence of the Boltzmann distribution and because a neutron is heavier than a proton by about 1.293 MeV making it energetically easier to convert into protons in interaction with leptons as the temperature drops below about 1 MeV. At temperatures below ~100 keV (10^9 K), deuteron became stable against photodisintegration and the nucleo-synthesis in the universe started converting protons and neutrons into deuterons. This started roughly 1 min after the big bang and could go on only up to about 20 min since free neutrons decay with a half-life of 10.3 min. The absence of any stable nucleus with a mass number of $A = 5$ mainly constrained the nucleo-synthesis up to helium (^{4}He), produced through a sequence of fusion reactions, given in Table 1.1. After about 20 min the universe was left with a huge number of protons and helium was about 6% of the total number of protons (apart from ^{2}H, ^{3}H and ^{3}He that are produced in much lower amounts, down by more than four orders of magnitude), resulting in a mass fraction of about 0.24. Experimental observations on the abundances of helium in extremely metal-poor old stars, which are more likely to have element distributions more closely representing the big bang nucleo-synthesis, result in values of relative mass fraction very close 0.24, providing strong evidence in support of the big bang and the nucleo-synthesis that happened during the first few minutes. Despite the gap at $A = 5$, the big bang nucleo-synthesis also produced some heavier elements: Li, Be and B in trace quantities. The synthesis of elements during the first 20 min or so after the creation of the universe is often referred to as Big Bang Nucleo-synthesis (BBN). The rest of the elements in the universe are made in the stars formed out of the seed materials (primarily protons and helium nuclei) left after BBN.

TABLE 1.1

Big Bang Nucleo-Synthesis up to ^{4}He

$p + n \rightarrow d + \gamma$

$d + p \rightarrow {}^3He + \gamma$

${}^3He + n \rightarrow {}^4He + \gamma$ or ${}^3He + d \rightarrow {}^4He + p$

${}^3He^3 + n \leftrightarrow {}^3H + p$

$d + n \rightarrow {}^3H + \gamma$

${}^3He + p \rightarrow {}^4He + \gamma$ or ${}^3He + d \rightarrow {}^4He + n$

For stars to form, the universe needed to first cool down to a temperature much below 13.6 eV, the ionization energy of hydrogen, to allow formation of neutral hydrogen atoms by the electron capture of protons. Once the photons, even in the high-energy tail of the Maxwell distribution, had too little energy to scatter electrons, the radiation was decoupled from matter. Since the universe became transparent to photons, the photons would continue with energy distribution at the time of the decoupling. As the universe continued to expand the photons became stretched (or red shifted) by many orders of magnitude to become 2.7 K Cosmic Microwave Background Radiation (CMBR) today. The radiation-matter decoupling took place roughly 380,000 years after the big bang. The stars came into existence from hydrogen nebula much later, about 200 million years after the big bang and galaxies came into existence later, after about 1 billion years. CMBR has no direct contribution to nucleo-synthesis but fluctuations of the order of 1 in 10^5 observed in the CMBR spectrum (using a Cosmic Background Explorer (COBE) and then a Wilkinson Microwave Anisotropy Probe, (WMAP)) is a consequence of density fluctuations in the early universe which offers an explanation for the formation of stars and galaxies.

1.3.2 Nucleo-Synthesis in Stars up to Iron

As already mentioned, the BBN mainly produced hydrogen and helium and heavier elements like Li, Be and B in trace quantities. All other heavy elements found on Earth, in the Sun, meteorites and in other stellar objects must therefore have been synthesized in stars in nuclear reactions. Any theory on nucleo-synthesis and the stellar environments in which it took place leading to the formation of heavier elements should be able to explain the observed abundance of elements in the universe. The abundance, however, could be measured most exhaustively (both in quality and quantity of the data) only for our solar system and the sources comprise of analysis of spectroscopic measurements of the Sun, analysis of terrestrial rock samples, lunar samples and meteorites. It should be mentioned that cosmic abundance is not the same as solar abundance, although the two do not appear to be very different. This is because our sun is from a later generation of stars born out of a supernova explosion. A cursory look at the observed relative abundances in our solar system reveals a number of key features:

(a) Relative abundance of elements varies over 12 orders of magnitude from hydrogen (by far the most abundant element in the universe) to uranium, which has natural abundance of about one trillionth of that of hydrogen. After hydrogen the most abundant element is helium, which has roughly 1/15 of the abundance of hydrogen. Thus, for hydrogen and helium, relative abundance has changed only a little since the big bang. Together, hydrogen and helium account for more than 98% of visible matter in the universe and all the heavy elements together contribute to the remaining 2%.

(b) The abundance decreases with atomic number: steeply up to $Z \sim 35$ and then gradually. There is a strong odd–even fluctuation in Z; the yields of even Z elements are always higher than the odd Z neighbor.

(c) The elements Li, Be and B have abnormally lower (orders of magnitude) abundance than their neighbors.

(d) There is a strong abundance peak for iron in which elements from Mn to Ni contribute.

(e) The abundance as a function of atomic number A reveals a number of additional interesting trends such as a higher abundance for light "α nuclei" up to ^{40}Ca and broad (sharp) peaks at $A = 80$ and (88); $A = 130$ and (138) and $A = 195$ and (208). The solar abundance of elements from around iron and above is shown in Figure 1.9.

In stars, gravitational collapse is balanced by radiation pressure and the source of radiation is the fusion of light elements forming heavier elements. Fusion of two light nuclei has positive Q value only up to a certain extent when binding energy per nucleon reaches the maximum. Beyond $A \sim 60$, the binding energy starts decreasing and thus fusion cannot prevent gravitational collapse once ^{56}Fe is formed inside the stellar interior. The high abundance of ^{56}Fe is a consequence of the dependence of binding energy per nucleon on the mass number. It should be noted that for fusion of two nuclei, the Coulomb Barrier (CB) must be overcome. The CB for fusion of, say, two helium nuclei is around 1.26 MeV, which corresponds to a temperature of 1.26×10^{10} K. Such high temperatures are not reached even in explosive stellar environments. In fact, the temperature in the stellar interior lies typically between 10^7, in normal stars like our sun, to a few times 10^9 K in explosive events (like supernovae). Thus, fusion in the case of ^{4}He nuclei can only take place far out in the high energy tail of the Maxwell energy distribution. A small fraction of He nuclei with sufficiently high energy can therefore participate in the fusion reaction. The reaction rate is decided by the product of two exponentials: one rapidly falling tail of the number of nuclei with the right energy for fusion and the other a rapidly increasing barrier penetration probability. The product gives rise to a broad peak in the reaction probability with the peak energy many times higher than kT, where k is the Boltzmann constant. This peak is called the Gamow peak (Figure 1.10) which decides the effective temperature zone for fusion in stars, which lies below the CB. The Gamow peak is given

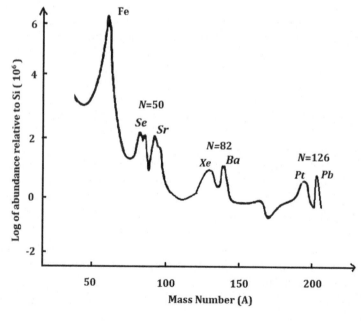

FIGURE 1.9
The solar abundance (relative to silicon) as a function of mass number for elements around iron and heavier elements up to lead.

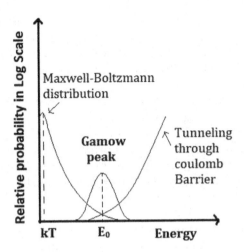

FIGURE 1.10
The Gamow peak resulting from the product of two exponentials. The charged particle capture reactions below the Coulomb barrier are mainly confined in the narrow energy window around E_0.

TABLE 1.2

Hydrostatic Burning Stages in a Massive Star Leading to ^{56}Fe

---

H burning: $4p \rightarrow {}^4He + 2e^+ + 2\nu + 26.7$ MeV

He burning: $\alpha + \alpha \rightarrow {}^8Be$;

$^8Be + \alpha \rightarrow {}^{12}C + \gamma + 2\nu + 7.2$ MeV

C burning: $^{12}C + {}^{12}C \rightarrow {}^{23}Na + p + 2.23$ MeV

$\rightarrow {}^{20}Ne + \alpha + 4.6$ MeV

$^{12}C(p,\gamma){}^{13}N(e^+,\nu){}^{13}C(\alpha,n){}^{16}O$

Ne burning: $^{20}Ne + {}^{20}Ne \rightarrow {}^{16}O + {}^{24}Mg + 4.5876$ MeV

O burning: $^{16}O + {}^{16}O \rightarrow {}^{31}P + p + 7.68$ MeV

$^{28}Si + \alpha + 9.59$ MeV

$^{31}S + n + 1.5$ MeV

O burning: $^{28}Si + \alpha + {}^{25}S + \gamma$

$\alpha \rightarrow {}^{56}Fe$

---

by $E_0 = (0.998 Z_1 Z_2 \, \mu^{1/2} \, kT/2)^{2/3}$, Z_1, Z_2: atomic number of the interacting charged particles and μ is the reduced mass.

In massive stars, fusion can proceed to much heavier elements compared to less massive ones since the temperature at the core is much higher (owing to the stronger gravitational pull) allowing heavier nuclei to tunnel through the CB and fuse. It is estimated that for fusion to proceed all the way up to iron-nickel, the initial mass of the star should exceed about 10 to 12 solar masses. The fusion reactions (hydrostatic burning) in a massive star leading to synthesis of iron, starting from hydrogen are given in Table 1.2 (Woos 02 and the references therein). After each burning stage, the gravitational pull first heats up the core to raise its temperature and density suitable for the next phase of burning, and it is the next phase of burning that keeps the star in hydrodynamic equilibrium until that phase of burning is over and the sequence continues. From H to Si burning, the temperature of the core increases from a few times 10^7 to a few times 10^9 K and the density from approximately a few g/cm^3 to ~10^7 g/cm^3. In this way the sequence of burning as listed in Table 1.2 leads ultimately to a Fe-Ni core.

This produces an onion skin type (also termed as concentric shell type) structure as shown in Figure 1.11. Since hydrogen burning (the p - p chain) leading to formation of ^{4}He proceeds through weak interaction, it is the slowest of all the burning processes. A star spends about 90% of its life in the hydrogen burning phase, which is more than several million years in a massive star (in the Sun the hydrogen burning phase is much longer; more than 9 billion years). The time for He burning is dictated by a cross-section of the 3 α reaction leading to ^{12}C. ^{8}Be (= $\alpha + \alpha$) is unbound but resonantly captures an α particle forming ^{12}C at its 7.654 MeV, 0$^+$ excited state, popularly known as the Hoyle state. The excited state decays mostly through 3 α emission but a small fraction decays into the ground state of ^{12}C, synthesizing carbon and is the next, longer, burning phase, generally an order of magnitude less in duration compared to that of hydrogen burning in a massive star. The time thereafter decreases sharply further by about three orders of magnitude (compared to the He burning phase) for carbon burning. The trend continues and the typical time span for Si burning is about a day.

After the formation of ^{56}Fe, fusion can no longer generate energy and huge gravitational pull results in what is called core collapse. The electron degeneracy pressure can prevent collapse but only for stars with a mass up to 1.4 times the solar mass. There is thus nothing that can stop the core collapse in a massive star until the density of the core reaches nuclear matter density (~10^{14} g/cm^3). The core collapse proceeds very fast, a typical time scale being a few seconds. The collapse increases the density and the temperature of the core and leads to the break-up of all nuclei by photodisintegration, predominantly the break-up of ^{56}Fe into 13 α particles since this takes away a lot of energy. Also, the electrons would be captured by nuclei to reduce the electron degeneracy pressure and the neutrinos created will escape carrying a lot of energy which would accelerate the collapse. The details of the core collapse are rather complex and readers may refer to excellent review articles (for example, Bethe 90, Burr 07) to find out more about the present understanding of the collapse scenario. The collapse ultimately leads to neutronization of matter, resulting in an extremely high density of free neutrons (>10^{20} neutrons/cm^3) and n-rich nuclei. As the density of the n-rich matter reaches nuclear density the collapse would come to a halt due to finite compressibility of the nuclear matter. Further compression would lead to a bounce back, which creates a shock wave that traverses outward triggering what is known as type II supernova explosions or Core Collapse Supernovae (CCS). In the explosion, except in a

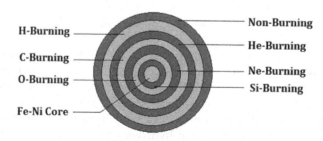

FIGURE 1.11
Onion skin/concentric shell-type structure of a pre-supernova massive star.

part of the core, most of the matter is thrown out into the interstellar medium. Eventually a new next-generation star is born from the condensation of these materials, maintaining the continuity of the evolution cycle. The new star is obviously richer in heavier elements than the first-generation star born out of the debris of the big bang. The elemental composition of a star is thus a strong indicator of whether it is a first-generation star formed out of debris of the big bang or a second- or higher-generation star born out of supernova explosions. The portion of the core that is not ejected cools down to form a neutron star or sometimes a black hole in the case of more massive stars. The amount of energy released during CCS explosions is in the form of energetic neutrinos and is enormous, about 10^{53} ergs. It is believed that the elements Na, Mg, Si and Ca are mainly synthesized in CCS. Also, CCS is considered to be a suitable site for the rapid neutron capture (r) process nucleo-synthesis that produces about half of the elements beyond iron.

1.3.3 Synthesis of Elements Heavier Than Iron: The S, R and P Processes

The nuclei that are heavier than iron are mostly produced in neutron capture processes initiated by capture of neutrons by seed nuclei already present in the star. The production of heavier nuclei in the fusion of charged particles involving two heavy ions is not possible since that requires too high a temperature in the stellar interior (because of the Coulomb barrier) which would lead to destruction of the seed nuclei. Two types of neutron capture processes are needed to explain the abundance of most of the heavy elements and their relatively more abundant isotopes. These two processes were identified long ago in the seminal works of M. Burbidge, G. Burbidge, W. Fowler and F. Hoyle, working together and independently by A. Cameron (Burb 57, Cam 57). One is a slow neutron capture process, called the s process, where s means slow and the other a rapid neutron capture process, abbreviated similarly as the r process. The term slow or rapid requires some qualification. If the neutron capture rate is slower than the β-decay half-lives of the nuclei that lie in the path of the process, it is termed as slow; in the opposite case it is termed as rapid. Thus, in the s process the β-decay always takes place first, creating the next higher Z element and restraining the process to follow a path very close to the stability line (Figures 1.12 and 1.13). In the r process (Figures 1.12 and 1.14), on the contrary, neutron capture proceeds very fast resulting in successive neutron captures producing very exotic n-rich nuclei and the r process path thus lies much further away from the β-stability valley, typically 10 to 30 neutrons more than the number of neutrons in the most β-stable isotopes (Figure 1.14). Such neutron capture rates require very high neutron flux which is only possible for a brief period (~1 s) during explosive stellar scenarios. Once the neutron flux falls below a threshold the r process stops, and the unstable n-rich nuclei produced in the process undergoes successive β decays to produce stable isotopes.

1.3.3.1 The S Process

The slow neutron-capture s process accounts for synthesis of nearly half (the other half being accounted for by the r process) of the elements and relatively abundant isotopes heavier than iron. Readers may refer to review articles by Kapp 11 and Reif 14 and references therein. The s process needs seed nuclei and ^{56}Fe is the main seed nucleus where the s process begins. If the star contains nuclei heavier than iron (from a previous star where the s process produced them), then those nuclei would also act as seed nuclei but in view of the high abundance of iron, ^{56}Fe remains the main seed nucleus for the s process. Also, the s process, being a slow process, cannot continue nucleo-synthesis beyond ^{209}Bi because

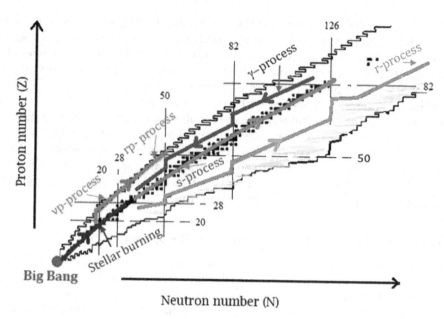

FIGURE 1.12
The various nucleo-synthesis processes in stars under the normal and explosive scenarios and their approximate paths. The r process path is shaded up to the neutron drip line since the path might extend up to the neutron drip line depending upon the actual conditions (temperature, neutron flux and matter densities) prevailing in the explosive stellar environments.

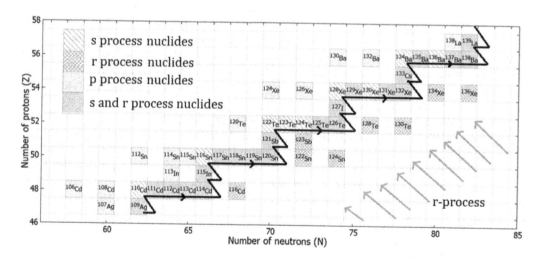

FIGURE 1.13
Typical s process nucleo-synthesis path for elements from Ag to La. The isotopes that are produced in only s, only r, both by s and r and p processes are shaded.

thereafter neutron capture would result in much shorter-lived nuclei compared to the capture rate. Since the s process needs seed nuclei, it can only occur in a second or later generation star that has formed out of interstellar matter that contained debris of a star which had evolved earlier (for example, the debris of a supernova explosion), where elements up to iron would already be present during the star's hydrogen and helium burning stages.

Rare Isotope Beams—The Scientific Motivation

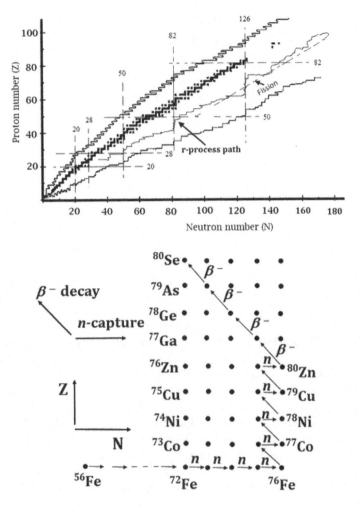

FIGURE 1.14
(Top) Simulated r process nucleo-synthesis path showing fission recycling that replenishes the environment with mid mass (A ~ 125) nuclei. (Bottom) A portion of the r process path expanded to show a tentative chain of reactions leading to production of ^{80}Se from ^{56}Fe as the seed nucleus.

Observationally, it is the asymptotic giant branch (AGB) stars that are identified to be the site for the s process. The s process seems to take place in the helium burning shell of stars or more precisely in the He-rich inter-shell region separating the He and H burning shells, at a temperature ~1.5 to 3.5×10^8 K. The neutron density at which the s process synthesis can take place is about a few times 10^8 neutrons/cm^3. However, even up to a neutron density of 10^{13}/cm^3, the s process more or less proceeds along the line of β stability. It should be obvious that for synthesis to proceed through neutron capture the presence of seed alone is not enough; exothermic nuclear reactions that would be capable of supplying enough neutrons for the s process also need to be identified. The most probable source of neutrons is a ^{13}C (α, n) ^{16}O reaction that needs a temperature of about 10^8 K. The ^{13}C becomes available for the reaction because of the CNO cycle in the previous hydrogen burning phase. It should be mentioned that it is the CNO cycle that becomes the predominant process in the hydrogen burning phase in the second- and later-generation stars. The end result of CNO

TABLE 1.3

Hydrogen Burning in Second and Later Generation Stars—The CNO Cycle

---

$^{12}C + p \rightarrow {}^{13}N + \gamma + 1.944$ MeV

$^{13}N \rightarrow {}^{13}C + e^+ + \nu_e + 2.220$ MeV

$^{13}C + p \rightarrow {}^{14}N + \gamma + 7.551$ MeV

$^{14}N + p \rightarrow {}^{15}O + \gamma + 7.297$ MeV

$^{15}O \rightarrow {}^{15}N + e^+ + \nu_e + 2.754$ MeV

$^{15}N + p \rightarrow {}^{15}C + \alpha + 4.965$ MeV

---

cycle is the same as with the *p–p* chain: the conversion of four protons into an α particle along with two positrons, two neutrinos and an energy release of 26.7 MeV (Table 1.3). Another reaction ^{22}Ne (α, n) ^{25}Mg becomes an important source of neutrons in larger AGB stars with a mass exceeding three solar masses. This is because this reaction requires a temperature of at least 3×10^8 K. Once again, the source of ^{22}Ne is ^{14}N (the main product of the CNO cycle) through the reaction sequence ^{14}N (α, γ) ^{18}F (β, ν) ^{18}O (α, γ) ^{22}Ne, that is two α captures and one β$^+$ decay. The s process in AGB stars can explain the observed abundance of elements heavier than $A \sim 90$ quite well. However, for lower masses ($56 < A < 90$), the s process in low mass AGB stars underestimates the observed abundances. Additional contribution from a second, rather weak s process operating in massive stars, (which ultimately ends in a type II supernova explosion) during the He burning phase (and later during C burning) can reproduce the observed abundances (Arn 85; Rai 91).

The s process path for a few nuclei is shown in Figure 1.13. It can be seen that mostly, the stable isotopes and the unstable isotopes which have one neutron more than the last stable isotope participate in the s process. Since the neutron capture cross-sections are very small at the shell closures, some accumulation occurs at the magic numbers corresponding to $N = 50$, 82 and 126 leading to rather sharp abundance peaks at $A \sim 88$ (Sr), 138 (Ba) and 208 (Pb).

As already mentioned, the s process is terminated at ^{209}Bi, since beyond that *n*-capture leads to isotopes that are much shorter lived compared to the typical neutron capture rates in the s process. Thus, to explain the existence of Th and U (having seven and nine more protons respectively than Bi), the existence of the r process nucleo-synthesis must be assumed, which should be a fast neutron capture process with a time scale of ~1 s or less, shorter than the half-lives of intermediate β (also α) unstable short-lived nuclei. The short timescale would allow the process to run through these intermediate isotopes by successive capture of a large number of neutrons to produce ^{232}Th and ^{234,235,238}U.

1.3.3.2 The R Process

Theoretical estimations show that unlike the s process which needs only moderate neutron densities, the r process needs an environment of very high neutron densities exceeding about 10^{20}/cm^3 so that neutron capture can proceed much faster than β decay. Such high neutron fluxes, however, could only be available for a brief period (approximately a few seconds) in explosive scenarios such as CCS and the merger of neutron stars and are always accompanied by a high temperature exceeding about 10^9 K. Once the neutron density decreases by a few orders of magnitude, say, below $10^{18}/cm^3$ or the matter density decreases beyond a critical value, the r process stops. The highly n-rich unstable nuclei that are produced in the r process undergo thereafter successive β decays producing a large number of β-stable isotopes, especially the n-rich stable isotopes of heavier elements including those of most precious metals Pt, Au and Ag.

The r process path (and hence the abundance) is believed to be decided, either through equilibrium between neutron capture and the inverse photo-disintegration reactions, that is equilibrium between $(n, \gamma) \leftrightarrow (\gamma, n)$ reactions (the equilibrium is assumed to be reached instantly, much faster than the capture or β-decay rates) as in the case of a hot r process or for a cold r process through the interplay of neutron capture and β-decay rates. In the former case, when the temperature is high enough for a hot r process to be operative, the path is mainly decided not by the neutron capture rates (assuming capture rates to be much faster than β-decay rates) but by neutron separation energies. The successive neutron captures of the seed nuclei would produce more and more n-rich nuclei along an isotopic chain and with each neutron capture the neutron separation energy decreases on average. In the hot stellar environment, the inverse process, photodisintegration, would become more and more significant as the neutron separation energy decreases (there will be enough high energy gamma photons at the Maxwellian tail at temperature of ~10^9 K to induce photodisintegration as S_n decreases to a value of <4 MeV). Finally, when the separation energy for a n-rich nucleus falls to a value of ~2 MeV, the inverse photodisintegration reaction rate would become equal to the neutron capture rate and subsequent neutron capture has to await β-decay.

The β-decay, considered as a leakage from the equilibrium, would produce a less exotic isotope with higher neutron separation energy and the capture can proceed again till a new equilibrium point is reached for an isotopic chain with a higher atomic number. The equilibrium for each isotopic chain is governed by the Saha equation and is dependent only on neutron separation energy, neutron density and temperature (assuming these remain constant during the r process). So, for any isotopic chain the r process would proceed along a path of constant neutron separation energy (S_n ~ 2 MeV) and there will be accumulation of matter at points where $(n, \gamma) \leftrightarrow (\gamma, n)$ equilibrium is reached. Since β-decays are much slower than the capture rates, these points are termed as "waiting points" and the β-decay rates of these waiting point nuclei decide the abundance pattern in the hot r process nucleo-synthesis. Neutron capture rates are, however, important for the cold r process which operates at lower temperature where photodisintegration does not play a significant role. In this case, the neutron capture reaction rates and the β-decay rates together decide the r process path and consequently the abundance pattern.

The r processes are thought to proceed in dynamic environments created in the explosive stellar scenarios. In such an environment the neutron density and temperature change very fast (in a time much shorter than 1 s) during the period that the r process operates. Thus, during its course the r process will change its path to a different constant neutron separation energy path from about S_n = 4 MeV to about 1 MeV and for very high neutron flux ($10^{30}/cm^3$ or so) and matter density S_n may even be equal to zero (neutron drip line) and the r process path proceeds along the neutron drip line. The r process nucleo-synthesis path is therefore often represented (in the N–Z chart) by a band extending from about midway between the stability and the drip line, to all the way up to the drip line (Figure 1.12).

It should be noted that both in the cold and hot r processes there will be accumulation of matter far in the n-rich side for closed neutron shells at N = 50, 82 and 126. This is because at closed neutron shells the neutron separation energies of the isotope resulting from capture of a neutron are small and the r process will have to move for quite a while along these magic neutron numbers (which has relatively longer β-decay half-lives ranging from a few tens of ms to a few hundreds of ms), resulting in a piling up of abundance. Also, the neutron capture cross-sections are much smaller for closed-shell nuclei. Thus, these closed-shell n-rich nuclei represent a special set of waiting points. The piling up results in the observed solar abundance peaks at $A \sim 80$ (elements Se, Br, Kr), 130 (elements Te, Xe) and 195 (elements Os, Pt, Au). These peaks, however, appear at lower mass numbers compared to those in the s process since the r process runs through n-rich nuclei and therefore the neutron shell closures are reached for much smaller proton numbers summing up to lesser atomic numbers. There is, however, some doubt whether the peak at $A \sim 80$ is purely from the r process since it is not well separated from the abundance of lighter nuclei but the other two peaks at $A \sim 130$ and 195 are considered as prominent r process abundance peaks. It should be noted that these abundance peaks cannot be explained by any other nucleo-synthesis process. It is of course not known whether the shell gaps or the magic numbers will remain the same for very n-rich nuclei or changes their character as has been observed experimentally for lighter n-rich nuclei. In fact, simulations tend to indicate that the observed r process abundances are better reproduced if the shell gaps are assumed to be quenched rather than pronounced for n-rich nuclei participating in the r process synthesis (Pfe 97; Dil 03). Once again, a clear picture will emerge only after some of these n-rich nuclei have been subjected to experimental studies in the third-generation RIB facilities and we have more reliable theories to predict masses, reaction rates, etc.

The r process is responsible for synthesis of about 50% of the relatively abundant isotopes heavier than iron that are not synthesized in the s process. Since heavy elements between Fe and Bi are synthesized both in the s and r processes, the r process abundance needs to be determined after subtracting the s process abundance (which can be more reliably calculated since it involves stable and long-lived isotopes whose properties are well known) from the total observed abundance. There are, however, a number of stable isotopes which are purely produced in the s process since those are shielded from the r process β decay by stable isobars; for example, ^{128,130}Xe ($Z = 54$) are pure s process isotopes that are shielded by stable ^{128,130}Te ($Z = 52$). This also means that ^{128,130}Te are purely r process isotopes. There are, in fact, many such pure r process even–even isotopes (for example, ^{82}Se, ^{96}Zr, ^{110}Pd, ^{116}Cd, ^{122,124}Sn, etc.) apart from all the isotopes beyond ^{209}Bi. The abundances of these isotopes become very useful to test r process theoretical predictions.

Unlike the s process, the r process is primary in nature since it does not depend upon the presence of any particular seed material like ^{56}Fe. This is borne out by the observed fact that the nature of the heavy-element abundance (or the ratios of abundances) in very old stars of low metallicity, in which the s process hardly contributes to element synthesis, quite closely match the solar r process abundance pattern (Cow 97). The observation also means that the r process became operative in the cosmos much earlier than the s process.

The termination of the r process is a topic of great interest. There is sufficient indication that the r process proceeds beyond uranium. The existence of ^{244}Pu (half-life $\sim 8 \times 10^7$ y) in the early solar system (the present age of the solar system is $\sim 4.5 \times 10^9$ y), as evidenced from the observation of its decay products in meteorites, and production of einsteinium ($Z = 99$) and fermium ($Z = 100$) in terrestrial nuclear explosions (that create a high neutron density environment necessary for the r process) clearly provide support for the r process nucleo-synthesis proceeding much beyond U. Under some explosive stellar scenarios, it might even be possible for the r process to continue up to the predicted super-heavy island around $Z \sim 114$ and $N \sim 184$ (Pan 16). However, since spontaneous and neutron-induced fission probability increases as nuclei become heavier and heavier, it is generally taken that the r process can reach roughly up to $A \sim 270$ beyond which it undergoes fission (Figure 1.14). The fission would result in replenishing the environment with lighter mid-mass nuclei ($A \sim 125$) which would then participate in the r process nucleo-synthesis leading to an increased production of heavier (> 125 or so) species. This process, termed fission recycling, is included in r process calculations but it lacks accurate information on fission barrier heights and the fragment distribution.

The CCS and the neutron star mergers have long been considered to be the potential sites for the r process nucleo-synthesis. The simulation studies indicate that the observed r process abundances cannot be explained by assuming a single site of constant temperature and neutron density (Kr 93). Thus, a site where conditions can change very fast during the period that the r process is operating is needed. Both CCS and neutron star mergers can present such dynamic scenarios and have been considered as potential candidates to host the r process. The CCS has long been considered to be a favored r process site but comparatively recent simulations (Arc 07, Mar-Pin 12) tend to show that it is difficult to achieve the full-scale r process in CCS. The alternate scenarios involving ejection of matter from a neutron star due to tidal disruption created by a companion black hole (Latt 74; Latt 76, Sym 82) or due to merger of two neutron stars (Eic 89; Fre 99) gained prominence because of their natural n-rich environment (crust of a neutron star contains n-rich matter and free neutrons) that has at least no problem with the existence of high neutron density required for the r process. The r process nucleo-synthesis apparently takes place in the material ejected during the last seconds of the merger that contains the seed nuclei, which rapidly capture neutrons available in the high-density free-neutron environment. However, these events being infrequent, doubts have been raised whether such mergers can explain the observed cosmic abundance of r process material, which in turn depends on the rate of mergers and on the amount of r process materials synthesized per merger. However, simulations have shown that the merger of two neutron stars could explain cosmic abundance of r process material (Eic 89; Fre 99).

The detection of gravitational waves in the merger of two neutron stars (GW 170817) at LIGO/VIRGO collaboration (LIGO and VIRGO 17) on August 17, 2017 was a landmark event for two important reasons: first, because it detected gravitational waves for the first time, and second, because it led to confirmation of at least one site where the r process indeed operates. The gravitational wave data could specify the location of the event and the mass of the binary system that produced it. Subsequent optical/spectroscopic measurements of the event for several weeks by different groups (the observations led to quite a large number of publications, all published in 2017; refer the article of Hotokezaka et.al, (Hoto 18) for a comprehensive review that also contains all the original references), using ground as well as space based telescopes found the spectroscopic evidences for the presence of r process elements in the material ejected as a result of the merger. The long duration of the afterglow observed for several weeks could only be due to heating provided by radioactive decay in the r process synthesis at the waiting points as was predicted earlier in the kilonova (the kilonova refers to a short-lived near infrared weak supernova-like signal originated from the heating produced by radioactive decay of n-rich nuclei in the ejected material) model (Li 98, Met 10). The afterglow is found to have two components. One is a hot blue component resulting from decay of light r process elements with $A < 140$; this component decayed within a few days. The other one is a cooler red component lasting much longer than the blue component. This component is thought to have resulted from decay of heavier ($A > 140$) r process elements, especially the lanthanides. The presence of lanthanides, because of their complex atomic structure, results in an increase in the opacity of the stellar environment by more than an order of magnitude. The opacity delays the peak time and consequently shifts the color of the peak towards red. The opacity due to the presence of lanthanides in the ejected material and its effect on the electromagnetic spectrum has also been predicted earlier (Kas 13; Bar 13). Thus, the spectroscopic observations following binary neutron star merger GW 170817 were all in line with the earlier theoretical predictions, confirming the neutron star merger as a site for the *r* process. It is estimated that the *r* process matter amounting to ~1/20 of solar mass was synthesized in the merger. So, despite the merger event being rare, the yields of the *r* process matter per event are sufficient for such events to contribute in a major way to the observed cosmic abundance of *r* process elements (Hoto 18).

It is estimated that about 5000 nuclei, all the way from the stability to the neutron drip line, participate in the *r* process nucleo-synthesis. Any *r* process network calculation therefore needs accurate information of their masses, neutron capture rates, β-decay half-lives, β-delayed neutron emission probability, and as the *r* process reaches the actinide region the fission barriers, the *n*-induced, spontaneous and β-delayed fission probabilities, and the fission-fragment distributions. The majority of these participant nuclei, being heavy (*A* > 80) and very *n*-rich, are not, so far, amenable to experimental study. We have only either insufficient or no information on them. Sensitivity studies show that the *r* process abundance simulations are highly sensitive to these inputs; for example, in the case of neutron star mergers nuclear neutron capture rates within a factor of 2 or so (Lid 16) need to be known so that the validity of a particular model or the astrophysical conditions it assumes can be checked. The situation is the same for other inputs as well. One, therefore, looks forward to third-generation RIB facilities to produce enough data on these exotic *n*-rich nuclei for the development of reliable theoretical models to predict these properties. The less the uncertainties of the input nuclear data, the more would be the constraint on the astrophysical models dealing with the evolution of these explosive scenarios. This would help us to understand these scenarios much better and to explore if there are other stellar sites where the *r* process can operate in part or in full.

1.3.3.3 The P Process Nucleo-Synthesis

The *s* and *r* processes together can account for synthesis of most of the isotopes heavier than iron but not for all stable naturally occurring heavy isotopes. There are about 35 neutron-deficient stable isotopes with low natural abundances listed in Table 1.4, at least 30 of which cannot be produced either by the *s* or by the *r* processes. The abundances of these isotopes are 10 to 1000 times lower than the abundances of other stable isotopes that are considered to be produced via the *s* and the *r* processes. These isotopes, in their respective isotopic chains, are the most neutron-deficient ones (e.g., ^{74}Se, ^{78}Kr, ^{84}Sr, etc.) or the last two most deficient ones (e.g., ^{92,94}Mo and ^{96,98}Ru). Apart from low abundances, the other unique feature of these isotopes is that they are shielded from production either via the *s* or *r* process. The *s* process path cannot produce these nuclides through β^- decay because of a lack of unstable isobars of lower Z elements in the isobar chain and the *r* process cannot feed all the stable isotopes through β decay because of stable isobars of a lower Z element coming in between to truncate the decay chain (Figure 1.13). These isotopes are therefore left to be fed by some other process, called the *p* process. These isotopes are thus often referred to as pure *p* nuclides. Five of the *p* nuclides (^{113}In, ^{115}Sn, ^{152}Gd, ^{164}Er and ^{180}Ta) shown in italics in Table 1.4 are not pure *p* nuclides since these can have significant contributions from the *s* process too. The *p* process actually needs to include a number of processes to explain the presence of isotopes that are heavier than iron from ^{74}Se to all the way up to ^{190}Pt and ^{196}Hg. These processes are rapid proton capture or the "*rp*" process, the gamma (γ) process and

TABLE 1.4

β-Stable Neutron-Deficient Isotopes with Low Natural Abundances. Except for the Five Isotopes Listed in Italics, All the Other Isotopes Can Only Be Produced in Stars in P Process Nucleo-Synthesis

Isotope	% Contribution	Isotope	% Contribution
^{74}Se	0.89	^{132}Ba	0.101
^{78}Kr	0.355	^{138}La	0.08881
^{84}Sr	0.56	^{136}Ce	0.185
^{92}Mo	14.53	^{138}Ce	0.251
^{94}Mo	9.15	^{144}Sm	3.07
^{96}Ru	5.54	*^{152}Gd*	0.2
^{98}Ru	1.87	^{156}Dy	0.056
^{102}Pd	1.02	^{158}Dy	0.095
^{106}Cd	1.25	^{162}Er	0.139
^{108}Cd	0.89	*^{164}Er*	1.601
^{113}In	4.29	^{168}Yb	0.123
^{112}Sn	0.97	^{174}Hf	0.16
^{114}Sn	0.66	*^{180}Ta^m*	0.01201
^{115}Sn	0.34	^{180}W	0.12
^{120}Te	0.09	^{184}Os	0.02
^{124}Xe	0.0952	^{190}Pt	0.012
^{126}Xe	0.089	^{196}Hg	0.15
^{130}Ba	0.106		

the neutrino (ν) process (Figure 1.12). For a recent review of production of p nuclides and the astrophysical sites suitable for synthesis of these nuclides refer, among others, to the article by T. Rauscher (Rau 13).

The first obvious reaction process that can produce isotopes on the neutron-deficient side of stability is the rapid proton capture or the *rp* process. The successive proton capture of low Z nuclei can produce higher Z-unstable nuclei beyond iron and these unstable isotopes would undergo β^+ decay to synthesize the stable p nuclides. However, because of suppression due to the Coulomb barrier, the proton capture cannot take us to the heavier region unless the temperature of the site is very high and the proton density is also high (the proton capture takes place at the high-energy tail of the proton distribution corresponding to the energy of the Gamow peak). But at higher temperatures (approximately a few times 10^9 K) the inverse processes (γ, p) reactions become faster than proton capture reactions and would prevent further proton captures. At such high temperatures the sequence of photodisintegration reactions such as (γ, n), (γ, α) reactions on relatively more n-rich stable or unstable heavy isotopes would lead to neutron-deficient nuclei which would then decay by β^+ decay to produce the p nuclides (Woos 78; Wall 81). In this process a fraction of heavy nuclei produced in the s and r processes are transformed into p nuclides. Thus, for heavier p nuclides the γ (gamma) process, that is γ-induced reactions on seed nuclei, becomes the most important reaction leading to production of heavier p nuclides, while for lighter p nuclei, the dominant contribution is from the *rp* process. In particular, the abundances of light p nuclides such as ^{92,94}Mo and ^{96,98}Ru, which are very high and comparable with neigboring nuclides produced via the s and the r process, cannot be explained in the framework of the γ process and the *rp* process needs to brought in to explain their high abundances.

It should be noted both the rp and the γ processes involve neutron-deficient nuclei away from stability and the exact path is determined by the temperature of the site and the nuclear properties of the nuclei that take part in the reaction sequence. Quite often both the processes occur together and contribute to the synthesis of p nuclides. However, unlike the rp process that flows from lighter to heavier elements and traces a path quite close to the proton drip line involving some drip line nuclei as well, the γ process path flows downwards in mass and also runs comparatively closer to the line of β-stability. The masses, as well as the β-decay half-lives of most of the nuclei that participate in the γ process, have been measured while the same is not true for the nuclei participating in the rp process. It is estimated that more than 2000 p-rich or neutron-deficient nuclei participate in these nucleo-synthesis processes (rp and γ) connecting more than 40,000 reaction paths. An accurate input of nuclear properties (masses, half-lives, decay modes including β^+ delayed proton emission probabilities, proton capture rates at sub-barrier energies, photodisintegration reaction rates, etc.) are therefore of key importance in determining the reaction paths and thereby to constrain the possible astrophysical sites in which these processes can take place.

The p processes can only happen in explosive scenarios such as type 1a novae, X-ray bursts, CCS, etc. This is because the proton capture process (the rp process) needs a high proton flux in a high temperature environment so that the proton capture can dominate despite the inverse photodisintegration or (γ, p) reactions. However, the timescale of the process should be small enough and the temperature should not be too high to avoid a rapid erosion of heavy s or r process seed nuclei to produce tiny amounts of p process nuclides. These considerations limit the temperature range between ~0.4 to 3×10^9 K and the time to a few tens of seconds. Also, the site should have enough heavy seed material present to effect production of α and p nuclides through the γ process. These conditions could be met in the sites mentioned above and, of them, the X-ray bursts are the most frequent events in which thermonuclear burning (explosions) of hydrogen and helium takes place on the surface of a neutron star. The bursts, having a typical timescale of a few seconds or a bit longer take place when a neutron star accretes hydrogen- and helium-rich matter from the envelope of its companion star that forms a close binary system. The accumulated matter, once it gets dense and hot enough to trigger instability leads to a thermonuclear runaway because the nuclear reactions heat up the layer, further accelerating, in turn, the reaction process. The X-ray bursters provide an ideal environment for the sequence of proton captures (rp process) because the peak temperature lies in the range $2-3 \times 10^9$ K.

The seed nuclei for the rp process are produced in helium burning that includes the breakout from the hot CNO cycle. Some of the most probable breakout reactions are: $^{15}O(\alpha, \gamma)^{19}Ne$, $^{19}Ne(p, \gamma)^{20}Na$; $^{14}O(\alpha, p)^{17}F$, $^{17}F(p, \gamma)^{18}Ne$, $^{18}Ne(\alpha, p)^{21}Na$. It should be mentioned that for light nuclei both (p, γ) and (α, p) reactions remain operative in explosive scenarios and both the reactions lead to the production of heavier isotopes. As Z increases the proton capture and its inverse process (γ, p) reactions become the main reactions and successive proton captures within a short time take the seed isotope to an isotope either on the proton drip line or close to the drip line. The time scale for the rp process is typically a few tens of seconds and the process can last up to about 100 s. Because of the rather long (compared to the duration of the rp process) β-decay lifetime, some of the very p-rich isotopes, such as ^{64}Ge ($t_{1/2} \sim 64$ s), ^{68}Se ($t_{1/2} \sim 36$ s) and ^{72}Kr ($t_{1/2} \sim 17$ s), which lie on the rp process path, are thus potential waiting-point nuclei. Proton capture of any of these exotic nuclei would only lead to a nucleus where the last proton is weakly bound and thus can easily be removed by the inverse (γ, p) reaction (the exact extent of weak binding requires accurate knowledge of the mass or the proton capture Q value, which is presently not experimentally determined for

^{68}Se and ^{72}Kr). Thus, the reaction flow has to wait quite a considerable amount of time for the β decay before it moves to heavier isotopes. The waiting points decide the burst light curves, since energy generation is higher at the waiting points. Under favorable conditions that include high temperatures in the range 2 to 3×10^9 K, the *rp* process can produce elements up to the Sn–Sb–Te region. The limitation imposed by the proton drip line isotope for Sb ($Z = 51$), short-lived α emitters of Te ($Z = 52$), e.g., ^{104,106}Te, and the natural cycle like the CNO that is likely to operate between the participating isotopes of Sn–Sb–Te restrict the flow of the rp process to below $A \sim 110$ (Scha 01). The typical nature of the *rp* and the γ process paths or the reaction networks involving a small number of isotopes are shown schematically (more to give the readers a mental picture of these nucleo-synthesis processes) in Figures 1.15 and 1.16 respectively. It should be mentioned that in type 1a novae where the accretion occurs on a white dwarf and the peak temperature reached is about 4×10^8 K (much lower compared to the peak temperature reached in X-ray bursts), a rather mild *rp*

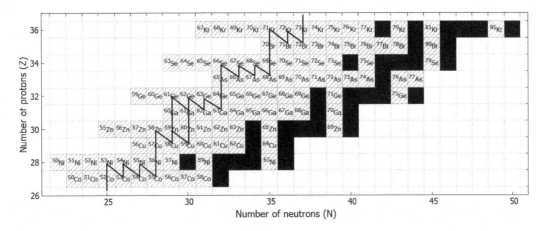

FIGURE 1.15
Typical expected path of *rp* process nucleo-synthesis in a limited zone involving neutron-deficient isotopes of Co to Kr.

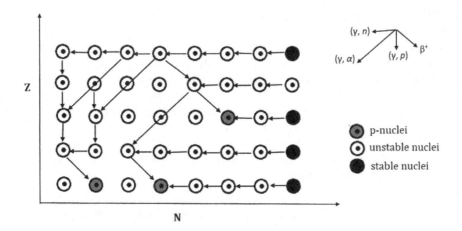

FIGURE 1.16
Typical reaction network in γ process nucleo-synthesis.

process becomes operative. The process uses neon as the seed and proceeds rather close to stability. This mild rp process in type 1a novae can typically produce isotopes up to Ca. This process thus does not contribute directly to p nuclide synthesis. However, it is highly probable that nuclei such as ^{18}F, ^{22}Na and ^{26}Al are synthesized in novae through proton and α capture reactions on lighter neutron-deficient seed nuclei and these nuclei are of special importance since these are the sources of galactic γ-rays. Lastly, it should be mentioned that the neutrino-induced reactions or the ν process (Woos 90; Heg 05) seems to provide a solution to explaining the natural abundances of two of the lowest abundant p nuclides, ^{138}La and ^{180}Ta, through neutral current (ν, n) and charged current (ν_e, e^-) reactions.

To summarize this section, it should be emphasized once again that a very large number of exotic nuclei, both neutron rich and neutron deficient, take part in nucleo-synthesis in different types of explosive stellar scenarios. A proper understanding of these processes and the sites ideally requires experimental determinations of masses, reaction rates, half-lives and decay schemes of all these nuclei. However, this is an impossible task, and it is hoped that these properties can be studied for a sufficient number of exotic nuclei in the RIB facilities, existing or coming up in the future (pushing/trying to continuously push the intensity frontier to higher and higher intensities of RIBs) so as to aid the development of more reliable theories capable of predicting all these properties of exotic species with the required accuracy.

1.4 RIBs and the Test of Fundamental Symmetries of Nature

The Standard Model of particle physics has so far been one of the most elegant and successful theories, which accounts for almost all subatomic phenomena. However, even ignoring the fact that the standard model cannot account for gravity and it has about 20 adjustable constants, there are certain other facts that clearly indicate the existence of physics beyond the standard model. These are: (a) we have a matter or baryon universe and the excess of matter over antimatter (about one part in a few billion) as reflected in the measured baryon to photon ratio, is too high (by many orders of magnitude) for the standard model to accommodate; (b) neutrinos have been discovered to have mass that requires adjustment by introducing new mass terms to the model; and (c) the model describes only 5% of the visible universe and provides no clue for the other 95%.

The availability of a huge number and variety of radioisotopes with sufficient intensities in RIB facilities offers the exciting possibility of studying some of the fundamental symmetries and the violation of these at a minute level using "low energy" nuclear phenomena. ^{60}Co stands as a classical example where the study of radioactive decay led to the discovery of parity violation in weak interaction. With the new isotopes available in RIB facilities, measurements can now be carried out to obtain information (carry out stringent tests) on other symmetries as well, such as the violation of time-reversal symmetry (or CP violation since CPT has to be conserved), atomic parity violation and possible deviations from the conserved vector current (CVC) hypothesis and the unitarity of Cabibbo–Kobayashi–Maskawa (CKM) matrix.

1.4.1 The Electric Di-Pole Moment in Atomic Systems and the CP Violation

The permanent electric di-pole moment or EDM arises from the spatial separation of opposite charges along the axis of a particle's angular momentum. The possible existence of

EDM in particles and nuclei was suggested long ago and it was pointed out that EDM, being a polar vector, should align with the axial angular momentum (which specifies the system's orientation) and in the process would violate both time reversal and reflection symmetry (Pur 50), as schematically explained in Figure 1.17. It should be noted that the electric di-pole moment will be zero if there is a reflection symmetry ($x \rightarrow -x$) or time reversal symmetry ($t \rightarrow -t$). Thus the very existence of EDM is clear proof of CP violation.

CP violation has been considered as a necessary condition for generating matter–antimatter asymmetry (Sak 67) if an initial condition of equal matter and antimatter is assumed. CP violation has been experimentally observed in the flavor-changing decays of neutral K mesons (Chr 64). The Standard Model accounts for CP violation through the phase in the CKM matrix and also through the gluon field in QCD. However, neither of these mechanisms can explain the magnitude of CP violation required to explain the matter–antimatter asymmetry that has presumably taken place moments after the big bang leading to the matter universe. It should be noted that the Standard Model predicts values for EDM that are very low: 10^{-38} e.cm for electrons, 10^{-33} e.cm for atoms/nuclei and 10^{-31} e.cm for neutrons. The present experimental upper limits for EDM for electrons, neutrons and nuclei are 8.7×10^{-29} e.cm, 2.9×10^{-26} e.cm and 7.4×10^{-30} e.cm respectively, which are still many orders of magnitude higher than the values predicted by the Standard Model and will be difficult to reach experimentally in the foreseeable future. However, the extensions of the standard model or "beyond the standard model" theories such as the supersymmetric theories (SUSY) have additional sources for CP violation and consequently predict EDM values that are much higher that are just below the present experimental limits and can thus be reached in the coming generation of RIB facilities.

The present discussion would be limited to EDM of atoms/nuclei that are more sensitive to chromo-EDMs of the quarks and other CP-violating mechanisms in the strong interaction. The most sensitive measurement conducted so far was on ^{199}Hg (Gra 16) that sets an upper limit of EDM (^{199}Hg) $< 7.4 \times 10^{-30}$ e cm. It should be noted that in an atom the nuclear EDM does not get canceled by rearrangement of electrons due to the presence of relativistic

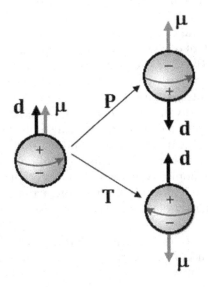

FIGURE 1.17
Existence of EDM and CP violation.

effects and the resultant moment, called the Schiff moment, increases as Z^n, where n lies between 2 and 3. Thus Hg with $Z = 80$ has been a very good probe for testing the possible existence of EDM. Radon nuclei ($^{221-226}$Rn) are now considered to be much better candidates for EDM than ^{199}Hg. Radon is six units higher in Z than Hg but the real gain comes from the enhancement of the Schiff moment by a factor of 100 to about 1000 due to permanent octupole deformation (pear shape) of n-rich radon nuclei. Recent studies at HIE-ISOLDE, however, revealed that even ^{224}Rn and ^{226}Rn nuclides although undergoing octupole vibration do not possess a static pear-shaped ground state, thereby disqualifying these nuclides as potential candidates for EDM research (But 19). The odd nuclei such as ^{223,225}Rn (But 96, Spe 97, Flam 03, But 16), however, are likely to have very closely spaced (energy difference ~50 keV) parity doublets (a manifestation of the pear shape), which would lead to significant enhancement in the Schiff moment. The parity doublet states (+ and −) enhances the sensitivity for test of EDM since T and P violating interactions mix the parity doublet and the T, P violating admixture coefficient is inversely proportional to the energy difference between the doublets. Thus, compared to ^{199}Hg, where there is no enhancement of Schiff moment due to octupole deformation, the sensitivity of EDM measurements can be improved by orders of magnitude in the case of odd radon nuclei. It should be mentioned that radon isotopes (half-lives of approximately several minutes) can potentially be produced with a considerable yield in a high-energy proton-induced spallation reaction using a ^{238}U target. However, since the reaction would predominantly produce fission products a clean separation of radon isotopes would be a challenging task. Experimental programs to search for atomic EDM in spin-polarized radon isotopes are being pursued at present at various laboratories including at the ISAC facility, TRIUMF (Tar 13).

1.4.2 Atomic Parity Violation

Parity violation in an atom leads to electric di-pole transition between two atomic states of the same parity which is strictly QED or electromagnetically forbidden. The parity violating exchanges of the Z^0 boson (neutral current interaction) between the electrons and nucleons (up and down quarks) in the nucleus gives rise to atomic parity violation (APV) or the parity non-conserving (PNC) effects. The PNC effect adds coherently for all nucleons. The parity non-conserving di-pole transition amplitude is proportional to the weak charge Q_W (which is dependent on standard model coupling constants) and a factor called the atomic structure factor. The Q_W for atomic systems can be calculated theoretically from the standard model and for heavy atomic systems ($N/Z > 1$), it roughly goes as: $Q_W \sim -N$, where N is the number of neutrons. This standard model value needs to be compared with the experimentally determined value for the same system. As such the electro-weak PNC effects are too small to be detected. However, because of the short-range nature of the interaction, the PNC effect or the transition amplitude in atomic systems roughly grows as $\sim Z^3 R$, where Z is the atomic number of the atom and R is a relativistic enhancement factor. This enhancement in heavy atomic systems makes this otherwise small effect detectable in principle in atoms of high Z (Bou 74; 82). The interaction causes a mixing of S and P states in an atom and the parity violating (PV) electric di-pole transition between two S or P states needs to be detected.

However, this requires not only a precise determination of the transition amplitude experimentally but also an accurate theoretical calculation of the structure factor. Thus, experimental determination of weak charge (Q_W) becomes subject to both experimental and theoretical uncertainties. Calculation of the atomic structure factor, among other complications, needs very high computational power that increases with the number of electrons. Usually, the theoretical calculations are fine-tuned by requiring it to reproduce the measured experimental values of various properties of the same atom.

The most accurate determination of the PV di-pole transition amplitude has been carried for the transition between 6S and 7S levels of ^{133}Cs atom (Wood 97). Although Cs, with $Z = 55$ is definitely not the element with the highest atomic number available for study, it was chosen as a best compromise since among the higher Z atoms Cs, having a single valence electron outside an inert core, has a relatively simpler atomic structure allowing its structure factor to be calculated quite accurately. The continuous improvements/refinements in the theory, together with availability of new experimental data on other properties of heavy atoms, have led to a substantial reduction in the uncertainty of the atomic structure factor calculations down to about 0.4% and even less. An important source of uncertainty lies in nuclear structure effects, especially the correction due to the neutron skin that changes the neutron distribution. A great deal of effort is now going on to precisely measure the neutron skin in ^{208}Pb and in other heavy nuclei. The new experimental inputs would certainly help in future to reduce the uncertainty related to nuclear structure effects. In any case, with the present level of accuracy achieved in the structure factor calculations, it is generally accepted that the experimentally determined weak charge of Cs agrees reasonably well with the standard model prediction of weak charge for Cs, which is Q_w (S.M) = −73.16 (Dzu 12). This agreement constrains the masses of additional gauge bosons predicted in new physics models (or beyond the standard model theories) to energies higher than the TeV scale.

The future of atomic parity violation measurements lies in the determination PV transition amplitudes in francium (Fr) atoms. Francium has very high Z (= 87) and like Cs its atomic structure is also simple. The PNC effect in Fr is expected to be about 18 times larger than that in Cs (Dzu 95). Francium does not have any stable isotopes and therefore was not amenable to earlier studies. However, with the advent of RIB facilities, production of a long chain of Fr isotopes in sufficient numbers has now become feasible. The measurement on a long chain of isotopes would help to reduce the uncertainties due to the nuclear structure. These isotopes can be produced with reasonable yields in a proton-induced spallation reaction using a ^{238}U target. Production experiments at ISAC and at ISOLDE predict that francium isotopes in the range of ^{206}Fr to ^{228}Fr could be produced with good yields (Aub 03). However, APV studies on these isotopes would require a further increase in the yield, long beam time and also improvement of experimental techniques to reach the desired experimental precision.

There is another atomic PNC effect which arises due to nuclear spin-dependent interactions between the nucleus and the orbital electrons. Due to inter-nucleon weak interactions within the nucleus, a parity non-conserving but time-reversal conserving (therefore charge conjugation non-conserving) electromagnetic moment arises, and this moment is known as the nuclear anapole moment. The interaction of anapole moment with the atomic electrons would allow di-pole transitions within states of the same hyperfine manifold; say, in between hyperfine states of the 7S level in francium, which lies in the microwave range (about 47 GHz in the case of 7S hyperfine levels in francium). Such a transition is otherwise parity forbidden in the absence of the anapole moment-induced "PNC hyperfine interaction."

The anapole moment-induced PNC effect increases in heavy atoms such as $Z^{8/3}R$ (Flam 84). The nuclear spin-dependent parity violation in francium is expected to be 11 times larger than Cs (Gom 06). Thus, precise determination of the anapole moment in a string of Fr isotopes would enable the probing of the inter-nucleon weak interactions in the nucleus and minimize the uncertainties due to the nuclear structure. Such measurements would therefore put constraint on weak nucleon–nucleon isovector and isoscalar couplings. Dedicated programs for anapole moment measurements in Fr isotopes have been undertaken at various laboratories, including at ISAC, TRIUMF (Aub 12, Aub 03).

1.4.3 The CVC Hypothesis, Nuclear Beta-Decay and the Unitarity of CKM Quark Mixing Matrix

1.4.3.1 The CVC and the Nuclear Beta Decay

At the quark level, β decay of a neutron (proton) into a proton (neutron) means conversion of a down (up) quark into an up (down) quark associated with the emission of a virtual W^- (W^+) boson that decays into an electron (positron) and an anti-neutrino (neutrino). The Conserved Vector Current (CVC) hypothesis (Feyn 58), a basic tenet of the standard model, proposes that the weak vector coupling constant G_v does not get normalized in the nuclear medium, that is, it should remain the same in all nuclei. In nuclear β decay, all super-allowed $0^+ \to 0^+$ Fermi transitions (in different nuclei undergoing β decay) between nuclear analogue states ($J^\pi = 0^+$; $T = 1$) take place purely via the vector current of the weak interaction since axial current cannot contribute in $0^+ \to 0^+$ transitions. Thus, according to CVC, G_v should not depend on the particular nucleus undergoing decay and remain constant in all $0^+ \to 0^+$ Fermi transitions. The coupling constant G_v is connected to Ft (which is obtained after suitably correcting the experimentally measured β-decay ft value for iso-spin symmetry breaking effects arising from a Coulomb interaction among protons and transition-dependent radiative effects), through a simple relation given by: Ft = $K/2G_v^2(1 + \Delta_R)$, where K is a constant and Δ_R is a transition-independent radiative correction. Thus, experimental determination of ft values for a number of $0^+ \to 0^+$ transitions would allow the testing of CVC through constancy of the corrected ft, that is, the Ft values. Thus, it is necessary to measure ft for a number of different β-decaying nuclei undergoing super-allowed β-decay and theoretically calculate the correction factors to compute the Ft values. If all the Ft values come out to be the same, the CVC hypothesis is confirmed. And if the CVC holds, G_v can be calculated from the Ft (if CVC does not hold, it is meaningless to extract G_v since it does not remain constant).

The experimental determination of the ft value involves determination of total transition energy, branching ratio of the transition and the half-life of the parent state. Experimentally determined ft values from 13 different nuclei (^{10}C, ^{14}O, ^{22}Mg, ^{26}Al^m, ^{34}Cl, ^{34}Ar, ^{38}K^m, ^{42}Sc, ^{46}V, ^{50}Mn, ^{54}Co, ^{62}Ga and ^{74}Rb) from $A = 10$ (^{10}C) to $A = 74$ (^{74}Rb; the super-allowed β-emitter with the largest Z), when corrected for the structure factors that are calculated theoretically, agree with each other; the average value of Ft being 3072.27 (72) s (Hard 15). So, the CVC hypothesis holds good to an uncertainty level of about 0.03%. However, further reduction of the uncertainties would allow more stringent tests of CVC. This requires efforts both in theory and in experiment. The theoretical inputs of radiative and iso-spin symmetry breaking corrections contribute significantly to the uncertainties. Since the iso-spin symmetry breaking corrections scale as Z^2, the calculations for this term can best be tested for higher Z nuclei, for example ^{74}Rb. It should be noted that 14 nuclei in the range $6 \le Z \le 37$ on which the measurements of β-decay properties have been carried out are all neutron-deficient exotic nuclei with short half-lives, in some cases as short as a few tens of ms. There are other super-allowed β-emitters, such as ^{42}Ti, ^{46}Cr, ^{50}Fe, ^{54}Ni, etc., for which experimental studies have recently been initiated (Zha 17). Experimental determination of the radial extent of the charge distribution using laser spectroscopy (Mane 11) of the parent and daughter nuclei is also essential for improving theoretical calculations of iso-spin symmetry breaking correction factor. This study also critically depends on the intensity of the unstable nuclei. The next generation of RIB facilities will be able to produce all these nuclei with much more intensity allowing more precise measurements, for example on the β-decay branching ratios of the super-allowed transitions which is, at the moment, one of the major contributors to the experimental uncertainties (Zha 17).

1.4.3.2 Unitarity of CKM Matrix

In the standard model the quark-mass eigen states are not the same as the weak-interaction eigen states and these two are connected through the Cabibbo–Kobayashi–Maskawa (CKM) quark mixing rotation matrix (Cabb 63; Koba 73), which must therefore be unitary. The unitarity of CKM quark mixing matrix is thus an integral part of the standard model and any test of its unitarity would be a test of the validity of the standard model itself. In particular, even a small deviation from the unitarity would suggest the existence of new physics beyond the standard model. The standard model by itself does not determine the values of the elements of the CKM matrix. Thus, to check the unitarity the matrix elements need to be determined experimentally. One of the most demanding tests on the unitarity is that the sum of the squares of the three top-row elements of the 3 × 3 CKM matrix should be equal to one, that is, $|V_{ud}|^2 + |V_{us}|^2 + |V_{ub}|^2 = 1$ (refer to Tow 10 for a detailed discussion on this topic). In this expression V_{ud}, V_{us} and V_{ub} are the up-down, up-strange and up-bottom quark mixing elements respectively.

The first top row matrix element V_{ud} is given by $V_{ud} = G_v/G_F$, where G_F is the weak interaction constant for purely leptonic muon decay which is known with good precision. Thus, precise experimental determination of ft values for super-allowed $0^+ \to 0^+$ Fermi decays for testing of CVC also allows precise determination of V_{ud}, through G_v. The precision of V_{ud} is mainly determined by the precision of Ft and Δ_R. Presently, the value of V_{ud} is $|V_{ud}| = 0.97417$ (21), that is, it is known at about 0.04% uncertainty level. Since $|V_{ud}|^2$ accounts for about 95% of the sum, precise determination of this term is of utmost importance for the test of unitarity. V_{us}, the second element of the first row, is determined from the decay of kaon and its present value is $|V_{us}| = 0.0.22521$ (94). The third element, V_{ub}, which represents the contribution from the third generation of quarks is not so precisely known but it is small enough ($|V_{ub}| \sim 4 \times 10^{-3}$) to have any significant impact on the unitarity test, unless a much better level of precision is achieved for the first two dominant terms to make precision of this third term important. The result is that the sum of the squares: $|V_{ud}|^2 + |V_{us}|^2 + |V_{ub}|^2$ satisfies the unitarity condition at a level of 0.06% precision and the standard model passes the test subject to this uncertainty level. The level of uncertainty or precision puts constraints on any theory that attempts to extend the standard model. An improvement in the precision, which would require theoretical as well as experimental efforts, is therefore of great importance to put tighter constraints on new theories beyond the standard model. Experimental programs for accurate determination of masses, charge radii and β-decay properties of super-allowed β-emitters will therefore remain a priority area in the next generation RIB facilities for tests of CVC as well as of unitarity of CKM.

1.5 RIBs and Condensed Matter Physics

Condensed matter physics (which includes soft condensed matter, chemistry and biology) has greatly benefited from the use of nuclear facilities: reactors as well as accelerators.

The neutrons from reactors, synchrotron radiation from electron storage rings, radioactive sources produced in reactors/accelerators, accelerated light ions such as proton and alpha particles and implantation of stable heavy ions: all these have been extensively used in condensed matter/material science studies including device developments. The use of various radioactive ion probes is the latest addition to this list. In fact, close to 20% of the beam time in major RIB facilities all over the world, especially in the ISOL-type facilities such as ISOLDE at CERN and ISAC at TRIUMF, is being used by condensed matter researchers and the demand on the beam time is ever increasing.

Radioactive ion beams of selected species are doped inside the host matrix, in the majority of cases, using ion implantation. The advantage of implantation is that the ions can be implanted at the chosen precise depth (with a small spread around its range in the host) by varying the incident beam energy and therefore short-lived ions can be implanted, unlike in diffusion doping where long-lived probes need to be doped. The energy of the implanted ion can be a few tens of keV (in which case the implantation depth is only about 10 nm or so) or higher up to about a few MeV if deeper penetration in the bulk is required. Also, the implantation energy can be varied to implant ions over extended regions in the bulk. The implanted or the probe ion comes to rest and occupies either an interstitial position in between the host atoms or a lattice site by replacing one of the host atoms. The slowing down of the implanted heavy ion might give rise to defects in the host matrix but these are sometimes corrected by annealing induced by the probe beam itself. Any defect that remains after self-annealing can of course be removed by conventional thermal annealing after ion implantation. The nuclear techniques are highly sensitive and to extract relevant information, it is usually enough to implant a relatively small number of ions (not exceeding 10^{10} atoms in most cases; <1 ppm concentration) and thus the probe hardly interferes with the properties of the host matrix. The radiation emitted by the probe atom is utilized to extract information on the host material.

The radioactive probe atom, after coming to rest, interacts with the local magnetic and electric fields (in the host) through its nuclear magnetic moment and the electric quadrupole moment respectively. Local fields thus perturb the nuclear spin which is manifested through the splitting of the energy levels (studied via γ transitions) and in the angular distribution/anisotropy of radiation (γ/β) emitted in the radioactive decay. The hyperfine interaction techniques: Mossbauer Spectroscopy (MS), Perturbed Angular Correlation (PAC) and the β-Nuclear Magnetic Resonance (NMR), all utilize one or the other of these manifestations and have been extensively used in RIB facilities to study the local magnetic and electric fields in the host. The low density of the probe atoms, the ability to implant ions at various depths in the host from the surface to deep inside and high sensitivity of the nuclear hyperfine techniques make them ideally suitable for the study of low-dimensional systems such as the quantum dots, well and wire-like structures, clusters, surfaces and interfaces. Also, the wide choice of radioactive probes usually available in a versatile RIB facility adds flexibility to the type of measurements. Thus, condensed matter studies using radioactive probes offer some unique advantages, such as high sensitivity, subatomic spatial resolution, etc., which are exploited/made use of in RIB facilities; measurements that are not possible using other competing techniques can often be carried out.

1.5.1 Mossbauer Spectroscopy

The Mossbauer Effect (ME) is the recoilless nuclear resonance absorption of γ-rays. The resonance line width is small enough (~10^{-9} eV) to allow study of the splitting of the energy levels (~10^{-6} eV) in the Mossbauer probe nucleus due to hyperfine interactions of the probe with the local magnetic di-pole and electrical quadrupole fields. A good Mossbauer source that satisfies the conditions of recoilless emission/absorption and a small enough line width should have γ-ray transition energy below 100 keV and a lifetime of the γ-decaying isomeric state ~1 ns. ^{57}Fe* is by far the most popular Mossbauer source that comes to the ground state from the decay of a 142 ns isomeric state emitting a 14.4 keV γ-ray. In conventional Mossbauer spectroscopy (MS) measurement, the source or the emitter, say, ^{57}Fe*, is placed outside the sample and the sample is doped with ^{57}Fe in its ground state. The ^{57}Fe in the sample or host absorbs the γ quanta emitted by the source in resonance and any splitting of levels due to local hyperfine fields in the sample are observed in the absorption spectrum as a function of source velocity (typically ~1 cm/s). This type of measurement is therefore called absorption-type MS. Alternately, the sample may be doped with ^{57}Fe*, that is a probe atom in its excited state, and the γ quanta emitted through decay of the excited isomeric state are studied in comparison with a standard absorber containing the probe atoms in the ground state in a matrix that has no magnetic di-pole field and quadrupolar electrical field. Since in this method any splitting is observed in the emission spectra, it is called emission-type MS. In RIB facilities, the MS studies are mostly carried out in the emission spectroscopy configuration which has a higher sensitivity than the conventional absorption spectroscopy configuration. At ISOLDE ^{57}Fe* is produced through β^- decay of ^{57}Mn (instead of from a conventional long-lived ^{57}Co source with a half-life of ~271 days which has been mostly used as a Mossbauer source in table-top MS set-ups). ^{57}Mn can be produced in a pure form with intensity exceeding 10^8 atoms/s and since it undergoes β^- decay with a much shorter (half-life of ~1.5 min, it allows recording of up to about 300 Mossbauer spectra in a single day. Further, a RIB facility allows implantation of other suitable Mossbauer probe nuclei. For example, a ^{119}Sn* probe has been extensively used at ISOLDE for studies on semiconductors. The ion implanted was ^{119}In which undergoes β^- decay to ^{119}Sn* and comes to ground state by emitting a 24 keV γ-ray. The 24 keV state in ^{119}Sn is the Mossbauer state. The MS studies with RI probes at ISOLDE have produced a wealth of new information on various advanced materials including semi-conductors and dilute magnetic semi-conductors (John 17 and the references therein).

1.5.2 Perturbed Angular Correlation

The PAC technique makes use of a cascade of γ-ray transitions in the probe nuclei where the intermediate state is an isomeric state with a mean-life in the range ns to μs. The probability of γ-ray emission in a particular direction depends on the orientation of the spin. Thus, an isotropic angular distribution of γ-rays is observed when an ensemble of unpolarized (randomly oriented) probe nuclei comes to rest inside a host matrix and undergoes γ decay. However, the direction of emission of the first γ-ray in the cascade sets a reference direction and with respect to that the angular distribution of the second γ-ray, resulting from the decay of intermediate isomeric state to the ground state, would be anisotropic if the spin of the intermediate state is greater than or equal to one. During the decay of the isomeric state, the angular correlation between the two γ-rays might be affected by the hyperfine interaction of the probe nucleus with the local magnetic field and the electric field gradient caused by the surrounding electronic charge distribution and nuclei. This results in a time-dependent correlation function which can be determined experimentally by recording γ–γ coincidence spectra as a function of time. The hyperfine magnetic field and the electric field gradient can be extracted from the time-dependence of angular correlation. It is imperative that all the information about the probe nucleus is known in advance: energy levels, multi-polarity of transitions, spins, magnetic moment and electric quadrupole moment. ^{111}In is the most extensively used isotope for the PAC studies, although there are also quite a few other isotopes (such as ^{111m}Cd, ^{181}Hf, ^{199m}Hg) that are used for PAC measurements. PAC requires quite a high intensity of the probe and high level of purity, which are demands that only a few facilities (e.g., ISOLDE, ISAC) can meet at the moment. The PAC technique has been extensively used in condensed matter studies that include phase transitions in multi-ferroic materials that possess both ferromagnetism and electric polarization (Lop 08, Joh 17), and in chemistry and biochemistry (Janc 17).

1.5.3 β--NMR

The β-NMR or the β-detected NMR is a unique tool for studying dilute and low-dimensional systems that utilize the asymmetric (as weak interaction violates parity) emission of β particles when spin-polarized nuclei undergo β decay. In this technique, unlike in an NMR, the monitoring of the polarization of an ensemble of implanted polarized nuclei as it interacts with the local magnetic and electric field and its time evolution is done through measurement of asymmetric β emission. Also, unlike in conventional NMR where a small degree of initial polarization (~10^{-6} to 10^{-4}; since in thermal equilibrium the degree of polarization depends upon $\mu B/kT$, symbols having the usual meaning) can only be sustained by applying a strong external magnetic field, in β-NMR a high degree of polarization (~70%) can be achieved in the probe beam before implantation, through optical pumping using a circularly polarized laser (Lev 14). Thus, in β-NMR, the techniques for achieving the initial polarization and detection are both different than in conventional NMR and are independent of the external magnetic field.

The β-NMR is about 10 orders of magnitude more sensitive than conventional NMR and is one of the most sensitive techniques for determining the electric quadrupole and magnetic di-pole moments. The figure of merit for β-NMR experiments is given by the product P^2Y, where P is the magnitude of polarization and Y is the production yield of the probe isotope. The resultant increase in the sensitivity is about ten orders of magnitude in favor of β-NMR compared to conventional NMR. It also provides the advantage of studies of near-surface phenomena as a function of depth by varying the implantation energy.

As in NMR, the probe nuclei are placed in an external magnetic field to maintain the polarization and an oscillating RF field is applied. At resonance, the RF frequency (which depends on the nuclear spin and the magnetic di-pole moment) matches with the Zeeman splitting and induces transition between adjacent Zeeman levels, thereby altering the nuclear magnetic sublevel population. This will result in a loss of polarization and consequently in a loss of the asymmetry in β emission. When there is no RF field the loss of polarization is caused by the fluctuating magnetic field. Experimentally through β counting in two detectors placed 180 degrees apart, the nuclear polarization and its time evolution can be deduced. Thus β-NMR can be used to study any property (local magnetism, spin lattice relaxation, etc.) in the host that causes a change in the resonance frequency or in the relaxation rate. The high sensitivity of β-NMR makes it an ideal tool for the studies of surfaces, interfaces and other lower-dimensional systems like quantum dots and well- and wire-like structures. The β-NMR has been used for a wide variety of studies, a number of them of a pioneering nature in material science research that includes magnetism, superconductivity, ultra-thin structures, etc. (Mor 14 and the references therein). The technique also has great potential in the fields of chemistry and biochemistry (Janc 17).

The suitability of a probe isotope for β-NMR is decided by presence of a large β asymmetry factor, enough β-decay energy, ability to produce with enough intensity, degree of polarization achievable and the half-life. ^{8}Li is by far the most popular probe for condensed matter studies and the β-NMR facility at TRIUMF has already produced a wealth of new information using this probe.

Apart from nuclear hyperfine interaction techniques, a number of other techniques have been used extensively, utilizing radioactive isotopes for condensed matter studies, mainly from the ISOLDE facility. These include diffusion, emission channeling and Photoluminescence Spectroscopy, and readers might refer, among others, to the review articles (Haa 96, Dei 02, Joh 17) covering the studies carried out at ISOLDE using these techniques.

1.6 RIBs: Medical Physics and Applications

Radioisotopes have a long history of successful application in cancer therapy and diagnosis. For example, ^{60}Co has been used extensively in the treatment of cancer using radiation (γ) therapy and ^{99m}Tc is the most widely used isotope in medical imaging. ^{99m}Tc is the most popular isotope because of its convenient half-life of 6 hrs (the half-life should be long enough to allow for injection into the body and short enough to decay away quickly), it emits γ-rays of sufficient energy to come out of the body for detection (in contrast to charged particles that have much lesser range) and because it can be abundantly and cost-effectively produced in nuclear reactors from fission of ^{235}U. Actually, ^{99}Mo that decays into ^{99m}Tc with a half-life of 66 hrs can be produced in a reactor. The relatively long half-life of ^{99}Mo allows for transport of long distances without significant loss in activity. ^{99m}Tc is used for imaging a large number of organs including liver, spleen, kidney, lung, thyroid and brain. ^{131}I is another isotope used widely both for diagnosis and treatment of thyroid disorders and in fact has revolutionized the treatment of thyroid malfunction. Like ^{99}Mo, ^{131}I is also produced in a nuclear reactor through fission of ^{235}U. Among the isotopes produced in an accelerator (cyclotrons, linacs), ^{18}F, a positron emitter with half-life of about 110 min, is the most widely used isotope for imaging. It emits two 511 keV γ-rays through positron-electron annihilation, which are used for diagnosis. Usually the β^- and the α-emitting isotopes of suitable half-lives are used for in-vivo therapy because of the short range of the radiation and γ-emitting isotopes for diagnostics.

RIB facilities can potentially play a key role in future radioisotope production for medical purposes. There are two ways. The first one is to produce isotopes like ^{99}Mo and ^{131}I in a cost-effective way through fission of ^{238}U (or through some fusion reactions) with some appropriate target using high intensity light ion beams from accelerators that are used for RIB production. The proton accelerators (proton-induced fission) and the electron accelerators (γ-induced fission) seem to be the two best choices for this purpose. This is important because most of the reactors that supply these isotopes are old and some are in the process of being shut down. A huge shortage of these isotopes is anticipated in near future. This shortage can be taken care of in light ion-based RIB facilities in a variety of ways including intercepting the beam with an isotope production target placed downstream of the RIB production target (about 80% of the beam survives after the RIB target) before it reaches the Faraday cup. Apart from being a source of conventional isotopes like ^{99}Mo, ^{131}I, etc., for which uranium has to be used as the material for the isotope production target, a lot of new and promising (in terms of their suitable half-lives and radiation they emit) isotopes can also be produced for pursuing the R&D required for their use in diagnostics or in therapy. The production of these isotopes needs a uranium target and also other targets. For example, a number of radiopharmaceuticals, especially the α emitters: ^{225}Ac, ^{223}Ra, ^{211}At, ^{213}Bi, ^{149}Tb, etc., which have suitable half-lives and α particle energies, are potential candidates for cancer treatment (in-vivo therapy). All these radioisotopes and a number of other promising radio lanthanides can be produced efficiently using the proton beam at ISOLDE. Thus, the beam which would otherwise have been wasted in the Faraday cup can be put to excellent use and a dedicated program has been undertaken at CERN, ISOLDE for isotope production using this technique (Augu 14). Similar programs for isotope production are also being conducted at all other major RIB facilities. It is also imagined that the production of radioisotopes in RIB facilities will provide a more perceptible justification (to taxpayers at large) for the high cost of the installation and running of these facilities.

The other potential use of RIBs is in cancer therapy through irradiation of cancer cells by energetic heavy ion beams. Therapy with γ-rays and even with high energy X-rays produced by compact linear accelerators have the shortcoming that because of the characteristic nature of energy loss, the radiation also kills the healthy cells around the malignant cells. Cancerous cells also have a low oxygenation rate requiring a higher energy deposition to kill them. This is achieved in X-ray/γ therapy by applying a larger dose which once again is harmful for the patient. Use of charged particles like protons and, more so, heavier ions like carbon, oxygen, neon, etc., offer a much better scenario. The heavy ions lose a large fraction of energy at the Bragg peak that allows a much higher linear energy transfer, requiring a lower dose (with respect to X-rays) to kill the cancer cells. Also, since its energy loss is much lower before reaching the Bragg peak, the normal cells are less affected by the radiation. By changing the beam energy the Bragg peak can be spread out to paint the entire tumor or the volume to be irradiated.

Cancer therapy with energetic ion beams was pioneered at the Lawrence Berkeley Laboratory (LBL) in the USA. After the initial proposition in 1946, the treatment with proton and α beams was initiated in the 1950s. Later, in the 1970s and 1980s treatments with high energy (~670 MeV/u) ^{20}Ne beams were started. The combination of high relative (with respect to X-rays) radiobiological effectiveness (RBE) and low relative oxygen enhancement ratio (OER), made the treatment very effective. The closure of Bevalac in 1992 brought an end to neon radiation therapy. Thereafter, a dedicated facility for ^{12}C therapy has been installed at the Heavy Ion Medical Accelerator (HIMAC), National Institute of Radiological Science (NIRS), Chiba, Japan. HIMAC started in 1994; at present there are eight centers in the world for heavy ion therapy, five of which are in Japan. The number of patients treated in these facilities has exceeded 15,000. Among the many excellent review articles on heavy ion therapy, readers may consult the review article of Schardt, Elsasser and Schulz-Ertner (Schar 10).

Use of an unstable heavy ion in place of a stable heavy ion offers an important potential advantage. Intense R&D activities have therefore been conducted to study the feasibility of using RIBs like ^{10}C, ^{11}C, ^{14}O, ^{15}O, etc., for cancer therapy. The advantage in using these unstable beams lies in the fact that they are themselves positron emission tomography (PET) isotopes and thus can be used simultaneously for real-time dose mapping as well as killing of cancer cells. The main physics uncertainty in heavy ion therapy comes from the range uncertainty. The target (the tumor) is inhomogeneous and the dose distribution is complex due to fragmentation of the beam particle and inelastic collisions inside the tumor. Fragmentation of the primary beam (say, ^{12}C) as it slows down in the tumor produces a number of radioactive and stable fragments including some positron-emitting fragments like ^{10,11}C. After fragmentation, the lighter fragments traverse longer distances than the ^{12}C beam, that is, beyond the Bragg peak for ^{12}C. Due to varied reaction products and their different ranges, the dose distribution delivered to the tumor becomes complex and in-vivo PET imaging with closed-loop control of treatment becomes very important.

In-vivo range verification by mapping the positron-emitting fragmentation products like ^{11}C produced in the reaction (fragmentation of ^{12}C) can of course be tried. This has indeed been tried but the signal was found to be too weak as these fragments are only produced in small numbers. Moreover, in this case only the regions where ^{11}C are (is) stopped get mapped. However, when the projectile itself is a positron emitter, the range could be more accurately pinpointed by measuring its emitting radiation since the beam intensity is much higher and a significant fraction of incident ions would survive until the ions come to rest at the Bragg peak location. However, theoretical simulations on fragmentation of these RIBs (^{10}C, ^{11}C, ^{14}O, ^{15}O, etc.) inside an equivalent biological target are necessary for complete knowledge of the total inventory of isotopes produced in fragmentation and other reactions. It should be noted that the disappearance of a fraction of the projectile would mean that these projectiles would not deposit their energy at the Bragg peak.

The additional dose that the patient might receive in a treatment using positron-emitting unstable radioactive beams is also a matter of concern that needs to be accurately evaluated. Studies at NIRS in Japan indicate that heavy ion therapy using ^{12}C could be combined or replaced with ^{11}C beams accelerated to the same energy (Hoj 05). A rather recent theoretical simulation carried out using GEANT4 code has also estimated and compared the relative radio biological effectiveness (RBE) of the positron emitting ^{10}C, ^{11}C and ^{15}O compared to their stable counterparts, ^{12}C and ^{16}O in heavy ion therapy (Chac 19). The simulations have shown that RBE remains more or less the same for stable and radioactive beams and the additional dose received by the target volume and the surrounding tissues in the case of RIB irradiation is negligible. Thus, the use of RIBs turns out to be clearly advantageous because these offer much accurate imaging of the treatment volume.

The main challenge of using RIBs for therapy lies in the need to obtain sufficiently high intensity (~10^8 pps) of these beams with the stringent purity required for clinical use. It has been felt that rather than using in-flight fragmentation for production of these unstable species, a higher yield and the required beam purity can be obtained by producing the isotopes using intense low-energy light ion beams, ionizing and separating them using the ISOL technique and then accelerating the mass-separated beam (say, of ^{11}C) using a synchrotron to about 400 MeV/u. At NIRS the production of ^{11}C is done using a compact cyclotron with 10–20 MeV protons using the ^{14}N (p, α) ^{11}C reaction. ^{11}C is selected (mass separated) in an ISOL system, accelerated in a linear accelerator and then injected into the synchrotron to accelerate the radioactive ^{11}C ion beam further to about 400 MeV/u. At ISOLDE, production of ^{11}C has been done using the high energy (1.4 GeV) proton-induced spallation reaction on molten fluoride salt targets (NaF:LiF eutectic) followed by separation and acceleration (Mend 14; Augu 16). However, the pulsed proton driver is not a suitable machine for ^{11}C production, and using a 70 MeV (or even higher), 1 mA cyclotron could be considered. Overall, the R&D on ^{11}C production and acceleration shows that it should be possible to achieve the desired purity and the intensity required for treatment. It is probably not long before a ^{11}C beam is actually used in radiation therapy.

2

Production of Rare Isotope Beams: The Two Approaches

2.1 Introduction

A Rare Isotope Beam (RIB) facility aims at producing the widest possible range (in N and Z) of ion beams of unstable nuclei spanning the nuclear chart. This means producing a significant fraction of 7000 particle-stable nuclei lying between the n and p drip lines, most of which are exotic nuclei. To meet the demands of experiments designed to identify these nuclei and to measure their properties, the ion beams need to be produced with sufficient beam intensity spanning a large range of energies from a few keV/u to a few GeV/u. This is needed to carry out different types of measurements on exotic nuclei and to study reactions induced by RIBs. The other requirements are the purity of the beam (no or minimum contamination from other isotopes) and good spatial and timing characteristics. Table 2.1 lists the desirable features—the so-called wish list of a versatile RIB facility.

Achieving this wish list is, however, extremely challenging and the reasons are:

- The production cross-section of unstable nuclei falls off sharply with their distance from the valley of beta-stability—more exotic the isotope, the less is the cross-section of its production. The nuclei close to drip line are produced with cross-sections typically in the range of a few pico-barns (1pb = 10^{-12} b; 1 b = 10^{-24} cm^2), reaching often as low as femto-barn (fb) for some drip line isotopes, compared to a few tens of millibarns (mb) for species close to the valley of beta-stability. The low production cross-section makes it extremely difficult to produce exotic RIBs with sufficient intensities. For example, a 1 fb cross-section roughly corresponds to a RIB production rate of only a few particles per day for 1 pµA of beam current (6 × 10^{12} beam particle per sec; pps) with 1 g/cm^2 of target thickness

- Most of the nuclear reaction routes found suitable for production of exotic species also produce a huge number of other species, often with orders of magnitude higher cross-sections making separation of a particular exotic isotope of interest an extremely difficult task

- The exotic n-rich and p-rich nuclei are short lived. Being close to drip lines the half-lives of these nuclei are often less than a few tens of ms (in some cases even smaller), allowing little time for preparation of beam for experiments

TABLE 2.1

Wishlist of a RIB Facility

Primary Beams

High intensity: up to 10 mA; limited by the power handling capacity of the target

Beam energy: a few MeV/u to a few GeV/u

Beam type: all from proton to uranium including beams of isotopes of low natural abundance such as ^{18}O, ^{48}Ca, ^{70}Zn, ^{36}Ar, ^{74}Se, ^{112}Sn, etc.

Cost of the driver accelerator/accelerators: should not be prohibitively high

Comments: Design and construction of accelerators capable of delivering high intensity beams of light/heavy ions at higher energies are technically challenging

RIBs/Secondary Beams

Production of a large number of Radioactive Beams; number exceeding 1000

Production of short-lived exotic isotopes on or very close to neutron and proton drip lines

Pure beam of single isotope

Small momentum (< a few %) and angular spread (<1 degree)

Beam intensity: ~1 to 10^{10} pps; the lower limit applies to near drip-line isotopes which are produced with very small cross sections

Beam energy: a few keV/u to about 1 GeV/u. Lower beam energy for spectroscopy studies on RIBs; higher beam energies for studying nuclear reactions using RIBs and for facilitating production of very exotic isotopes using reactions with RIBs

Beam transport time from target to experimental area: <1 ms

Comments: To produce RIBs with sufficient intensity, thick targets capable of handling the large beam power of the ion beam from the primary accelerator need to be used. Design of thick targets and its thermal management are technologically challenging. Further, production of pure a RIB of a single isotope is also a challenging task since both light ion and heavy ion induced reactions usually produce hundreds of isotopes

- Achieving sufficient intensity of RIBs needs intense beams of quite a large number of primary projectiles starting from proton to heavy ions up to uranium and energies at least up to 1 GeV/u. This together with the requirement of wide energy range puts a huge demand on the accelerator design and the cost often tends to become prohibitively high for securing funding

The intensity of a particular RIB beam is given by:

$$I_{RIB} = I_P \times N \times \sigma \times \epsilon \tag{2.1}$$

where I_P is the primary beam intensity or the number of primary beam ions per sec, N is the number of target atoms per cm^2, σ is the production cross-section and ϵ is the efficiency factor which is the ratio of intensity of the RIB delivered for experiment to that produced in the reaction at the production target.

It is clear that to produce RIBs with sufficient intensity, the following will be needed:

- Primary beam intensity as high as possible
- Target as thick as useful
- Optimum reaction route to maximize production cross-section of the nuclide of interest by choosing appropriate projectile and energy
- High efficiencies of different processes/stages necessary for preparation of the beam for experiments. Beam preparation always involves a magnetic separation stage that allows selection of the reaction product of interest. It is imperative that beam preparation time should be fast enough so that the loss of short-lived exotic nuclei through decay is minimized

Production of Rare Isotope Beams

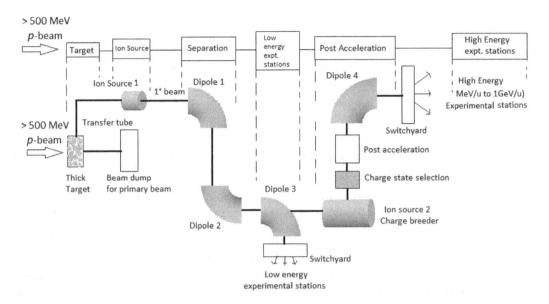

FIGURE 2.1
Schematic diagram showing the ISOL post-accelerator-type RIB facility.

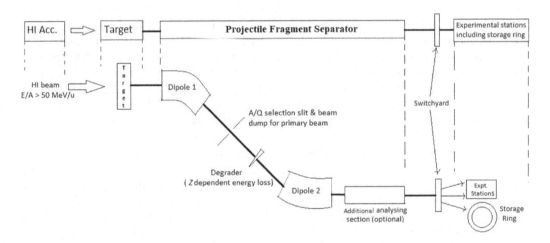

FIGURE 2.2
Schematic representation of the Projectile Fragment Separator (PFS) approach for producing RIBs.

There are two approaches to produce RIBs: the Isotope Separator On-Line (ISOL) post-accelerator technique (Figure 2.1) and the other is the Projectile Fragment Separator (PFS) or In-Flight (IF) technique (Figure 2.2). In both approaches, a stable ion beam (often called the Primary beam to distinguish it from the RIB which is quite often termed as the Secondary beam) from a driver accelerator bombards a target to produce unstable nuclei through a nuclear reaction and the reaction products are transported from the target station (production site), where the radiation background is too high to carry out any measurement, to a distant (usually the distance is more than 10 m) low background detection area where spectroscopic and other measurements are carried out on the reaction products. The most popular reaction routes such as spallation, fission and fragmentation produce a large number of isotopes, requiring an elaborate separation scheme to select the RIB of interest. The separation of reaction products for selecting the isotope of interest needs to be carried out in the transport beam line connecting the target and the experimental stations and this is where the ISOL and PFS approaches follow completely different schemes.

2.2 The ISOL Post-Accelerator Approach

The main components of an ISOL-type RIB facility are schematically shown in Figure 2.1. In the ISOL post-accelerator method a thick target is used to maximize the production rate of the RIB. Thick target implies that the target is thick enough either to stop the beam or to utilize the entire range of beam energy useful for the reaction. In the latter case only a large fraction of beam power, and not the total beam power, is deposited in the target. In either case, the reaction products are all stopped within the thick target and then come out of it by diffusion. The reaction products coming out of the thick target are in the form of neutral atoms. The neutral radioactive atoms are transported through a small tube to an ion source by the process of effusion in which the neutral atoms undergo a series of absorption and desorption cycles with the inner wall of the transport tube to reach the ion source. A small pressure gradient between the target chamber and the ion source aids the transfer process. The target and the ion source are both in vacuum (~10^{-4} torr or below) but relative pressure in the ion source is a bit lower than that in the target chamber. The neutral atoms are ionized in the ion source mainly to charge state $q = +1$. The ions are then extracted with a few tens of KV potential (usually in the range of 20 to 60 KV) and separated in an isotope separator comprising of a momentum analyzer di-pole magnet (and other optical elements for focusing, etc.) as per their charge to mass ratio and this helps to select the RIB of interest. Since the charge state is +1 for ions of all reaction products, mass separation is achieved after the di-pole magnet. It is important, therefore, for efficient separation that the ion source produces only ions of a single charge state (+1in this case). Reactions like spallation, fission, etc., produce a huge number of isotopes and one separation stage is usually not enough. The ions, after the first separation stage, are often cooled in an emittance cooler (not shown in Figure 2.1) to reduce the ion beam emittance and then extracted and fed into a second analyzing section to achieve higher resolution so that only isotopes of interest can be selected using a mass selection slit after the second di-pole magnet. At this stage, the energy of the RIB ions of interest is typically a few tens of keV, which is ideal for carrying out measurements of mass and spectroscopic properties. For further acceleration post-accelerators are used to increase the beam energy to the energy required for experiments, typically up to a few MeV/u or even higher. However, efficient post-acceleration needs a higher charge state and that is achieved by a second ion source, usually called the charge breeder. After the charge breeder a particular high charge state appropriate for the downstream post-accelerators is chosen using a di-pole magnet (not explicitly shown in Figure 2.1 but the process is indicated as charge state selection). The high charge state ion beam is then injected into the first post-accelerator. Usually, acceleration to the final energy is achieved in a number of stages using a number of accelerators in tandem. Usually either cyclotrons or linear accelerators are used for post-acceleration.

2.3 The PFS Approach

In PFS/IF method, schematically shown in Figure 2.2, peripheral in-flight fragmentation or fission reactions employing intermediate to high energy heavy ions (from about 40 MeV/u to a few GeV/u) as primary beams are used. Except for heavy projectiles ($A > 200$), the reaction is dominated by projectile fragmentation in which the nuclei of the primary beam split into projectile-like fragments (one projectile-like fragment per reaction) in a peripheral nuclear reaction with the target and the fragments move forward almost with the beam velocity in a narrow forward angle. Quite a few vibrant RIB facilities in the world are at present using projectile fragmentation reaction with high and intermediate energy heavy ions followed by projectile fragment separators (PFS) for the production of RIBs. This is why the term PFS technique is quite often used and accepted in place of IF technique.

In cases where the projectile is heavy and fissile, say uranium, fission and fragmentation-fission reactions take place in the target, increasing the spectrum of n-rich nuclei produced. In both cases the reaction products move with the velocity of the beam with a small momentum spread of a few percent and with an angular spread of a few degrees that becomes even smaller at higher beam energies. At relativistic energies close to 50% of the reaction products are in the form of fully stripped ions. The reaction products are fed directly into a Fragment Separator comprising typically of a number of analyzing (dipole), focusing (quadrupole) and corrector magnets that select and transport the RIB of interest with high efficiency. The first di-pole magnet separates the fragments according to their A/q ratios, but this separation is hardly enough. So, an energy degrader is used after the first di-pole magnet to introduce a Z dependent energy loss in the fragments which are then separated further in a second di-pole magnet positioned downstream of the degrader. The focal plane after the second di-pole still contains a number of fragments of other nuclei in addition to the fragment of interest. This makes the identification of exotic nuclei on or close to the drip line difficult since those are produced with much lesser cross-section than other nuclides. For identification of such nuclei, it is often advantageous to add an additional analyzing section (comprising of a number of di-poles and focusing elements). This enhances the A/q resolution and allows accurate tagging of each of the fragments leading to unambiguous identification of new exotic isotopes that are usually produced with very low intensities of about a few per day.

In the PFS/IF method the fragments are already of high energy. Obviously, no post-acceleration is needed in this case. The targets used in this technique are often referred to as "thin" in the sense that unlike in the ISOL method, the reaction products are not stopped in the target but are allowed to come out of the target with high enough velocity to ensure good kinematic focusing. The target in practice could be quite thick, usually about 1 g/cm^2 for projectile energy of 1 GeV/u, and with this thickness the projectile loses only a small fraction of its energy (<10%) in the target. But even this thickness is about two orders of magnitude less compared to the target thickness used in spallation reactions, justifying the nomenclature "thin" in a comparative sense.

The production of RIBs using low energy heavy ion reactions ($E/A < 10$ MeV/u) in inverse kinematics is also a kind of IF technique. In this, a low energy heavy ion projectile is allowed to incident on a lighter target. Because of the low energy of the projectile, the target in this method is thinner by two to three orders of magnitude compared to target thickness used in in-flight fragmentation or fission reactions with intermediate and high energy heavy ion projectiles. In this technique, the rejection of the primary beam poses the main challenge since the number of reaction products produced in low energy heavy ion reactions is not very large. Apart from being a useful technique for production of light and not-too-exotic RIBs ($A < 40$), this method is often suitable for studying nuclear reactions of astrophysical interest (Ha 00).

2.4 Comparison between the ISOL and PFS Approaches

To discuss relative advantages and disadvantages of the two approaches, the ISOL and IF, it may be noted that the inherent delay in the diffusion process in the target and hold-up time in the ion source are serious limitations of the ISOL approach. The times involved in these processes are ~10 ms even in most ideal cases and thus the ISOL method is often not suitable for RIBs that have very short half-lives; half-lives in the range of 1–10 ms being the limit in the most favorable cases. On the other hand, the flight time of fragments in fragment separators depends on the projectile energy (velocity), target thickness that determines the mean velocity of the fragments and the length of the separator but is usually less than a few hundred ns. Thus, with the PFS technique, in principle, all the isotopes with half-lives greater than 100 ns can be studied. This means the PFS technique would allow studies of almost all the isotopes between the proton and neutron drip lines, if drip lines are defined as the loci of S_{2p} and S_{2n} equal to zero. This is a clear advantage of the PFS approach over the ISOL approach in respect of studying exotic isotopes with half-lives shorter than a ms (in as much as it would be possible to produce them in enough numbers for identification and study!).

In the ISOL approach, as already mentioned, a target that is more than two orders of magnitude thicker than that in the IF method can be used, particularly if the primary projectile used for production of RIBs in the ISOL method is protons with energy of about 1 GeV. A1 GeV proton has a long range inside the target, allowing an effective target thickness of more than 100 g/cm^2. Also, if we assume a usable beam intensity (beam current limited not by the accelerator but by power handling capacity of the production target) of 1 mA for a 1 GeV proton and an achievable beam intensity of 1 µA for a 1 GeV/u ^{238}U heavy ion beam (the highest beam intensity so far achieved for ^{238}U is about 10^{11} particles/s at the RIKEN RI Beam Factory for beam energy of 345 MeV/u (Kam 19)), the relative gain in the yield on intensity account (assuming the same production cross-section) is a factor of 10^3 in favor of the ISOL approach. Overall, with target thickness and beam intensity considered together, there is an advantage of about five orders of magnitude in the yield of isotopes at the target position in the ISOL approach compared to the IF approach.

However, in the ISOL method the processes (diffusion, effusion, ionization, etc.) involved in converting the radioactive atoms produced inside a thick target to ion beams extracted from the ion source depend on the chemical nature of the isotope. These processes (effusion and ionization) are not highly efficient and result in some loss of beam. Further, these are slow processes and usually take more than a few tens of ms. For nuclei with half-lives in the range 1–100 ms, the delay leads to further losses by the way of radioactive decay. In fact, these two stages constitute the most technologically challenging part in an ISOL-type RIB facility and the combined efficiency of these two stages even in most favorable cases varies between 1 to 10%. This efficiency is, however, several orders of magnitude less for refractory elements with high melting points (for example, for isotopes of elements with atomic numbers Z = 40 to 46), compared to other elements possessing favorable chemical properties. This is because isotopes of these elements do not come out of the target easily (poor effusion efficiency) and are also difficult to ionize. It is thus difficult to get appreciable intensity for isotopes of refractory elements in the thick target ISOL technique. Of course, for isotopes of refractory elements thin targets (thickness limited by the recoil range of the reaction products) can be used and the reaction products recoiling out of the target can be collected in a gas and then ionized in an ion source. He-Jet recoil transport and ion guide-based ISOL systems (Macf 69; Arje 81) fall into this category and these systems have indeed been extensively used for spectroscopy on exotic isotopes of some of these refractory elements. However, since the target thickness in this case is typically about four orders of magnitude less compared with thick targets used in spallation reactions, the thin target ISOL case, although advantageous in certain specific cases, is not included for the comparison between the two approaches. Coming back to the thick target ISOL method, in addition to the losses in the processes of diffusion, effusion and ionization, there are also losses associated with the subsequent processes of magnetic separation of the singly charged ion beam of interest from ions of other reaction products. Quite often, one separation stage is not enough and two separation stages are required, with a beam emittance cooler stage in between to achieve desirable separation. The combined efficiency for 1+ low energy (a few tens of keV) mass separated ion beam hardly exceeds about 1% in favorable cases.

Post-acceleration of the selected ion beam often requires a second ion source (charge breeder) at low energy to increase the charge state for higher acceleration efficiency, and charge breeding efficiency (less than 10% in most cases) reduces the beam intensity further. Further, the acceleration efficiency in the post-accelerator(s) is usually less than 50%. Usually, in addition to charge breeding, charge stripping is also needed, especially for acceleration to high energies that result in further loss of the ion beam intensity. The combined efficiency on account of post-acceleration is usually about 1%. All these efficiencies put (multiplied) together may be termed as the beam preparation efficiency that varies widely in the ISOL method depending on the chemical nature of the element. However, for the sake of comparison between the two techniques beam preparation efficiency may be taken as 1×10^{-4} (1×10^{-2} without post-acceleration) for the ISOL method.

In the PFS/IF method, on the other hand, the reaction products are fragments moving in the beam direction with high velocity. There is practically no loss of intensity on account of decay in this technique. Also, since the fragments are in the form of highly stripped ions there is no need for an ion source and thus the beam preparation efficiency in this technique is independent of the chemical nature of the reaction product. However, even at relativistic energies of several hundred MeV/u, all the fragments are not fully stripped and there is a charge state distribution comprising of about four charge states including the fully stripped one. It is necessary to select the fragment of a particular charge state which means a loss in intensity. Also, the fragments have considerable momentum and angular spread, but any fragment separator only accepts a limited momentum and angular spread. The finite acceptance leads to further loss of intensity. Also, the momentum acceptance often needs to be limited further so as to allow fewer reaction products at the final focal plane where measurements are carried out. So, experimental requirements also add up to the intensity loss. Overall, the beam preparation efficiency can be taken to be about 10% for the purpose of comparison, which is about 1000 times higher compared to the ISOL post-acceleration technique. This brings the relative advantage in yield down to about 100 times in favor of the ISOL technique. However, it should be noted post-acceleration of ISOL beams is not required for the study of ground state properties of nuclei (such as mass, spin, nuclear moments, half-lives, etc.) as well as for that of excited states populated in β-decay.

These studies can best be done with low-energy beams from an ISOL facility and thus for these studies the relative advantage in yield is about 10^4 times in favor of an ISOL-type facility. It should be stressed that these numbers for relative advantages (10^4 times, etc.) are only indicative and may vary widely depending upon the isotope and the kind of measurement to be carried out.

The superiority of the PFS technique is of course unquestionable for the study of very short-lived isotopes with half-lives lower than about 1 ms and for isotopes of refractory elements that are difficult to ionize. Also, in the PFS technique a new isotope is identified by its Z and A, using online particle identification techniques rather than the decay spectroscopy techniques employed in the ISOL method. Since the online detection techniques are very efficient (detector efficiency close to 100%) and have less problem from background radiation (such as those originating from cosmic rays and other radioactive sources in the surroundings), the requirement on RIB intensities for the identification of a new exotic isotope (and also for various other measurements) is often two to three orders of magnitude less in the IF technique compared to the ISOL technique. For example, in the IF technique RIB intensity of a few particles per day or even per week may be enough for an unambiguous identification of a new exotic isotope produced with a very small cross-section of say 1 fb.

In the ISOL method, on the other hand, it is usually possible to get a much purer beam (less contamination from other isotopes produced in the same reaction) and also a beam of much better quality (smaller energy width and angular divergence) compared to the PFS method. The better beam quality of a low energy ISOL beam before post-acceleration allows very high resolution spectroscopy and precision mass measurements using Penning traps, for example. Low energy and relatively high intensity beams from the ISOL facilities have the potential to resolve a good number of issues in nuclear structure and in stellar nucleo-synthesis. Another advantage in the ISOL technique is the decoupling of the production and acceleration. The ISOL beam can be accelerated to any energy by suitable post-accelerators irrespective of its production route. This allows all kinds of experiments to be carried out using RIBs accelerated to various energies from sub-barrier to Fermi energies and above. The fact that many spectroscopic measurements need the beam to be slowed down/stopped make the PFS technique not an ideal one for carrying out accurate measurements of many of the properties of exotic species. The lack of purity of the beam creates further difficulty. However, precise determination of mass to about 100 keV accuracy is possible by injecting the high energy beam into a storage ring (for example, Experimental Storage Ring (ESR) at GSI). Table 2.2 summarizes the characteristic features of the two types of RIB production approaches and their relative advantages and disadvantages.

A great deal of R&D effort is at present going on to overcome the technological challenges that limit the RIB intensity and purity in both ISOL and PFS-type facilities. In fact, over the years many of the shortcomings could be at least partially overcome. The improvements allowed identification of a host of new exotic isotopes thereby extending the region of known nuclei towards the proton and neutron drip lines and also precise measurements of the nuclear properties of exotic isotopes. The ISOL method of producing RIBs is being used at, among other places, Louvain-la Neuve (Belgium; Huy 11), SPIRAL at GANIL (France; Jar 12), ISOLDE at CERN (Geneva; Cath17), ISAC at TRIUMF (Canada; Bal 16)—to name a few. The major laboratories where the PFS method of producing RIBs is used at present are GANIL (France; Thom 17), RI Beam Factory at RIKEN (Japan, Yan 07), GSI (Germany; Gei 92), and NSCL at MSU (USA, Gad 16). A detailed account of worldwide RIB facilities is presented in Chapter 8.

TABLE 2.2

Characteristic Features of the ISOL- and PFS-Type Facilities

ISOL-Type Facilities	PFS-Type Facilities
Reaction products, produced in light or heavy ion reactions, are stopped in the target and then allowed to come out of the target by diffusion. The reaction products in the form of neutral atoms are then ionized, extracted, mass separated and accelerated. These beam preparation steps, especially release from the target and ionization, are technologically complex and are considered to be the most challenging parts in an ISOL-type RIB facility.	Reaction products are produced in the peripheral collisions of high energy heavy ion projectiles with target. Reaction products (fragments) move with the projectile (beam) velocity and then separated in-flight. Technologically, beam preparation steps are much simpler compared to the ISOL technique, which is considered a major advantage of the PFS-type facilities.
RIB can be accelerated to any energy (from low to Coulomb barrier to Fermi energy and beyond) by using suitable post-accelerators.	RIB energy is above the Fermi energy and has the same E/A as the beam or the projectile after the target. Thus, the PFS technique does not allow in a straightforward way the study of Coulomb barrier physics, direct measurements of capture cross-sections for interest in nuclear astrophysics, etc.
The highest RIB intensity is achievable for isotopes with half-lives more than 10 ms and possessing favorable chemical properties. Beam intensity depends on the chemical nature of the isotope because the release efficiency from the thick target and the ionization efficiency both depend on the chemical properties. This limits severely the production rate of shorter-lived isotopes of refractory-type elements, making those difficult to study using the thick target ISOL technique.	Primary beam intensity and target thickness are both lower than the ISOL case and therefore RIB intensity is usually lower (often by orders of magnitude). Products are fully or almost fully stripped; no need for ionization and thus beam intensity is independent of the chemical nature of the isotope. Isotopes of all elements, including those of refractory elements, are therefore available for various kinds of studies.
The ISOL method can provide RIBs of high purity and better beam quality.	Compared to the ISOL method both beam purity and quality are poorer in the PFS technique.
RIBs from ISOL facilities are typically suitable for spectroscopic studies, high precision (within a few keV) mass measurements in traps, and studies on the sub-barrier and around Coulomb barrier nuclear physics.	RIB energy, as such, is too high for Coulomb barrier physics and detailed spectroscopic measurements. However, Coulomb excitation studies are possible using high energy fragment beams and also precision mass measurements (better than 100 keV) are possible in storage rings. Coulomb barrier physics and spectroscopy studies are possible by slowing down and stopping the fragment beam.
Identification of a new isotope using spectroscopic techniques usually requires higher intensities of RIBs (compared to the PFS technique) because of low beam preparation efficiency, low efficiency of radiation detectors and background radiation from cosmic rays and the surrounding structures.	Identification of a new isotope requires much less intensity (a few per day is sometimes enough) of RIBs because of online detection techniques that have very high detection efficiency.
Ionization, extraction and separation followed by post acceleration limits the half-life of the nuclei produced to about 1–10 ms in favorable cases and to about 100 ms in other cases. This would probably not allow study of some n-rich nuclei close to drip line with half-lives shorter than a few ms.	Beam preparation time is about 100 ns or less. So, neutron drip line isotopes can be identified in principle, if those can be produced with intensities exceeding about a few per week.

2.5 The Combined Approaches

Many of the existing/upcoming RIB facilities are trying to combine both ISOL and PFS approaches to produce the widest range of exotic nuclei with optimum yields. One way to combine is to start with a PFS facility and add an ISOL-type facility to it, as shown in Figure 2.3. The high energy heavy ion driver accelerator and the projectile fragments separator, together with experimental stations for experiments using the high energy fragment beams, constitute the PFS facility. The fragments can however also be directed to a different station where they are first slowed down by means of metal degrader plates and a metallic wedge and then are captured/thermalized in a gas cell of pure helium (Sa 03). The exotic ions pick up electrons in the process of thermalization to become singly charged ions in pure helium, since the ionization potential of helium is very high. The ions are guided to a small exit hole of about 1 mm diameter by DC and RF electric fields and are then extracted, mass separated and delivered to various experimental stations for precision spectroscopic measurements. Alternately the ions can be accelerated to higher energies, usually from a few keV/u to about 10 MeV/u for various reaction studies of importance to nuclear astrophysics and nuclear structure physics. The thicknesses of the energy degrader plates can often be adjusted to get rid of most of the unwanted contaminants (fragments other than the one of interest) present in the fragment beam from the fragment separator and thus almost pure exotic beams can be extracted from the gas cell. The delay time in the gas cell is expected to be about a few tens of ms but unlike in an ISOL facility, has no chemical dependence. The PFS method is thus combined with the ISOL method through the stopping and re-acceleration of the projectile fragments. All kinds of experiments at low energy and after acceleration around Coulomb barrier energies can thus be conducted. The original proposal for the Rare Isotope Accelerator (RIA) facility project in the USA contained this idea although the RIA proposal also included the provision for a standalone ISOL-type facility by using a 1 GeV proton beam that can be obtained from their versatile superconducting Linac driver accelerator that was designed to provide both an intense beam of 1 GeV proton and uranium beam of 400 MeV/u. The facility for Rare Isotope Beams (FRIB), a truncated version of RIA, coming up at MSU aims to deliver unprecedented high power beams of all heavy ions from C to U to be used for

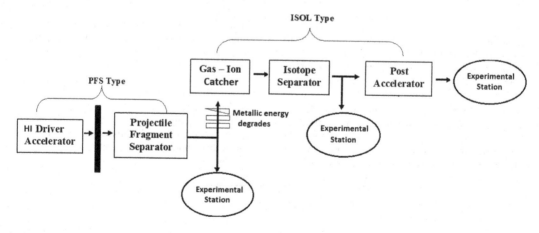

FIGURE 2.3
The schematic presentation of the PFS-ISOL combined approach.

Production of Rare Isotope Beams

studying RIBs using the PFS approach and also to use a gas catcher system to study exotic ions using the ISOL technique. In general, slowing down of high energy projectile fragments to thermal energies is now employed for example, at RIKEN RI Beam Factory (Wa 04) so that the techniques of atomic and nuclear spectroscopy including those for precision measurement of mass developed at low energy ISOL type facilities (ISOLDE, Louvain la Neuve, ISAC, etc.), can be used.

The other way to combine the ISOL and PFS-type facilities is to produce RIBs using the ISOL technique and accelerate the RIBs in stages up to, say, 100 MeV/u or more to effectively utilize the PFS technique (Figure 2.4). This approach, in principle, allows production of very exotic isotopes through projectile fragmentation and deep inelastic reactions with accelerated RI or secondary beams, initially produced and mass separated using the ISOL technique. Depending upon the chosen acceleration scheme, RIBs from a few keV/u to about 7–12 MeV/u can be routed to various experimental stations. This is the ISOL part. The RIB (7–12 MeV/u) can then be accelerated further to 100 MeV/u or higher. Projectile fragmentation reactions using a secondary target can then be utilized to produce and study, after separation in a fragment separator, very exotic reaction products resulting from the fragmentation of RIBs. Theoretical estimation of production of exotic n-rich nuclei in the "cold" fragmentation reaction of n-rich secondary beams (Hel 03) and experiments conducted at GSI using ^{132}Sn beams (Per 11) have indeed shown great potential for producing very exotic n-rich nuclei using secondary n-rich beams. Since the ISOL method can produce RIBs with the maximum possible intensity, there is a clear advantage of producing these secondary n-rich beams in the ISOL method, using say, fission of uranium. In the ISOL technique employing a thick target, the fission of uranium can produce exotic n-rich radioactive isotopes of considerable intensity that can in principle be as high as 10^{10} pps in favorable cases. For example, n-rich RIBs such as ^{92,93}Kr, ^{132}Sn, ^{144}Xe, ^{146}Cs, ^{149}La, etc., can be produced with quite high yield using fission of ^{238}U. Projectile fragmentation (also deep inelastic transfer collisions in combination with stable and unstable n-rich targets) of these n-rich secondary beams offer an exciting prospect of producing very n-rich nuclei lying on the r process nucleo-synthesis path; and for some lighter elements drip line can hopefully be reached. Projectile Fragmentation of RIBs, produced using the ISOL technique, would also allow production and study of isotopes of very exotic isotopes of refractory elements, otherwise not possible in an ISOL-type facility. On the neutron-deficient side, n-deficient RIBs with the highest possible intensity can be prepared, which can be used to produce and study p drip line nuclei using either projectile fragmentation, deep inelastic transfer or

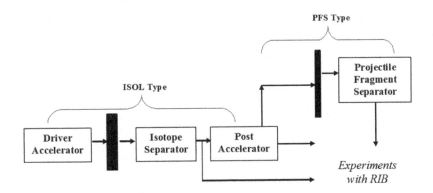

FIGURE 2.4
The schematic presentation of the ISOL-PFS combined approach.

compound nuclear fusion reaction. It is also obvious that the same accelerators that accelerate RIBs can also be used to accelerate stable heavy ions by injecting stable ions from a separate ion source into the same accelerator module(s). The ANURIB facility in Kolkata (Cha 07) has been configured to derive these advantages of a combined ISOL-PFS-type facility. The RAON project in Korea also has been planned following the same approach of combining the ISOL and in-flight methods to produce very exotic isotopes through projectile fragmentation of beta-unstable RIBs that are to be produced using the ISOL technique with a proton cyclotron as the driver (Je 16).

As mentioned already, the major challenge to the RIB community is to produce accelerated beams of a few thousand exotic isotopes with sufficient beam intensity to allow study of their properties. These would make possible the development of reliable theoretical models capable of predicting the properties of the rest of the exotic nuclei with the desired precision. The proton drip line is closer to the beta-stability line compared to the neutron drip line and the drip line or near drip line neutron-deficient isotopes of all the elements up to lead are produced in the laboratory. However, determination of properties of many of these nuclei would be possible only if they can be produced with larger yields, which requires building the next generation of RIB facilities comprising very powerful driver accelerators. On the n-rich side, the drip line is only reached up to neon (Ahn 19) and there are still more than 2500 exotic nuclei waiting to be produced. Also, the yields of isotopes already produced need to be increased to allow complete spectroscopic studies. To understand and appreciate the efforts that are presently being made to meet these challenges, it is necessary to discuss the factors of RIB production in some detail: the nuclear reactions, the production target, ion sources and accelerators for primary beam as well as secondary beam (RIB) acceleration. These factors together determine the ultimate yield of RIBs. Chapter 3 is dedicated to discussions on various nuclear reactions used for RIB production: spallation/target fragmentation using high energy protons/light ions in Section 3.1; projectile fragmentation and in-flight fission using intermediate and high energy heavy ions in Section 3.2; fission induced by thermal neutrons, protons and gamma rays in Section 3.3 and CN fusion and deep inelastic transfer reactions using low to medium energy heavy ions in Section 3.4. Following the reactions, we discuss the technical aspects of RIB production, namely, the production target, ion sources and accelerators in Chapters 4, 5 and 6 respectively. The separation of RIBs is also an important aspect of production, but this will be discussed in detail in Section 7.2 of Chapter 7, which is dedicated to discussing experimental techniques for measurements on RIBs/exotic nuclei.

3
Nuclear Reactions for Production of Rare Isotope Beams

3.1 Production of RIBs in High-Energy Proton-Induced Reactions (Spallation/Target Fragmentation)

3.1.1 Introduction

High-energy protons are used to produce RIBs at the ISOLDE facility at CERN, Switzerland, and at the ISAC facility at TRIUMF, Vancouver, Canada. The ISOLDE facility currently makes use of a 1 to 1.4 GeV proton beam from its Booster Proton Synchrotron (PSB) and the maximum average beam current is about 2 µA (about 10^{13} protons per second), whereas the ISAC facility uses a 500 MeV proton Cyclotron delivering 500 MeV proton beams but the beam current delivered to the target could be much higher and can reach up to 100 µA. ISOLDE has been the first extensive facility to be built for exotic nuclei physics and, along with the ISAC facility that was commissioned in the late 1990s, have made very important contributions to our knowledge of exotic nuclei physics.

When a proton or light particles (such as ^{3}He) of energies exceeding several hundred MeV interact with heavy target nuclei, it results in the emission of a large number of hadrons (mostly neutrons) and fragments of various masses. The reaction has been named "Spallation" following the patterns of the reaction products (fragments of various masses/sizes), first observed in nuclear track detectors in reactions with high energy cosmic rays. Much later, accelerators capable of accelerating protons to relativistic energies were developed, allowing much more detailed study of the reaction process. The Spallation reaction has been found to be one of the most optimum routes for the production of a wide variety of Radioactive Ion Beams. This reaction is also very important for the study of nuclear equation of state since it can deposit a large amount of energy inside a nucleus. Further, since an interstellar medium is always getting bombarded by high energy protons in the cosmic rays, a detailed understanding of the Spallation reaction is necessary to determine the original composition of interstellar media and how were they produced.

There were also other motivations that created wide-ranging interest in Spallation reactions. The reaction can produce a strong flux of neutrons of a very broad energy spectrum that can be suitably pulsed to make it highly suitable for the study of materials. It is also considered the most cost-effective and safe neutron source for Accelerator Driven Subcritical Systems (ADSS). In ADSS the nuclear reactor is designed to be sub-critical and is made critical by the additional supply of neutrons (that can be switched off quickly if required) from Spallation reactions of relativistic proton beams on heavy targets like Pb/Bi placed close to the reactor core. Also ADSS, suggested by Carlo Rubia and his collaborators

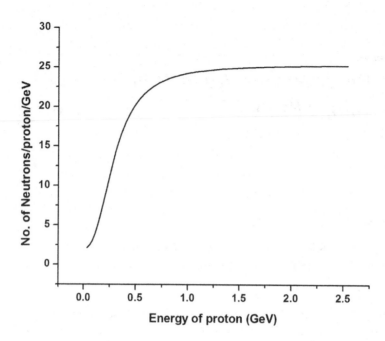

FIGURE 3.1
Spallation neutron yield as a function of incident proton energy.

(Rub 94) have the potential to provide a solution to the nuclear waste disposal problem by converting the long-lived radioactive waste products to shorter-lived ones. For ADSS applications the incident proton energy of around 1 GeV has been found to be cost effective and an optimum choice (Figure 3.1) for the neutron source. For proton energy of 1 GeV and using a Pb–Bi target, the number of neutrons produced per proton is more than 25 and this gives the best neutron economy among different options of producing neutrons using accelerated light and heavy ions. Incident energy of around 1 GeV has also been found to be a good choice for production of exotic nuclei at RIB facilities.

3.1.2 The Spallation Reaction Process

The Spallation reaction (in which a light ion-like proton of relativistic energies interacts with the nucleus of a heavy target, say lead) can be understood as a two-stage process: intranuclear cascade (INC) followed by statistical de-excitation (Se 47). This is schematically represented in Figure 3.2. The de Broglie wavelength of a proton even at 500 MeV is less than 1 fm (1 fm = 10^{-15} m), that is of the same order as the size of a proton or neutron. The reaction is thus dominated by individual nucleon–nucleon collisions rather than the projectile interacting with the entire target nucleus through the average mean field of the entire target nucleus. The high energy proton thus interacts with individual nucleons in the target nucleus and shares its energy mostly through elastic collisions. The presence of other nucleons limits the number of available quantum states after the scattering because of Pauli's exclusion principle which does not allow all classically possible energy transfers in the collision process. Thus, target nucleons not participating in the collision process make their presence felt indirectly by effectively reducing the cross-section of nucleon–nucleon scattering.

The mean free path of protons at these high energies is also of the order of a few fm. So, the incident proton usually interacts with only a few nucleons before leaving the

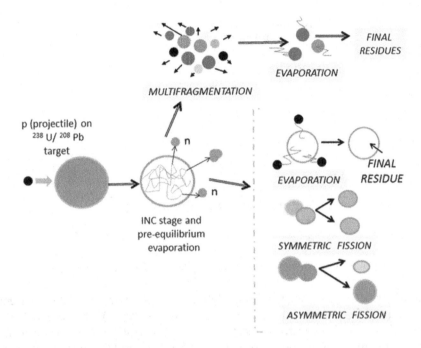

FIGURE 3.2
Schematic representation of the spallation reaction process.

target nucleus. In the process, it deposits a fraction of its energy in the target nucleus. The nucleons struck by the incident proton can have very high energy and interact with other nucleons in the same way as proton, resulting in an INC of fast nucleons—a sequence of two-body interactions that take place in a typical time scale of 10^{-22} s. During the INC process, fast nucleons, mainly neutrons (since protons and heavier clusters need to overcome the Coulomb barrier) but sometimes also light fragments, may be emitted from the target nuclei with high energies. These are called pre-equilibrium emissions. A majority of the cascade nucleons are absorbed in the target nucleus leaving the residual nucleus (at the end of the Cascade stage) in a state of high excitation energy. At energies in excess of the pion production threshold ($E_{CM} > 150$ MeV), the INC will also contain pions in addition to nucleons. Since the cross-section of the pion–nucleon interaction is higher than that of nucleon–nucleon interaction, the pions will have shorter mean free path and will be absorbed within the target nucleus resulting in an increase of the excitation energy of the residual target nucleus. The fast INC stage leaves a distribution of residual nuclei (cascade products) in (A, Z) and in excitation energy.

The statistical de-excitation of INC products (also termed as residual nuclei or prefragments) starts subsequently. This second stage is a slow process very similar to the statistical cooling of an excited compound nucleus and the timescale could be as long as 10^{-16} s. During this process, the excited residual nucleus loses energy predominantly by evaporation of neutrons (since proton and other light charge particles need to overcome the Coulomb barrier) and also by fission, which competes with evaporation. In the first case in which the de-excitation proceeds only through evaporation of nucleons, the end result is a target-like nucleus that extends from the target to 10 to 15 elements below. The de-excitation through fission, on the other hand, produces two medium mass nuclei since fission is a binary decay process.

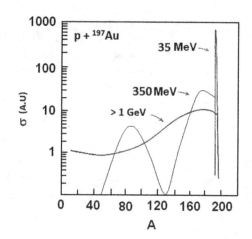

FIGURE 3.3
Nature of mass distribution in proton-induced reaction on ^{197}Au at different proton energies.

The mass distribution of products in spallation reaction changes as the incident proton energy is increased from a few hundred MeV to a few GeV (Fr 81). The basic nature of mass distributions that would result from reaction of high-energy protons with a heavy target like gold (Au) at (i) incident proton energy of about 350 MeV and (ii) energy exceeding 1 GeV are shown in Figure 3.3. The mass distribution for proton energy of 35 MeV is also included just for the sake of comparison since at 35 MeV, the reaction is entirely of the compound nuclear type producing a narrow band of masses close to the target mass. At 350 MeV, the reaction is dominated by spallation and a great number of isotopes characterized by two broad peaks, corresponding to spallation–evaporation and spallation–fission, are produced. The first peak closer to the target mass is typically about 10 to 15 mass units less than the target (spallation–evaporation products) and the other, much broader, is around a little less than the 50% of the target nucleus mass number (spallation–fission products). As the proton energy is increased beyond 350 MeV, the mass distribution pattern changes quite rapidly, presumably due to pion production. The pion–nucleon cross-section is much larger than nucleon–nucleon cross-sections and thus the pions have a much shorter mean free path. This leads to an increase in the deposition energy in nuclear volume during the INC stage that leads to an increase in the excitation energies of the INC residues. The higher excitation energy of the INC residues might result in an increase in the probability of INC residues decaying through a process called nuclear multi-fragmentation (Ran 81), which is a fast process following the INC stage leading to a break-up of INC residues into a number of clusters, as shown schematically in Figure 3.2. The clusters/fragments thus produced eventually cool down through slow statistical de-excitation, producing nuclei of a wide range of mass numbers (Intermediate Mass Fragments; IMF). It is observed, as a consequence, that at higher proton energy exceeding the pion-production threshold (300 MeV in a laboratory system) energy by a few hundred MeVs and especially at energies exceeding 1 GeV, the broad peaks corresponding to spallation–evaporation and the spallation–fission groups are replaced by a continuous distribution of products extending to light nuclei, as shown in Figure 3.3.

3.1.3 Production of Neutron-Deficient Exotic Nuclei Using Spallation–Evaporation Reaction

Spallation–Evaporation can produce very neutron-deficient nuclei because of a loss of neutrons in the INC stage and in the subsequent evaporation stage which is dominated by

TABLE 3.1A

Some of the Proton-Rich Nuclei Produced at ISOLDE (from ISOLDE Website)

Isotopes	Half-Life	Proton Energy (GeV)	Targets	Yields (ions/µC)
^{9}C	126.5 ms	1	CaO	2.0×10^3
^{17}Ne	109.2 ms	1	MgO	1.0×10^3
^{31}Ar	15.1 ms	1	CaO	3.4×10^1
^{35}K	190 ms	0.6	Sc2C3	2.5×10^5
^{70}Se	41.1 m	1.4	ZrO$_2$	2.0×10^5
^{73}Se	39.8 m	0.6	Nb	2.4×10^7
^{71}Br	21.4 s	1.4	Nb	7.5×10^4
^{75}Br	96.7 m	1.4	Nb	7.0×10^8
^{71}Kr	64 ms	1.4	Y$_2$O$_3$	10.0
^{75}Kr	4.3 m	1.0	SrO	8.0×10^5
^{82}Rb	1.273 m	1.4	UC$_X$	4.5×10^6
^{75}Sr	71 ms	1.0	Nb	2.5
^{83}Y	2.85 m	0.6	Nb	1.0×10^6
^{102}Ag	7.7 m	0.6	Ta	7.7×10^4
^{102}Cd	5.5 m	1.4	LaC$_X$	8.0×10^5
^{105}In	5.07 m	1.4	LaC$_X$	1.3×10^6
^{108}In	58 m	0.6	LaC$_X$	5.5×10^6
^{107}Sn	2.9 m	1.4	LaC$_X$	6.0×10^6
^{117}Te	62 m	1.4	ZrO$_2$	1.0×10^6
^{123}I	13.27 h	1.4	PbBi	8.5×10^8
^{118}Xe	3.8 m	1.0	La$_2$O$_3$	9.7×10^6
^{114}Cs	0.57 s	0.6	La	8.4
^{126}Cs	1.64 m	1.4	UC$_X$	4.6×10^5
^{160}Yb	4.2 m	1.4	Ta	3.1×10^7
^{167}Yb	17.5 m	1.4	Ta	2.5×10^8
^{188}Hg	3.25 m	1.4	PbBi	1.4×10^9

neutron evaporation. In fact, for production of highly neutron-deficient isotopes with mass numbers not very far away from the target, the spallation-evaporation reaction is considered to be of the most efficient routes. Some of the neutron-deficient nuclei produced at ISOLDE are listed in Table 3.1A. It can be seen, for example, that using a 600 MeV proton beam and La target, it was possible to produce ^{114}Cs with measurable yields in ^{139}La (p, 3p23n) ^{114}Cs reaction (Ra 79). It should be noted that ^{114}Cs, having a half-life of 570 ms, is a neutron-deficient proton drip line isotope (for cesium ^{133}Cs is the beta-stable isotope) and its production involved the removal of 23 neutrons and 3 protons from the p + La system. Similarly, 1 GeV and 1.4 GeV proton-induced spallation–evaporation of Ca and Y targets produced exotic species such as ^{31}Ar (2p and 7n removal from target nucleus ^{40}Ca; $t_{1/2}$ = 15.1 ms) and ^{71}Kr (3p and 15n removal from target nucleus ^{89}Y; $t_{1/2}$ = 64 ms) respectively.

In a spallation–evaporation reaction, the centroid of mass distribution of isotopes for any element is usually quite a few neutrons less than the neutron number of the beta-stable isotope of the same Z. The centroid, however, tends to shift towards the line of stability if a n-rich target like uranium is used. For example, it has been found in the case of the production of Cs isotopes that the centroid of production yield shifts from (A = 127) to almost on the beta stability line (A = 133) if a uranium target is used instead of a lanthanum target (Ra 79). So, for production of neutron-deficient species, it is always better to choose relatively (compared to other isotopes of the same element) neutron-deficient stable isotope targets. In fact, the spallation–evaporation reaction is most suitable for producing p drip line isotopes for relatively lighter elements. Owing to the bending of the beta-stability line towards neutron excess with an increase in atomic number, heavy stable isotope targets like ^{208}Pb, ^{197}Au, ^{193}Ir, ^{181}Ta, etc., have about 40 more neutrons than protons. These targets are obviously not the optimum choices for production of neutron-deficient nuclei using a spallation–evaporation reaction. However, in cases where the choices of other possible reaction routes are limited, the spallation reaction on heavy targets can be and has been used for production of proton-rich (neutron-deficient) nuclides. A good example is the production of a chain of Fr isotopes at ISOLDE, CERN, some of which are extremely neutron-deficient, using spallation of a ^{238}U target; for example, the spallation reaction produced p-rich ^{202}Fr which requires removal of five protons and 31 neutrons from the ^{238}U target. Lastly, it should be also mentioned, for the sake of completeness, that a spallation–evaporation reaction can also produce a few isotopes that are a little n-rich (lying close to the stability line on the n-rich side). This becomes possible due to large fluctuations in the excitation energy of the INC residues.

3.1.4 Production of n-Rich Exotic Nuclei in Spallation–Fission Reaction

In the case of de-excitation through fission (often termed the Spallation–Fission reaction), which becomes predominant for heavier targets like Pb and more so for the actinide targets such as Th, U, etc., it is expected that two n-rich fragments would be produced since fission fragments tend to retain the memory of N/Z ratio of the compound nucleus; in this case N/Z ratio of the residual nucleus after the INC stage. However, because of the loss of neutrons from the target nucleus during the INC stage, the neutron-richness is often limited compared to low-energy fission. Also, the fission products are in an excited state and de-excite through neutron evaporation further reducing the neutron-richness of the spallation-fission products. So, the medium mass n-rich nuclei produced as a result of spallation are often not as n-rich as in the low energy fission. In fact, a spallation–fission reaction also produces some p-rich medium mass species close to the beta-stability valley. However, when the incident high-energy proton undergoes peripheral collisions, a loss of neutrons from the target nucleus is much less and so is the excitation energy left by INC in the target nucleus volume. In such cases quite n-rich nuclei corresponding to low energy fission are produced (Be 94). It is the small contribution (about 5% of the total spallation cross-section) of low energy asymmetric fission that contributes to the production of n-rich nuclei in the range of $23 < Z < 65$. One noticeable difference between fission produced by low energy neutrons or light ions and the spallation-induced fission is that the width of the isotopic distribution is much broader in the case of spallation-fission, producing on average about 25 isotopes per element. This makes separation of the isotope of interest more difficult in the case of spallation–fission products because of the greater amount of isobaric contaminations compared with low energy fission.

To digress a little it could also be mentioned that a low energy fission reaction with proton beams of relativistic energies can be selected by using just a converter target like Pb to produce spallation neutrons. The spallation neutrons in turn can be used to induce fission in a uranium target and a fission rate as high as 10^{15}/s could be reached with 1 GeV and a few mA of beam current. This route has the definite advantage of producing n-rich nuclei in the range of $28 < Z < 60$ with higher cross-sections and less contamination. In fact, ^{82}Zn, an extremely n-rich nucleus, was produced in a neutron-induced fission of uranium at ISOLDE, where neutrons are produced by bombarding a tungsten converter with 1.4 GeV protons (Wolf 13). The production method allowed the measurement of the mass of ^{82}Zn for the first time. In general, however, use of a neutron converter limits the range of n-rich nuclei that can be produced using, say, a 1 GeV proton beam. In particular, as will be discussed in the following paragraphs, spallation–fission can produce light n-rich nuclei (Z < 23) through highly asymmetric fission and by multi-fragmentation, which cannot be produced in low energy fission.

Much of the detailed understanding of the spallation reaction has resulted from experimental studies employing inverse kinematics, carried out at GSI; that is, instead of a proton beam bombarding a heavy target, the heavy ion beam of comparable energy per nucleon (E/A) hits a proton target. The inverse kinematics allowed in-flight identification of all reaction products irrespective of their half-lives and also allowed study of reaction kinematics. This is in contrast to the direct kinematics studies performed earlier that allowed detection of mostly longer-lived nuclei.

Up to 1 GeV of incident proton energy, spallation–evaporation and spallation–fission are the most prominent de-excitation modes and these two reaction channels together account for most part of the total spallation cross-section of about 2 b. In the case of a 1 GeV proton beam bombarding a fissile target like uranium, the spallation evaporation residues typically account for less than 25% of the total cross-section and spallation–fission accounts for more than 75% of the reaction products (Ar 04).

For proton energy of 1 GeV and with a uranium target (actually, an incident beam of ^{238}U with energy of 1 GeV/u hitting a proton target in the inverse kinematics), three groups of products can be identified, a high Z group corresponding to spallation–evaporation products, a medium Z group corresponding to spallation–fission and a low Z group produced by spallation-asymmetric fission/target fragmentation (Ar 04). A substantial fraction of the light fragments produced in proton-induced reactions has been found to be n-rich as the proton energy is increased to 1 GeV and beyond. These fragments with Z < 23 belong to the third group of nuclei produced in spallation reactions. The isotopic distributions of these light nuclei peak towards the n-rich side of stability and that results in the formation of very n-rich light nuclei with cross-sections sufficient to allow spectroscopic measurements on some of these nuclei. For less fissile targets, multi-fragmentation, rather than asymmetric fission, is expected to be the dominant source of production of these light n-rich isotopes.

3.1.5 Highly Asymmetric Fission vs Multi-Fragmentation

Experimentally, it has been found that in a spallation reaction of a 1 GeV proton on a ^{238}U target, performed in inverse kinematics at GSI using a 1 GeV/u ^{238}U beam and a hydrogen target, that these light n-rich nuclei are produced predominantly from highly asymmetric fission of INC residues. The inference was drawn based on the kinematic properties of these light nuclides (Ri 06). However, in cases of more central collisions or at higher proton energies exceeding several GeV, INC residues might have much more excitation and the nucleus becomes unstable due to overheating in a short period of time. In such cases the binary de-excitation through fission, which is a slow process, is replaced by a fast decay process, in about 10^{-22} to 10^{-21} s, called multi-fragmentation (Ran 81). Multi-fragmentation is a simultaneous break-up process into a number of mass fragments/heavy clusters, which have different kinematic properties compared to the binary decay mechanism. It has been found that the kinematic properties of light nuclei produced in a 1 GeV proton on a ^{56}Fe reaction, performed in inverse kinematics using a 1 GeV/u ^{56}Fe projectile on a hydrogen target, can be explained assuming a prompt break-up process (Na 04). The multiple fragments are thought to arise from a nuclear liquid-to-gas phase transition. The overheated residual nucleus would vaporize and then cool down to condense into smaller droplets called Intermediate Mass Fragments (IMFs) (Hi 84). For heavy target nuclei with a lot more neutrons than protons, the IMFs tend to retain the memory of the neutron excess (A/Z ratio), resulting in formation of light fragments that are very n-rich. Multi-fragmentation is thus an important reaction channel that contributes to the production of very n-rich light nuclei especially when the proton energy is around or exceeds 1 GeV. It is believed that multi-fragmentation is the dominant reaction mechanism in the production of ^{11}Li and ^{11}Be which were produced in high-energy proton reactions with a Ta target. The same is true for productions of other n-rich isotopes such as ^{22}O and ^{16}C using a CaO target and production of ^{14}Be and ^{33}Na using a UC$_x$ target. In addition to contributing to the production of n-rich light nuclei, multi-fragmentation is also thought to be largely responsible for the formation of neutron-deficient isotopes in the fission product mass range (Beg 71). The de-excitation stage of the spallation reaction can proceed through multi-fragmentation followed by the cooling of fragments by evaporation or by a slow statistical process that includes highly asymmetric fission. Thus, apart from conventional fission both multi-fragmentation and highly asymmetric fission contribute to the production of intermediate mass fragments. In literature, especially while discussing the production of light n-rich exotic isotopes using high-energy protons, the terms "spallation reaction products"/"target fragmentation products"/"nuclear multi-fragmentation products" are all used because it is not always easy to determine whether a particular product has resulted from highly asymmetric binary fission or multi-fragmentation. In fact, both the reaction processes are quite likely to contribute to the production of a particular isotope. In general, the multi-fragmentation reaction becomes more and more important as the incident proton energy increases and contributes in a major way to the production of light n-rich nuclei from heavy targets.

3.1.6 Measured Yields of Exotic Species Using Spallation Reaction at ISOLDE

Spallation reaction produces a very large number of isotopes through spallation–evaporation, spallation–fission and spallation–multi-fragmentation reactions. The yields of individual isotopes are therefore often not large. The total spallation reaction cross-section for a 1 GeV proton on ^{238}U has been experimentally determined to be about 2 barn (1 barn = 10^{-24} cm^2). As is true for any other reaction, in this case also the production cross-section decreases sharply as the reaction product becomes more and more exotic. Thus, the more exotic the isotope, the lower its production cross-section, But the long range of high-energy protons in heavy targets allows the use of very thick targets that makes the ultimate yield of exotic nuclei good enough for many types of experiments. In fact, the spallation reaction has remained one of the most popular and effective routes of production of RIBs since their first use at ISOLDE in the early 1970s. More than 600 isotopes with half-lives down to milliseconds belonging to 70 elements (Z = 2 to 88) have been produced at ISOLDE with intensities in the range of 1 to 10^{11} atoms per microampere of proton beam (proton beam intensity of 1 µA = 6 × 10^{12}/s). The yields of some of the p-rich and n-rich exotic isotopes produced and studied at ISOLDE (taken from the ISOLDE web site) are listed in Tables 3.1A and 3.1B which clearly show the suitability of the spallation route for the production of RIBs, which has been exploited at ISOLDE and TRIUMF for a good number of novel measurements on exotic nuclei. It may be noted, for example, that substantial yields of very n-rich light nuclei such as ^{11}Li and $^{11,\,14}$Be can be obtained by using a 1 GeV proton beam from a PS Booster and a thick tantalum target consisting of 199 thin tantalum foils, each of 2 µm thickness (Ber 02). Similarly, extremely p-rich nuclei, such as ^{9}C, ^{17}Ne, ^{31}Ar, ^{35}K, ^{71}Kr, ^{75}Sr, ^{114}Cs, etc., could be produced with substantial yields.

TABLE 3.1B
Some of the Neutron-Rich Nuclei Produced at ISOLDE (from ISOLDE Website)

Isotopes	Half-Life	Proton Energy (GeV)	Targets	Yields (ions/μC)
^{8}He	119 ms	1	MgO	1.2×10^4
^{8}Li	838 ms	1	Ta	5.8×10^8
^{11}Li	8.5 ms	1	Ta	2.5×10^3
^{11}Be	13.8 s	1	Ta	3.4×10^6
^{14}Be	4.35 ms	1	UCx	4.0×10^0
^{16}C	747 ms	1	CaO	2.5×10^3
^{22}O	2.25 s	1	CaO	1.0×10^3
^{25}Ne	602 ms	1	TiO2	2.7×10^3
^{33}Na	8.2 ms	1	UCx	1.3×10^1
^{52}K	105 ms	1	UCx	5.6×10^2
^{85}Br	2.9 m	0.6	ThNb	5.7×10^6
^{90}Kr	32.32 s	1.4	PbBi	1.8×10^7
^{95}Rb	377.5 s	1.4	UC$_X$	5.8×10^3
^{102}Sr	69 ms	1.0	UC$_X$	30.0
^{95}Y	10.3 m	0.6	Ta	7.8×10^4
^{116}Ag	2.68 m	0.6	ThTa	5.9×10^6
^{118}Cd	50.3 m	1.4	UC$_X$	7.2×10^7
^{117}In	43.2 m	0.6	UC$_X$	3.5×10^8
^{137}Xe	3.818 m	1.0	ThC$_X$	7.5×10^8
^{201}Au	26 m	1.4	UC$_X$	4.0×10^4
^{202}Au	28.8 s	1.4	UC$_X$	1.4×10^3
^{206}Hg	8.15 m	1.4	Pb	1.5×10^8

3.1.7 Reaction Codes for Spallation Reaction

The importance of spallation reactions, for exotic nuclei production and as a strong neutron source for use in basic and technological research including energy applications (e.g., an accelerator-driven sub-critical system – ADSS), has made it a rather intense field of study both experimentally and theoretically. A huge effort has gone into developing theoretical codes to estimate the cross-sections of different product nuclei in proton-induced spallation/target fragmentation reaction. This has resulted in more than a dozen codes that deal with both the reaction stages—the INC of high energy nucleon–nucleon collisions followed by pre-equilibrium emission of energetic nucleons, followed by the statistical de-excitation stages where evaporation and fission are two competing processes. Following pioneering contributions of a number of researchers who applied the Monte Carlo stochastic technique to investigate INC (Gol 48; Bert 63; Che 68; Ya 79; Cu 87) a number of versatile codes have been developed based on the Monte Carlo method and using different models of spallation reactions. These codes also use the cross-section libraries of neutron-induced reactions as input. Some of the most often-used codes are: LAHET (Pr 89), Monte Carlo N-Particle Transport Code eXtended (Pe 05), CASCADE (Ba 00) and FLUKA (Fe 05). The improved version of the ABLA code (ABLA 07) includes binary asymmetric fission and multi-fragmentation and has been found to describe the de-excitation phase quite well. The simulated cross-sections often match reasonably well with the experimentally determined cross-sections for a wide range (in (A, Z)) of reaction products, at least for those product nuclei which are not very exotic.

3.2 Production of RIBs Using High and Intermediate Energy Heavy Ion Induced Projectile Fragmentation and In-Flight Fission Reactions

3.2.1 Introduction

Production of exotic nuclei in nuclear reactions involving intermediate and high energy heavy ion beams is one of the most efficient routes for the production of exotic nuclei. A number of world-class laboratories, including GANIL, GSI, MSU and RIKEN, have built extensive and state-of-the-art in-flight RIB facilities around accelerators delivering intermediate to high energy heavy ion beams. The beam energy of the primary beams (projectiles) at GSI is about 1 GeV/u, in RIKEN up to 350 MeV/u, and at GANIL and MSU in the range 50–100 MeV/u. These facilities study exotic nuclei produced through Projectile Fragmentation and In-Flight Fission reactions and used in Flight Separators, often called Projectile Fragment Separators (PFS).

The fragmentation of high energy heavy ions was first studied at LBL in the early 1970s at the BEVALAC accelerator (He 71). Subsequently, the nature of the fragmentation reaction was studied experimentally and the reaction was found to be suitable for the production of exotic nuclei in new regions of the nuclear chart and for carrying out studies on them using in-flight separators (He 72; Gri 75; Gol 74; Gol 78; Cun 78; Vi 79; Sy 79; We 79). These initial studies established the capability of a PF reaction for the production of exotic species of light elements. These studies followed by the discovery of the halo structure in light n-rich drip line nuclei at BEVALAC, LBL (Ta 85), gave a great boost to the utilization of PF reactions for the production and study of new exotic isotopes. In BEVALAC, the beam current was pretty low, and it was ultimately decommissioned in 1993. Studies of projectile fragmentation reaction and production of new exotic species using the projectile fragmentation route were subsequently taken up in several intermediate and high energy heavy ion facilities, in a major way at GANIL, MSU, RIKEN and GSI. The Ligned'Ions Super Epluches or Super Stripped Ion Line (LISE) spectrometer at GANIL (Duf 86) was the first fragment separator to implement a more effective separation technique and it was in experiments using LISE that many exotic isotopes, both n-rich and neutron-deficient, were identified for the first time, firmly establishing the usefulness of the Projectile Fragmentation reaction for the production of exotic isotopes. Subsequently fragment separators A1200 and A1900 at MSU (Sh 91; Mo 03), RIPS and Big RIPS at RIKEN (Ku 92; Ku 03) and FRS at GSI (Gei 92) were built, and these have been used extensively for new isotope research and studies on the properties of exotic nuclei. These facilities created great opportunities for studies in exotic nuclei and contributed significantly to our knowledge of exotic nuclei and their properties.

3.2.2 The PF Reaction Process

Like spallation, Projectile Fragmentation Reaction at relativistic energies can also be described as a two-step process, a fast process in the time scale of 10^{-23} to 10^{-22} s which is dominated by individual nucleon–nucleon interactions followed by a slow statistical de-excitation process that can have, depending upon the excitation energy, a timescale as long as 10^{-16} s. Projectile Fragmentation is considered as a peripheral collision between the projectile and the target nuclei in which during the first stage (the fast process) a few nucleons are removed from the projectile forming a single pre-fragment. This occurs in a typical time scale of about 10^{-23} s. The first stage can be described microscopically by intra-nuclear

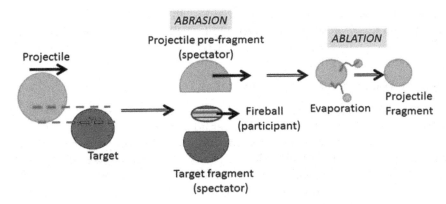

FIGURE 3.4
Schematic presentation of the Projectile Fragmentation process in the Abrasion–Ablation picture.

cascade models or macroscopically by the geometrical abrasion model. The main assumption of the abrasion model (Bow 73; We 76) is that in peripheral high-energy collisions, since the projectile velocity exceeds the Fermi velocity of nucleons, the interaction between the target and the projectile nuclei is mainly limited to the overlap zone. The projectile and the target nuclei make a clean cut through each other while remaining parts of the projectile (the excited pre-fragment) and targets outside the overlap region continue to move roughly with their initial velocities, as if they were spectators (Figure 3.4). Thus after abrasion the pre-fragment moves almost with the beam velocity and the target fragment remains almost at rest for a fixed target case, and both of these are only mildly excited. The overlap or the participant region in which the projectile and target nucleons interact forms a hot and dense fireball that decays through the emission of nucleons and light-charged particles. Accordingly, the abrasion process is essentially a spectator–participant picture and it has been found to be quite suitable for describing the peripheral PF reaction.

After the abrasion stage, the excited pre-fragment de-excites statistically through evaporation of nucleons, light clusters and gamma rays. Depending upon excitation energy the timescale for the second stage is in the range of 10^{-20} s at higher excitation energies to 10^{-16} s at lower excitation energies of a few tens of MeV. This second stage or the de-excitation phase of the pre-fragment is generally called the ablation process during which the pre-fragment loses all its excitation energy and becomes the projectile fragment observed in the experiment. The PF reaction is thus seen as an abrasion–ablation process as shown schematically in Figure 3.4.

As a result of nucleon removal, the pre-fragment is left in an excited state. This excitation energy can be seen as the excess surface energy in the highly deformed pre-fragment with respect to a sphere of equal volume. Calculations based on this picture (Gos 77; Me 83; Me 74) appear to reproduce close to the experimental results but there are some systematic deviations that are considered to have resulted from an underestimation of the excitation energy of the pre-fragment. Alternately, it is assumed that the fast timescale of the abrasion stage does not allow any re-arrangement of the orbits of the nucleons in the pre-fragment (spectator) and the pre-fragments keeps their geometrical shape and size (with reduced density) the same as that of the projectile and the excitation energy can be calculated as the sum of the energies of the vacancies or holes (with respect to Fermi level) created in the single particle levels by the removal of the nucleons (Ga 91). This approach results in higher excitation energy of about 13.5 MeV per abraded nucleon that results in a better fit of experimental cross-section data although not in all cases. In many cases the excitation energy requires a multiplication factor ranging from 0.6 to 2 to fit the experimental cross-section data.

3.2.3 Limiting Fragmentation and Factorization

The cross-section of the fragments produced in a reaction has been experimentally found to be almost independent of beam energy over a large span of energy from intermediate to high energy (Cun 78). This is known as the concept of limiting fragmentation. Also, for projectile energies of more than 100 MeV/u, there is practically no target dependence of the cross-section beyond the nuclear size effect (Cun 90). This is known as the concept of factorization. The validity of the concept of factorization, however, is questionable in intermediate energy PF reactions since it has been found that the cross-sections of production of very n-rich nuclei depend on the N/Z ratio of the target and is higher for heavier target nuclei (say for Ta target as opposed to Ni or Be targets) (Gu 83; Sc 02; No 07). This is probably because at energies lower than 100 MeV/u nucleon exchange between the target and the projectile during the abrasion process is not negligible. Thus, at energies in the range 40–100 MeV/u, projectile-like fragments observed in the experiments might not originate purely from the fragmentation mechanism as viewed in the participant-spectator picture but have contributions from the transfer reaction as well.

3.2.4 Momentum/Energy Width of the Projectile Fragments

The momentum distribution of projectile fragments gives important insight into the reaction mechanism. It is also of great practical importance for designing PF separators where the design aims to collect all the fragments of a particular exotic isotope in view of small production cross-sections of exotic isotopes, which is in the range of a few µb and below depending on the exoticity (how far away it is from the beta-stable isotope of the same element or how close to the drip line) of the final fragment. The momentum distribution of projectile fragments in the direction of the projectile (parallel or longitudinal momentum distribution) has experimentally been found to be Gaussian in the rest frame of fragments (He 72). For example, the momentum distribution of ^{10}Be fragments produced in the fragmentation of ^{12}C projectile at 2.1 GeV/u using a ^{9}Be target (Gri 75) was observed to follow a Gaussian distribution in the rest frame of the projectile closely. For ^{10}Be, the mean momentum was found to be (–)30 MeV/c, where the negative sign implies that the fragment velocity is a little less than the projectile velocity. The standard deviation was found to be 129 MeV/c. So, one standard deviation momentum spread of the fragment is less than ±0.5% of its parallel momentum at beam energy of about 2 GeV/u. The low momentum spread and high kinematic focusing of the fragments make it possible to collect the projectile fragments efficiently in an in-flight magnetic separator.

The Gaussian shape for fragment momentum distribution has in fact been found to be a common feature for all projectiles of different energies and for different fragments unless the projectile energy is below 100 MeV/u, in which case the distribution becomes asymmetric with a tail in the low momentum side. The tail is believed to be due to dissipative transfer in the projectile that starts contributing at lower projectile energies in the abrasion stage, affecting the momentum distribution of the pre-fragments.

The momentum width of fragments has been calculated by Goldhaber (Gol 74) based on a statistical model and on the assumptions that nucleons in a projectile can be described as free fermions in a volume and fragmentation is a process in which the nucleons are suddenly removed. As the total momentum of projectile nucleons in the rest frame of the projectile is zero, these assumptions allow the momentum width of the fragments to be calculated from the sum of the Fermi momentum of the removed nucleons. The Gaussian momentum width parallel to the beam is thus expressed as:

$$\sigma_{||}^2 = \sigma_0^2 \frac{A_f(A_p - A_f)}{A_p - 1} \quad (3.1)$$

where σ_0 is the reduced width which is a fraction of the Fermi momentum and A_f and A_p are the mass numbers of the fragment and projectile respectively. It is clear from Equation (3.1) that parallel or longitudinal momentum width increases for fragments that are away from the projectile and becomes maximum for $A_f = \frac{A_p}{2}$. The value of σ_0 has been estimated to be around 115 MeV/c, based on quasi-elastic electron scattering measurements of Fermi momentum (Mon 71) but a smaller value of about 85 MeV/c appears to fit the experimental data better.

One reason for observing a smaller momentum width might be Pauli blocking that limits the number of participating nucleons (Bert 81). It should be noted, however, that in Equation (3.1), A_f is strictly the mass number of the pre-fragment after the abrasion stage. Most often it is not the same fragment that we observe in the experiment which results from de-excitation of the pre-fragment. Equation (3.1) is thus expected to hold true more in cases when the final fragment is the same as the pre-fragment which can happen if the pre-fragment is formed in a cold condition with too little excitation energy for neutron evaporation in the statistical de-excitation stage. In general, for projectile fragmentation reactions where appreciable evaporation is expected at the second stage of the reaction, the empirical relation (Mo 89)

$$\sigma_{||}^2 = 150^2 \frac{(A_p - A_f)}{3} \quad (3.2)$$

appears to give a reliable estimate of the momentum spread induced by the reaction. However, for much lighter fragments, the observed momentum widths are much less than the values given by Equation (3.2). This is probably because the lighter fragments are more likely to be produced in more central collisions and originate from multi-fragmentation rather than in a peripheral projectile fragmentation reaction. The momentum width in any case remains more or less independent of beam energy from high energies down to about 50 MeV/u, below which the momentum width decreases because of competition from the direct transfer of nucleons (Eg 83).

The width of the momentum perpendicular (σ_t) to the beam or the transverse momentum distribution should be equal to $\sigma_{||}$ at high energies where the contribution from nuclear and Coulomb scatterings are small. This is consistent with isotropy of the angular distribution of the projectile fragments in the frame of the projectile. However, at projectile energies less than 200 MeV/u, the transverse momentum is better fitted by the expression (Bi 79):

$$\sigma_t^2 = \sigma_{||}^2 + \sigma_1^2 \frac{A_f(A_f - 1)}{A_p(A_p - 1)} \quad (3.3)$$

where σ_1 is approximately 200 MeV/u. It is important to note that the momentum widths of the fragments are but a small fraction of the average fragment momentum and since the widths are essentially independent of energy over a large energy range, the relative parallel momentum width ($\sigma_{||}$/fragment momentum) and the relative transverse momentum width (σ_t/fragment momentum) of a fragment decreases with beam energy. For example, at around 2 GeV the longitudinal fragment momentum spread is always less than ±0.5% irrespective of how far the fragment is from the target and the angular spread is less than ±0.5 degrees. The strong kinematic focusing and the conservation of the velocity of the projectile fragments are the two most attractive features of this peripheral reaction (PF), which have been exploited in the design of in-flight separators. At lower energies of about 50 MeV/u the relative widths increase, but the relative parallel momentum width remains within a few percent and the angular spread within a few degrees. So, PF reactions from intermediate to high energies allow efficient collection of the fragments in PF separators although the efficiency decreases at lower energies primarily because of practical limitations in the size of the di-pole magnets in the dispersive (usually horizontal) plane and the pole gap (usually vertical) that limit, respectively, the momentum and the angular acceptance.

3.2.5 Production of Exotic Species in PF Reaction

A Projectile Fragmentation reaction produces neutron-deficient fragments with much higher cross-sections compared to n-rich fragments. This is because the projectile is a beta-stable isotope and in the abrasion stage the probability is highest for abraded nucleons to have the same N/Z ratio as that of the projectile (Sc 93), thus producing a pre-fragment close to the beta-stability. Therefore, it is best to choose as the projectile (the heavy ion beam from the accelerator) the most neutron-deficient beta-stable projectile (say ^{40}Ca for the element Ca) for production of neutron-deficient exotic species (lying on the n-deficient or p-rich side of beta-stability), while for production of exotic n-rich species the choice is the most n-rich stable isotope of the same element (say ^{48}Ca) as the projectile. The pre-fragment then cools, predominantly by neutron evaporation, because the probability of neutron evaporation is a few orders of magnitude higher than proton evaporation for moderate excitation energies of a few tens of MeVs, favoring the production of a neutron-deficient final fragment close to beta-stability on the proton-rich side. However, the abraded nucleons may have large fluctuations in their N/Z ratio and in their excitation energies. These large fluctuations are thought to be a consequence of high kinetic energy of the projectile that by far exceeds the total binding energy of the reaction partners and is a typical characteristic of the peripheral PF reaction. The fluctuations make possible the production of n-rich pre-fragments with an N/Z ratio much different (higher) than the projectile and with relatively small excitation energies that allow evaporation of only a few neutrons from the excited pre-fragment leaving the final fragment very n-rich. The probabilities of such processes are of course very low, and these very n-rich nuclei are produced with several orders of magnitude smaller cross-sections, say a few nb or even lower compared to production cross-sections of a few mb for neutron-deficient products close to beta-stability. But even with such low cross-sections, the identification of new exotic isotopes is possible in experiments using in-flight separators.

3.2.5.1 Production of Neutron-Deficient Nuclei

As explained in the last paragraph, a PF reaction is very efficient for producing very exotic neutron-deficient nuclei on or close to the proton drip line nuclei. To produce neutron-deficient exotic nuclei, the most neutron-deficient beta-stable isotopes are usually chosen as the projectile. For example, projectile fragmentation of intermediate energy heavy ion beams of ^{28}Si, ^{36}Ar, ^{58}Ni, ^{78}Kr and ^{112}Sn have been used to produce a good number of exotic isotopes on or close to the proton drip line. For example, ^{36}Ar and ^{28}Si beams were used for the production of exotic isotopes of Si, such as ^{22,23,24}Si (Bl 96; Naik 01), ^{58}Ni beam was used to produce ^{45}Fe, ^{48,49}Ni (Bl 00), ^{78}Kr beam was used to produce ^{65}As, ^{69}Br, ^{75}Sr (Moh 91) and ^{112}Sn beam was used for the production of ^{98,99}In (Baz 08). Since the beta-stability line curves towards neutron excess slowly from Z = 20 onwards, projectile fragmentation is best suited to produce proton drip line isotopes only up to about Z = 50, beyond which compound nuclear fusion type reactions with low energy heavy ion beams and light ion induced spallation reactions are the most suitable reaction routes.

3.2.5.2 Production of n-Rich Nuclei

PF reaction has been extensively used for the production of light neutron-rich nuclei up to about A ~ 80, which cannot be produced in low-energy fission of actinides. In these experiments the beam (projectile) of the even–even beta-stable isotopes with the maximum neutron excess, such as ^{86}Kr, ^{82}Se, ^{76}Ge, ^{50}Ti and ^{48}Ca (A/Z = 2.388, 2.41, 2.375, 2.273, 2.4 respectively), were used to optimize the production of n-rich isotopes. A large number of experiments have been carried out using these beams at different times and at different laboratories leading to a proportional number of publications in journals. (Just to suggest a few, readers might go through the references: Ahn 19, Tar 13; Tar 09; Bau 07; No 97; No 96; Pf 95; Web 94; Ta 85; Mu 85; Sy 79; and We 79.) It is through these efforts that the neutron drip line is reached up to neon (Ahn 19) and nearly reached up to about Z = 3 (^{43}Al has been observed experimentally, while ^{45}Al has been predicted to be the last stable isotope on the drip line (Erl 12)) and a large number of exotic light n-rich nuclides could be produced in the laboratory and important nuclear structure effects are discovered. It was observed in a pioneering experiment involving projectile fragmentation of a 212 MeV/u ^{48}Ca beam using a ^{9}Be target (We 79) that the cross-sections of n-rich isotopes of all the elements from carbon to chlorine decrease roughly by an order of magnitude with each neutron added to a n-rich isotope (for example, the cross-section of ^{31}Na is a factor of 10 less than that of ^{30}Na). The decrease, however, was found to be much faster for n-rich oxygen isotopes, which was ascribed to structural effects.

It should be stressed that except for the very light n-rich nuclei, the production cross-section of isotopes lying on the n drip line may be quite small, often a few tens of fb (10^{-39} cm^2) or could be even less for drip-line nuclei of high Z elements (because the drip line moves further away from beta-stability towards more and more neutron excess). This makes the identification of these nuclei extremely difficult. For example, only three events of ^{40}Mg could be detected in continuous 7.6 days of beam run using fragmentation of a 140 MeV/u ^{48}Ca projectile (Bau 07). It appears that the production cross-sections of exotic isotopes (projectile fragments) follow an exponential dependence on the difference of the mass excess of the projectile and the fragment, which is independent of the target (Tar 07). Any deviation from this trend tends to indicate the presence of structural effects. This provides a simple way of estimating the cross-section by extrapolation if the masses are experimentally known (which is quite often not the case for n-rich nuclei with neutron excess exceeding about ten neutrons) or can be predicted with good accuracy using some mass model.

Figure 3.5 shows the exotic n-rich isotopes produced in the fragmentation of 132 MeV/u ^{76}Ge beams using a ^{9}Be target at NSCL, MSU (Tar 09). It can be seen that ^{76}Ge, which has

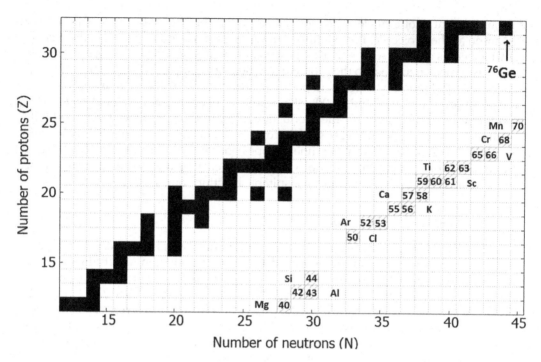

FIGURE 3.5
Production of very n-rich isotopes in the projectile fragmentation of a 132 MeV/u ^{76}Ge beam (Tar 09).

12 extra neutrons is very effective in producing a good number of new *n*-rich isotopes for elements ranging from Cl to Mn (such as ^{50}Cl, ^{58}Ca, ^{68}Cr, ^{70}Mn, etc.). It may be noted that production of ^{68}Cr requires removal of eight protons and no neutrons, while the production of ^{70}Mn requires removal of seven protons from the projectile and transfer of one neutron from the target. The cross-sections of production of these isotopes are understandably very low but it is these possibilities that need to be explored to reach the neutron drip line, perhaps through fragmentation of secondary exotic beams.

The most exotic *n*-rich nuclei are produced when a number of protons are removed from the projectile at the abrasion stage. It is also necessary that after proton removal, the pre-fragment is left in a state of low excitation energy that restricts neutron evaporation totally or only to a few. In cases where the pre-fragments are left, after proton removal, with excitation energies below the threshold for neutron emission (roughly 8 MeV), the final products have the same (N, Z) as the pre-fragment and the cross-sections are therefore determined by the first stage abrasion process. Such fragmentation reactions are often termed "Cold Fragmentation." The production of isotopes resulting from proton removal channels in projectile fragmentation has been studied experimentally at GSI using various projectiles: 0.8 GeV/u ^{136}Xe and 1 GeV/u ^{197}Au (Sc 92) and 0.95 GeV/u ^{197}Au (Ben 99). Productions of isotopes up to five proton removal channels were studied. The production cross-sections of these proton removal channels were determined along with their longitudinal momentum distributions. An analytical model of Cold Fragmentation was found to reproduce the experimental data with a factor of two (Ben 99).

It should be noted that a five-proton removal from ^{197}Au produces ^{192}W which is a quite *n*-rich isotope, having six more neutrons than the last stable isotope ^{186}W. The fact that *n*-rich isotopes with $A > 160$ are not accessible through fission of actinides makes the projectile fragmentation of heavy projectiles like [197]Au, [196,198]Pt, [208]Pb, [209]Bi, etc., the most efficient route, if not the only route, to produce n-rich nuclei with $A > 160$. Detailed spectroscopic studies on the properties of these nuclei would require production of these heavy projectiles with higher and higher intensities to compensate for the small cross-sections of production for the exotic n-rich species. Cold Fragmentation appears to be a promising reaction route to produce the most n-rich isotopes in various mass regions. However, a proper understanding of Cold Fragmentation will need further effort, both experimental and theoretical.

3.2.6 Production of n-Rich Nuclei in In-Flight Fission of [238]U

In-flight fission of relativistic projectiles like [238]U has been proved to be one of the most efficient ways of producing n-rich nuclei. Both Coulomb fission and abrasion fission reactions take place that can produce a wide range of isotopes starting from Z as low as 20 (Ca). The Coulomb fission proceeds from a low excitation energy of [238]U (~15 MeV) and is well suited for production of exotic n-rich nuclei. Fission produces two fragments, each having a kinetic energy of about 1 MeV/u in the rest frame of the fissioning nucleus, in this case the pre-fragment. This kinetic energy is much smaller compared with the kinetic energy of the projectile. The fission process thus does not appreciably hamper the kinematic focusing at relativistic energies. Usually, the fission fragments emitted in the forward direction in the rest frame of the fragment are collected by the in-flight separator. For identification of new n-rich exotic nuclei, fragments with a magnetic rigidity higher than the projectile rigidity are usually selected to ensure that only the exotic fission fragments produced in peripheral collisions with A/Z ratios higher than the projectile are collected (magnetic rigidity is proportional to momentum per charge state; the charge state is equal to atomic number Z for fully stripped ions which is usually the case at relativistic energies; so fragments of higher rigidity than the projectile mean that the fragments have higher A/Z value than the projectile).

The projectile fission of [238]U (A/Z for [238]U is 2.587) has been extensively used at GSI (Be 97) and at RIKEN (Oh 10) for new n-rich isotope search and has led to the discovery of more than 100 new n-rich nuclei of a wide range of elements starting from Ca ($Z = 20$) to Ba ($Z = 56$) with A/Z ratio reaching as far as 2.846 and 2.84 for [74]Fe and [71]Mn respectively in the RIKEN experiment. Since the production cross-sections are of the order of a few pb for the most exotic nuclei, higher intensity beams are required to produce these nuclei in enough numbers to allow spectroscopic studies. Figure 3.6 shows the new n-rich nuclei produced in projectile fission using a 345 MeV/u [238]U beam at RIKEN schematically on the N–Z chart. The calculated r process nucleo-synthesis path is also shown in the same diagram to show that many of these new nuclei indeed lie on the r process path. It is important to point out that the intensity of the [238]U projectile was higher by more than an order of magnitude in the RIKEN experiment (3.5×10^8 pps) compared to that in the previous measurement at GSI (2×10^7 pps). This is a major reason why more exotic species could be identified in the RIKEN experiment.

In Table 3.2, the measured cross-sections of some of the most exotic species produced through projectile fission of [238]U, compiled from the data of GSI (Be 97) and RIKEN (Oh 10) experiments are listed. It may be noted once again that in many cases the production cross-section of isotopes of an element decreases roughly by an order of magnitude with each neutron added. Experimental data tends to suggest that up to $Z = 20$ or a little higher, the projectile fragmentation of beams like [48]Ca, [76]Ge, [82]Se and [86]Kr are better suited for production of near drip line/drip line n-rich species. In the range of $20 < Z < 60$, abrasion

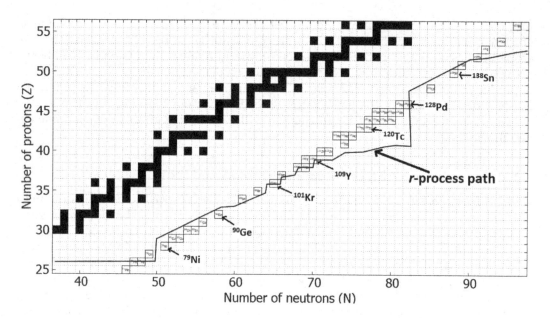

FIGURE 3.6
n-Rich isotopes produced for the first time at the RIKEN RI Beam Factory using in-flight fission of 345 MeV/u ^{238}U. It is noted that a good number of these new isotopes fall on or around the suggested r-process nucleosynthesis path.

fission (optimum target: Be) and Coulomb fission (optimum target: Pb) of ^{238}U beam are the two appropriate routes. For production of *n*-rich isotopes with $Z > 60$ and $A > 160$, fragmentations of heavy ion beams like ^{197}Au, ^{196,198}Pt, ^{208}Pb, ^{209}Bi, etc., appear to be the best option.

3.2.7 Choice of Target Thickness, Target and Projectile Energy

To maximize the yield of any exotic isotope produced in a projectile fragmentation reaction, higher energies for the incident heavy ion projectile need to be used, as long as the beam intensity remains the same. As the reaction cross-section is largely independent of energy above the Fermi energy domain, higher beam energies allow use of higher target thickness and the yield is expected to increase proportionally. Target thickness is however limited by two factors. The first one is the momentum spread of the reaction products (projectile fragments), which increases with the target thickness. The longitudinal momentum spread results from the slowing down of the projectile and fragments inside the target because of interactions with target electrons. The energy loss in the target is different for the projectile and a fragment, and the momentum spread of the fragment increases with the target thickness, as explained schematically in Figure 3.7 (Ta 89). A fragment with an atomic number (Z) lower than the projectile and produced at the first layer of the target will have higher momentum than the same fragment produced at the last layer of the target because of differential electronic energy loss of the projectile and the fragment inside the target. Since the longitudinal momentum acceptance of PF separators is practically limited to a few percent, increasing the target thickness beyond a limit is not expected to increase the yield. Target thickness also affects the transverse momentum spread due to multiple Coulomb scattering, but at higher projectile energies the effect is negligible compared to the increase in the longitudinal momentum and can be ignored.

TABLE 3.2
Cross-Sections of Some Very Exotic n-Rich Isotopes Produced in "In-Flight" Fission of ^{238}U at GSI and RIKEN

Nuclei	Cross-section (nb)	Projectile (Energy in MeV/u)	Nuclei	Cross-Section (nb)	Projectile (Energy in MeV/u)	Nuclei	Cross-Section (nb)	Projectile (Energy in MeV/u)
^{56}Ca	1	^{238}U (750)	^{79}Ni	5×10^{-3}	^{238}U (345)	^{115}Nb	3×10^{-3}	^{238}U (345)
^{57}Sc	10	^{238}U (750)	^{82}Cu	3×10^{-3}	^{238}U (345)	^{117}Mo	4×10^{-3}	^{238}U (345)
^{58}Sc	3	^{238}U (750)	^{85}Zn	2×10^{-3}	^{238}U (345)	^{120}Tc	2×10^{-3}	^{238}U (345)
^{60}Ti	10	^{238}U (750)	^{87}Ga	19×10^{-3}	^{238}U (345)	^{123}Ru	2×10^{-3}	^{238}U (345)
^{61}Ti	2.5	^{238}U (750)	^{90}Ge	6×10^{-3}	^{238}U (345)	^{124}Ru	0.6×10^{-3}	^{238}U (345)
^{63}V	6.4	^{238}U (750)	^{92}As	0.6	^{238}U (750)	^{125}Rh	11×10^{-3}	^{238}U (345)
^{64}V	0.3	^{238}U (750)	^{94}Se	2	^{238}U (750)	^{126}Rh	0.7×10^{-3}	^{238}U (345)
^{67}Cr	0.5	^{238}U (750)	^{95}Se	0.02	^{238}U (345)	^{124}Pd	32	^{238}U (750)
^{69}Mn	0.4	^{238}U (345)	^{98}Br	0.01	^{238}U (345)	^{128}Pd	12×10^{-3}	^{238}U (345)
^{71}Mn	4×10^{-3}	^{238}U (345)	^{100}Kr	0.5	^{238}U (750)	^{133}Cd	0.026	^{238}U (345)
^{74}Fe	1×10^{-3}	^{238}U (345)	^{101}Kr	0.01	^{238}U (345)	^{138}Sn	0.6	^{238}U (345)
^{76}Co	8×10^{-3}	^{238}U (345)	^{107}Sr	1×10^{-3}	^{238}U (345)	^{140}Sb	4.3	^{238}U (345)
^{77}Ni	1.4	^{238}U (750)	^{109}Y	4×10^{-3}	^{238}U (345)	^{143}Te	0.3	^{238}U (345)
^{78}Ni	0.2	^{238}U (750)	^{112}Zr	7×10^{-3}	^{238}U (345)	^{148}Xe	0.07	^{238}U (345)

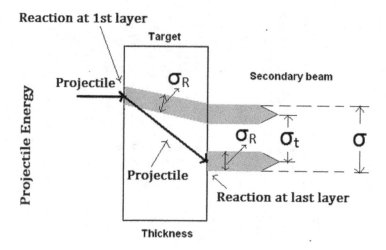

FIGURE 3.7
Momentum spread in projectile fragmentation reaction due to the reaction process and the finite thickness of the target.

The other limit from the target thickness results from the consideration of secondary fragmentation. Higher target thickness, while effective for producing more fragments of a particular isotope (A, Z), at the same time also reduces the chances of survival of the same fragments. This is because a fragment produced at a certain layer of the target might undergo further fragmentation while passing through the target layers downstream. So, beyond an optimum thickness the yield of a particular exotic isotope would not increase, because there would be no net increase in the production of that isotope because of secondary fragmentation reactions. This consideration restricts the target thickness to about a few tens of g/cm² (Ta 89) at relativistic energies. However, longitudinal momentum spread sets a much lower limit for the maximum usable target thickness. The square of the total longitudinal momentum spread is obtained by adding in quadrature the momentum spreads due to the reaction as given roughly by Equation (3.1) and due to the target thickness. The contribution to the spread from the reaction itself is a fixed value for a particular projectile–fragment combination. So, the target thickness is chosen such that the total momentum spread of the fragment of interest does not exceed the momentum acceptance of the PFS/in-flight separator. Based on these considerations, the optimum target thickness for 1 GeV/u projectiles typically comes to be about a few g/cm², which is an order of magnitude less than the limit set by the consideration of secondary fragmentation. The optimum target thickness in an experiment is thus decided in practice solely by the consideration of longitudinal momentum spread of the fragments. Also, as mentioned earlier, higher energies lead to stronger forward focusing and smaller fractional momentum spread, allowing more efficient acceptance by the fragment separator. By increasing the beam energy from about 100 MeV/u to 1 GeV/u, the production rate can be gained, roughly by an order of magnitude, by using thicker targets. Also, higher projectile energies lead to fragments that are predominantly fully stripped ions with little admixture from charge states $q < Z$. This is important for unambiguous identification of reaction products at the focal plane of the PF separator. These considerations together make 1 GeV/u to be a sort of optimum projectile energy for RIB production, at which reasonable beam intensity of heavy ions could also be obtained.

Although the PF reaction cross-section is almost independent of the choice of the target, the dependence of the cross-section on nuclear size makes it roughly proportional to $A^{2/3}$. The fact that the number of target nuclei for a given target thickness in g/cm² is proportional to $1/A$ (the Be target has about 20 times the atoms/cm² than a lead target) makes the production yield proportional to $A^{-1/3}$. Thus, light targets like ^{9}Be, ^{12}C, etc., are usually considered to be the optimum choice for PF reaction although for Coulomb fission (and for Coulomb excitation experiments) high Z targets like Pb, Au, etc., are chosen in place of Be/C, etc.

3.2.8 Reaching Closer to the Neutron Drip Line Using Fragmentation of Secondary RIBs

To reach the neutron drip line for heavier elements (say for $Z > 20$) through projectile fragmentation, or for that matter using any reaction, seems to be an impossible task at the moment. This is because for heavier elements the drip line drifts away towards more and more neutron excess as the atomic number increases; for example, the $2n$ drip line ($S_{2n} = 0$) isotopes for Sn (Tin; $Z = 50$) can be as far as ^{176}Sn (Erl 12). However, in the journey towards the neutron drip line, projectile fragmentation of n-rich unstable beams (say ^{54}Ca, ^{92}Kr, ^{132}Sn, ^{149}La, etc.) offers an exciting possibility of producing very n-rich nuclei with much higher cross-sections because of the favorable memory effect (higher N/Z ratio of the unstable (RIB) projectile compared to that of a beta-stable isotope). This is shown in Figure 3.8 (Bh 20) where the estimated cross-sections of rhodium (Rh) isotopes produced in the projectile fragmentation of three tin (Sn) isotopes of increasing neutron-richness are plotted. It can be seen that the n-rich isotopes of Rh are expected to be produced with higher cross-sections in the fragmentation of unstable ^{128}Sn and ^{132}Sn projectiles compared to the

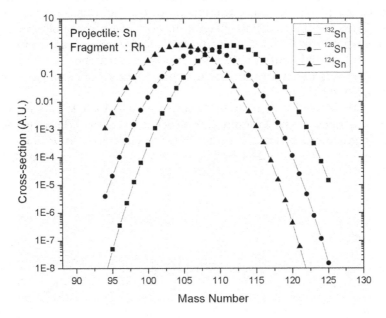

FIGURE 3.8
Simulated production cross-sections of rhodium (Rh) isotopes using projectile fragmentation of stable (^{124}Sn) and unstable (^{128,132}Sn) tin (Sn) isotopes.

fragmentation of stable ^{124}Sn and the relative gain increases with the neutron-richness of Rh isotopes. It can be seen that compared to the fragmentation of stable ^{124}Sn, the fragmentation of ^{132}Sn is expected to produce more n-rich isotopes of Rh with cross-sections that are higher by a few orders of magnitude.

The enhancement in the production cross-sections of n-rich exotic nuclei using unstable projectiles has also been seen experimentally. As mentioned in Section 3.1.4, the production of n-rich isotopes in the fragmentation of unstable n-rich ^{132}Sn has been studied (Per 11) at GSI in a pioneering experiment. In the experiment ^{132}Sn was produced using in-flight fission of ^{238}U beam (a Pb target was used to enhance the contribution of Coulomb fission). The separated ^{132}Sn was then allowed to impinge on a secondary Be target of thickness 1 g/cm^2 to produce n-rich fragments. The fragments produced were then separated using a second PF separator stage and were identified at the focal plane of the second separator. The experimental set-up used is shown schematically in Figure 3.9. Because of low intensity of the secondary ^{132}Sn beam, which was only about 10^3 per second, the experiment could detect only the n-rich isotopes produced with a cross-section greater than 1 μb. Nevertheless, it is interesting that in the fragmentation of ^{132}Sn exotic n-rich isotopes such as ^{120,121}Rh, ^{124,125}Pd, ^{128,129}Ag, ^{130}Cd, etc., are produced with cross-sections in the range of 1–100 μb. If we compare, for example, the production of a cross-section of ^{124}Pd, produced in this reaction (fragmentation of ^{132}Sn) with a cross-section of about 30 μb (Per 11), with that produced directly in in-flight fission of ^{238}U where it is produced with a cross-section of 32 nb (Be 97; Table 3.2), we clearly see an enhancement in the cross-section of about three orders of magnitude. The production of exotic species using fragmentation of unstable projectiles also has an additional advantage. Fragmentation in this case mainly produces exotic isotopes far from stability and thus at the focal plane of the fragment separator the relative intensities of unwanted species compared to the one of interest would be comparatively less compared to the case when the same nuclei are produced using fragmentation/in-flight fission of a primary beam (such as ^{238}U). This results in a better signal-to-noise ratio that helps in identification of new exotic nuclides and also in carrying out spectroscopic measurements on them. However, despite these advantages it is necessary to produce the exotic n-rich secondary beams with intensities exceeding 10^6 pps to achieve any substantial gain over the fragmentation using stable beams.

Using fission of ^{238}U (used as a target; see Section 2.4) in an ISOL facility to produce n-rich isotopes and subsequent post-acceleration of selected n-rich nuclides to energies suitable for projectile fragmentation, it might be possible in near future to produce intermediate energy ion beams of ^{132}Sn (A/Z = 2.64) and other n-rich nuclei such as ^{144}Xe (A/Z = 2.66) and

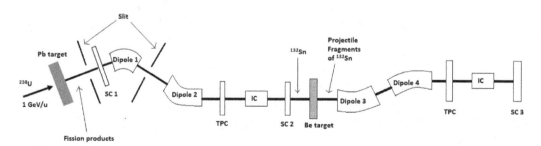

FIGURE 3.9
Schematic experimental set-up for a fragmentation study of a ^{132}Sn beam used at GSI.

^{146}Cs (A/Z = 2.65) with intensities of about 10^8 pps. Comparable intensities of these n-rich unstable projectiles can also be obtained using in-flight fission of ^{238}U if the primary ^{238}U beam can be produced at high energies with intensity of about 10^{13} pps. In the near future this also seems to be possible. Thus, production of new n-rich nuclei using fragmentation of unstable n-rich secondary beams is likely to become a thrust area in the near future for the production and study of new n-rich nuclei. Considering the fact that a nuclide as exotic as ^{129}Ag is produced with a cross-section of about 5 μb in the fragmentation of ^{132}Sn (Per 11), the fragmentation of these exotic secondary beams would potentially allow identification of a large number of new n-rich nuclides close to the drip line, especially for odd Z elements up to about $Z = 50$.

3.2.9 Theoretical Estimation of Production Cross-Sections in PF Reaction

Accurate predictions of projectile fragmentation cross-sections are required for working out optimum ways of approaching the neutron drip lines through appropriate selection of the projectile, and to examine the efficacy of using secondary fragmentation of n-rich projectiles to reach the neutron drip line. It is also very much needed for the planning of experiments where the details are worked out on the basis of the expected production cross-sections of the isotope of interest and other contaminant isotopes that are likely to be present in the focal plane of the separator. Obviously, the goal can only be reached through a better understanding of the heavy ion reactions at intermediate and high energies. Among the available computer codes for computing the cross-sections, EPAX (Su 00) and ABRABLA (Ga 91; Ri 03) are the two that are used most often.

EPAX is a semi-empirical code that gives reliable predictions of fragmentation cross-sections of all non-fissile projectiles up to ^{209}Bi, especially when the fragmentation products are neither very n-rich nor very neutron-deficient. The code assumes the validity of limiting fragmentation in the sense there is no energy dependence of its parameters and provides useful guidance in planning experiments using fragmentation reaction in IF/PFS separators. For very n-rich nuclei, the EPAX code usually overestimates the experimental cross-section by about 1–2 orders of magnitude although the improved version of the code, EPAX3 (Su 12), seems to reproduce the cross-sections of n-rich isotopes better (Tar 13). In EPAX, the N/Z dependence of cross-sections of projectile fragmentation has been accounted for through the introduction of the "memory effect." It is felt that this needs to be optimized further for better predictability of cross-sections of very exotic species. For n-rich nuclei, a very simple parametric expression involving the average binding energy of the fragments (Moc 07) appears to give quite accurate predictions and can be used beneficially for planning experiments aimed at searching for new n-rich exotic isotopes using projectile fragmentation of non-fissile projectiles. Although the parametric expression is extremely simple and very useful, it is not clear conceptually why such a simple dependence on average binding energy should work.

The ABRABLA code is a Monte Carlo simulation code suitable for peripheral and semi-peripheral collisions. The abrasion stage assumes a clean cut and the number of nucleons abraded out of the projectile depends purely on the impact parameter. For a given number of nucleons abraded out, protons and neutrons are removed randomly (assumption: no correlation between the nucleons) and the average excitation energy per abraded nucleon is taken to be 27 MeV, which is a factor of two higher than the calculated value of 13.5 MeV. This adjustment is necessary to fit most of the experimental cross-section data resulting from the fragmentation of very heavy ion beams like ^{197}Au. However, recent experiments for the production of relatively lighter nuclei show that excitation energy of 15 MeV (Tar 13) and for still lighter nuclei even 8 MeV (Th 16) per abraded nucleon is needed to fit the experimental data.

To include multi-fragmentation, a condition is put on the excitation energy in ABRABLA code. If the excitation energy exceeds about 3–4 MeV/u in the abrasion stage, thermal instabilities would lead to a break-up of the pre-fragment into clusters and nucleons. Otherwise, the pre-fragment undergoes statistical de-excitation, the ablation or ABLA stage, that is modeled using an evaporation code. The decay of the excited pre-fragment is considered as competition between the particle and fragment evaporation, gamma decay and fission. The ABLA code can be used to describe low energy compound nucleus type reactions or can be combined with an appropriate INC model for predicting cross-sections of spallation reaction products. Compared to EPAX, the ABRABLA code seems to reproduce experimental cross-sections for n-rich nuclei better in many situations. It is interesting to mention that simple and computationally less intensive approaches based on a statistical model of multi-fragmentation (the Canonical Thermodynamical Model), where multi-fragmentation is considered as a necessary intermediate stage after the abrasion, also appear to reproduce experimental data pretty well (Chau 13; Chau 07).

Another code, COFRA (COld FRAgmentation), which is a simplified analytical version of ABRABLA requiring much less computation time (Ben 99), seems to quite accurately reproduce the cross-sections of production of very n-rich nuclei formed in the fragmentation of stable and unstable projectiles, such as ^{197}Au and ^{132}Sn (Per 11) respectively, with a large neutron excess. The COFRA code considers cold abrasion of protons and neutrons and in the second stage considers only neutron evaporation from the cold fragment. Projectiles with large neutron excess have an enhanced probability of losing only neutrons by evaporation from the excited pre-fragments that are n-rich, having relatively smaller neutron binding energy. Thus, the COFRA code is typically suited for predicting production cross-sections of very n-rich nuclei from the fragmentation of stable and unstable beams with a large neutron excess. In summary, it can be said that while the various codes developed for predicting cross-sections of projectile fragmentation are quite reliable in certain domains, more experimental cross-section data, especially for the production of very exotic isotopes, are needed to fine-tune these codes for better predictability of the cross-section of nuclei closer to the drip lines.

It is important to note that fundamentally there is no difference between the spallation and the projectile fragmentation reactions; they just occur in different rest frames. In the projectile rest frame, projectile fragmentation can be considered in the same category as spallation or target fragmentation in the rest frame of the target. Both the reactions are high energy (~GeV and higher) reactions and are described assuming a two-step process; a fast process in the time scale of about 10^{-22} s dominated by individual nucleon–nucleon interactions that allow wide fluctuations in N/Z, corresponding to the statistics of nucleon–nucleon collisions followed by a slow statistical de-excitation process. Thus, the codes developed for spallation can be used for projectile fragmentation and vice-versa. The big difference is when it comes to the utilization of these two routes for production and to study exotic nuclei. For this, a completely different primary or driver accelerator (light ion accelerator for spallation and heavy ion accelerator for projectile fragmentation) and different technologies for the production of RIBs and for beam preparation are needed for experimental measurements. The measurement techniques for identification and study of the properties of exotic species are also different.

3.3 Fission Induced by Low-Energy Neutrons, Protons and Gamma Rays

3.3.1 The Fission Process

The production of *n*-rich exotic nuclei using fission of actinides (^{238}U) has already been discussed in connection to a high-energy proton-induced spallation fission reaction using ^{238}U (Section 3.1) target and also in in-flight Coulomb fission of high energy ^{238}U (Section 3.2) projectiles. Since the fission of actinides needs only low excitation energies (typically ~6 MeV), fission induced by low energy neutrons, protons and gamma rays offers equally potential routes for producing *n*-rich RIBs and a number of RIB facilities have been built/are being built where the fission of uranium induced by low-energy beams will be used as the principal mode of RIB production. This section, after a short introduction to the fission process, deals with fission induced by neutrons, light charged particles, and gamma rays at low and medium excitations (thermal to a few tens of MeV excitation of the fissioning nucleus) and their relative merits and demerits for production of *n*-rich RIBs.

Nuclear fission is a complex process and a complete understanding of the process that can explain all the experimental observations is yet to be reached. However, many of the basic features of the reaction are quite well-understood, at least qualitatively. Immediately after the discovery of fission, the phenomenon is described by treating the fissioning nucleus as a charged macroscopic liquid drop and by tracing the potential energy (deformation energy) of the liquid drop from its spherical (or slightly deformed stable shapes) ground state as a function of its deformation (Me 39, Bo 39, Fr 47). As the drop/nucleus deforms at constant volume, its surface energy increases while the Coulomb energy decreases. As long as the increase in surface energy is more than the decrease in Coulomb energy, the net effect is positive deformation energy, which is the summation of the surface and the Coulomb energy. In this situation there will be a net restoring force to bring the drop back to its stable ground state shape. However, beyond a critical deformation, called the saddle point (the highest point in the potential energy deformation diagram representing the potential barrier (Figure 3.10), the decrease in Coulomb energy overtakes the increase in the surface energy. The net effect is a decrease in the deformation

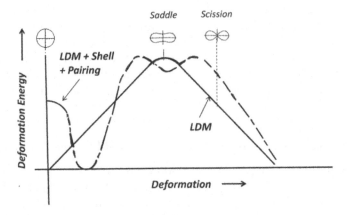

FIGURE 3.10
Potential energy of a liquid drop as a function of deformation: with and without shell correction. The configurations at the saddle and scission are also schematically shown.

energy. Beyond the saddle point, therefore, the potential energy decreases as the drop deforms further and further as there is no restoring force. Thus, if a nucleus is excited enough to reach the saddle point deformation, its ultimate fate is fission in which the nucleus breaks into two fragments. After crossing the saddle, the nucleus will deform continuously until in its final state it separates into two liquid drops (fragments/nuclei). The point on the potential energy–deformation curve at which the fissioning nucleus breaks into two fragments is called the scission point. At the scission point, two deformed fragments are in touching configuration. Because of large Coulomb repulsion between the two charged drops in contact, the fragments fly apart with kinetic energies. The liquid drop model predicts symmetric splitting into two fragments of equal sizes (masses) that leads to maximum energy release. The two separated deformed fragments in flight also quickly assume spherical or less deformed stable shapes and, in this process, the potential energy of deformation is converted into internal excitation energy of the fragments. The fragments then cool down by neutron evaporation and γ decay, resulting in the final fission products.

The simple charged liquid drop picture, although it describes the evolution of the shape of the excited nucleus leading to its break-up into two fragments quite well, is not adequate to correctly predict either the accurate barrier heights or the experimental observations, especially the observed marked asymmetry in the mass distribution of fission products in thermal neutron-induced fission of ^{235}U. The mass distribution shows that symmetric fission, most unexpectedly, is suppressed by a huge factor of about 600. To explain asymmetric splitting of the fragments leading to asymmetric mass distribution, the shell effects in the macroscopic liquid drop approach need to be introduced (St 67). The shell effect is essentially a bunching of nuclear levels or non-uniformity in the distribution of single particle states, like the liquid drop potential energy is a function of deformation. The shell effect introduces a correction to the liquid drop potential energy (the pairing correction term also needs to be included). The shell correction results in the modification of the fission barrier height and in the appearance of a second minimum. It also predicts a ground state that is slightly non-spherical (Figure 3.10). As the excitation energy of the compound nucleus undergoing fission (^{236}U in this case) is low, just above the fission barrier, shell effects play a decisive role in the splitting of the fissioning nucleus into two fragments. The formation of one of the nascent fragments is strongly influenced by the extra stability associated with a doubly magic shell closure at $Z = 50$ and $N = 82$ (^{132}Sn). The shell effect thus favors an asymmetric split in ^{236}U, with one fragment in the vicinity of $A = 132$. It may noted that actually the fissioning nucleus is itself quite deformed with deformation increasing in its journey from saddle to scission. In a deformed nucleus the magic numbers that would decide the splitting at the scission point are therefore different than the ones for spherical nucleus. To explain the experimental results it is necessary to consider the magic numbers corresponding to the deformed nucleus undergoing fission (Wi 76).

3.3.2 Production of *n*-Rich Isotopes in Fission Induced by Thermal Neutrons

The mass distributions resulting from thermal neutron-induced fission of ^{235}U and spontaneous fission of ^{252}Cf are shown in Figure 3.11 (Fr 81). It may be noted that in the mass distribution the position of the heavy fragment (heavier mass) peak remains the same in the two cases indicating the role of $Z = 50$ and $N = 82$ shell closure in the formation of the heavier fragment. The position of the lighter mass peak is then determined by subtracting the mass number corresponding to the heavy peak (roughly $A = 136$) from the mass number of the compound system undergoing fission (^{236}U, ^{252}Cf, etc.). The lighter mass peak

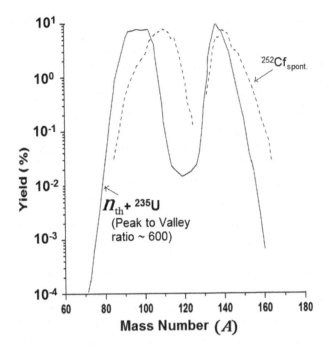

FIGURE 3.11
Mass distribution in thermal neutron-induced fission of ^{235}U. The dotted line shows the mass distribution for spontaneous fission of ^{252}Cf.

thus shifts towards higher mass for the fission of ^{252}Cf. The asymmetric split for ^{236}U results in heavy fragments, typically 1.4 times the mass of the light fragments. The mass yield curve shows that the distribution of yields is almost symmetric around the minima, corresponding to a symmetric split (A nearly equal to 118 in case of ^{236}U). The peak-to-valley ratio is nearly 600 and one consequence of this, from the viewpoint of the production of n-rich nuclei using fission of actinides (say ^{235}U), is that thermal neutron-induced fission is not the ideal choice for the production of nuclei in the symmetry region (say, $A = 107$ to 127) of the mass distribution.

Out of a total of 200 MeV energy released in the fission of one ^{236}U nucleus, on average about 170 MeV is taken away by the kinetic energies of the two fragments and about 30 MeV goes into internal excitation of the fragments. Internal excitation results, on average, in emission of about 2.41 neutrons per fission, apart from γ emission. Studies revealed that in thermal neutron-induced fission of ^{235}U, the fraction of neutrons emitted before the complete separation of fragments is usually a very small fraction (10–20%) of the total neutron emission. Thus, neutron emission primarily takes place from the two primary fragments which cool down by neutron evaporation forming the final fission products.

The yield of an isotope produced in fission is determined by three factors: the total fission cross-section, the mass distribution which decides what fraction of the total cross-section goes into the production of the particular isobaric chain and the charge distribution which determines what fraction of the chain yield goes into the production of the particular isotope of that isobaric chain. In thermal neutron-induced fission of ^{235}U (^{236}U undergoing fission) the total fission cross-section is as high as 580 b (Hy 64). The chain yield at the peak of the mass distribution is usually a few percent (about 4 to 5%) of the total fission cross-section, which is 580 b in this case.

During the fission process the N–Z equilibration appears to take place at an early stage, resulting in the highest probability for the emission of primary fragments with A/Z ratio very close to that of the compound system of ^{236}U (equal to A/Z ratio of the compound system, in this case 236/92, if there is no loss of neutrons in the pre-scission stage). Thus, for a given mass chain A, the charge (proton) number of the isotope that is expected to be produced with the maximum yield is given by $Z_p = 92 * A/(236-\nu_t)$, where ν_t is the total number of neutrons lost from the fissioning system prior to and after the separation of the two fragments, usually referred to as pre- and post-scission neutron multiplicities. The charge distribution for thermal neutron-induced fission has been found to be a narrow Gaussian around a most probable charge of Z_p with a Full Width at Half Maximum (FWHM) of about 1.5 charge units (Wa 62). If Z_p is decided assuming N–Z equilibration as explained above, the charge distribution is termed as the Unchanged Charge Distribution (UCD). The charge distribution has been determined for various mass chains and the width has been found to be almost independent of the mass number. The Z_p value for any mass chain has been found to be about three to four charge units smaller than the atomic number of its beta-stable isobar.

Thus, for an $A = 99$ mass chain, for example, a maximum yield for the isotope ^{99}Y ($Z = 39$) following the UCD prescription can be expected. It should be noted that ^{99}Y has A/Z ratio of 2.538 which is closer to A/Z ratio of 2.565 of the compound system (^{236}U) than the A/Z ratios of the other two neighboring isobars ^{99}Sr ($A/Z = 2.6$) and ^{99}Zr ($A/Z = 2.475$). Since the $A = 99$ mass chain also lies almost at the peak of the mass distribution, a very high yield of ^{99}Y is expected in the thermal neutron-induced fission of ^{235}U. Therefore, the production cross-section of ^{99}Y (and its complementary heavier fragment) is expected to be approximately as high as 10 b. The cross-sections for more exotic n-rich isobars of ^{99}Y (such as ^{99}Sr, ^{99}Rb) can be estimated on the basis of the charge distribution. It has been estimated that with the production target placed very close to the core of a modern high flux reactor (such as the ILL high flux research reactor in Grenoble or the Munich high flux reactor, FRMII), intensities of mass separated ions of very n-rich species like ^{132}Sn and ^{144}Cs can reach as high as 10^{11} ions/s (Kes 71). Such intensities are extremely difficult to obtain using any other reaction route. The PIAFE project in Grenoble (Kö 98), France, and the CARIF facility in China (Li 11) propose using the thermal neutron-induced fission of ^{235}U.

In spite of having the advantage of very high total fission cross-section, the number of RIB facilities planned around high flux reactors are so far very few. This is mainly because of constraints associated with radiation safety. Nuclear reactors are the best sources of thermal neutrons offering high flux exceeding $10^{14}/cm^2/s$ at the core. But safety-related issues get in the way of positioning the target chamber and ion source anywhere close to the reactor core where the flux is high. Also, for the integrated target ion source assembly to be positioned close to the core (in-pile position), radiation damage of the ion source materials, target holder, etc., due to high neutron and gamma flux poses a serious challenge (Kö 97). Placing the target chamber a few meters away from the core is generally considered to be safe but in that case the thermal neutron flux at that distance decreases by three to four orders of magnitude (hardly exceeding $10^{10}/cm^2/s$). So, the overall yield also decreases proportionally, neutralizing the advantage of very high cross-section. Also, as mentioned earlier, thermal neutron-induced fission is not optimum for production of n-rich nuclei in the symmetry region.

These factors together lead to the choice of alternate fission routes using accelerated charge particle beams that induce fission in actinide (usually ^{238}U) targets either directly or indirectly. Fission induced by protons bombarding a uranium target constitutes the direct route, while fission induced by energetic neutrons produced in the break-up of energetic deuterons and by Bremsstrahlung gamma rays produced via the stopping of energetic electron beams are the indirect approaches. The total fission cross-section in these approaches is more than two orders (three orders for gamma-fission) of magnitude less compared to the thermal neutron-induced fission of ^{235}U, but in this case safety issues do not limit the flux of the particles (p, n, γ) producing the fission, resulting in an overall gain in the yield. A number of upcoming RIB facilities are planning to use these routes to produce n-rich isotopes in the fission product range.

3.3.3 Production of *n*-Rich Isotopes in Fission Induced by Energetic Protons/Light Ions

In RIB facilities with low to medium energy driver accelerators fission may be induced by protons (or other light ions like ^{3,4}He) with energies up to about 70 MeV, energetic neutrons of a few tens of MeV resulting from the break-up of energetic deuteron beams, and by gamma rays in the range of 10–20 MeV where gamma rays are produced by the stopping of electron beams with energies in the range of 20–50 MeV. Heavy ion beams with energies around the Coulomb barrier also produce fission but are not considered an optimum route because of lower beam intensity and much lower usable target thickness compared to light ions. In all these reactions, as a consequence of higher excitation energy of the compound system, fission shows some characteristics different from compared to thermal neutron-induced fission. These are: (a) the symmetry region in the mass distribution is filled up with an increase in the incident projectile energy/excitation energy of the compound system because shell effects become less important at higher excitations; this results in enhanced production of the isotopes in the symmetry region; (b) the width of charge distribution increases rather slowly with excitation energy, presumably due to occurrence of multi-chance fission, which results in the production of a larger number of isotopes with measurable cross-sections; the wings of mass distribution also spread a little, resulting in the production of a wider mass range of n-rich nuclei, especially nuclei with $A \geq 150$ and $A \leq 80$ are produced with substantial cross-sections at higher excitations; (c) the peak of the charge distribution Z_p shifts, irrespective of the isobaric chain, towards higher Z values (fewer n-rich isotopes) because of more neutron loss from the fissioning system; and (d) the total fission cross-section decreases from 580 b to about 1–2 b or even less (around 1.6 b for 35 MeV protons/neutrons; for gamma-induced fission of ^{238}U it is about 160 mb at about 15 MeV).

It has been found that the peak-to-valley ratio, which is about 600 for thermal neutron-induced fission of ^{235}U reduced to about 6 for 14 MeV induced fission. This is presumably because the shell effect becomes less and less important at higher excitations and the valley becomes increasingly filled up at higher excitations since symmetric rather than asymmetric fission is favored at higher excitations. However, it has been found that the fission valley does not get filled up completely even up to excitation energy of 66 MeV (Cha 93, Hi 17, Ban 20). This is presumably because of multi-chance fission, where fission takes place from the initial compound nucleus as well as from other nuclei resulting from one, two and more neutron evaporations depending upon the excitation energy. The nucleus that undergoes fission after evaporation of a few neutrons is already cold enough for shell effects to play a dominant role and would split into two asymmetric fragments. Multi-chance fission thus allows for the survival of asymmetric fission to quite high excitations of the initial compound system. But the asymmetry at higher excitations is about a factor of two less compared to 14 MeV induced fission and the peak-to-valley ratio is usually in the range of three to four. Figure 3.12 shows the mass distributions at excitation energies of 45 and 66 MeV in alpha-induced fission of ^{232}Th.

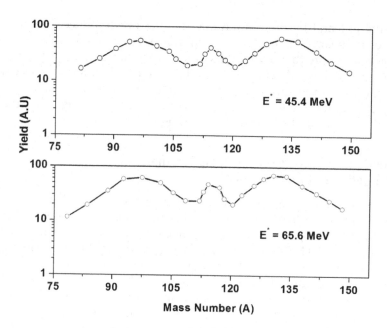

FIGURE 3.12
Mass distribution in ^{232}Th (α, f) reaction at excitation energies of 45.4 and 65.6 MeV of the compound nucleus (^{236}U).

As regards the charge distribution it has been found that for $A = 99$ mass chain the Z_p value at 23 MeV of excitation is around 39.4 which increases to 39.7 at excitation energy of 66 MeV (compared to $Z_p \sim 39.0$ in thermal neutron-induced fission for the same mass chain). The Z_p values in this excitation energy range follow approximately the UCD hypothesis. The width of the charge distribution is expected to increase slowly with the excitation energy due to an increase in the probability of multiple chance fission, that is, fission proceeding from different compound systems due to neutron evaporation prior to scission. FWHM in the range of 1.6–1.65 charge units seem to fit the experimental data up to about 40 MeV. At higher excitation energies, because of the shift in the Z_p value towards the beta-stability line, the more exotic isotopes of any isobaric chain are expected to have a lesser share of the chain yield. So, for the production of very n-rich nuclei lower excitation energy of the fissioning system is generally preferred. However, it might be noted that at higher excitations the width of the charge distribution also increases, which means cross-sections of exotic isotopes far from the Z_p would fall off less sharply. Thus, in cases where an increase in excitation energy leads to a higher total fission cross-section and a favorable mass distribution (which happens for the isotopes lying in the symmetry region and also for isotopes on the two edges of mass distribution because the wings spread a bit more at higher excitations) relatively higher projectile (excitation) energy might turn out to be a better choice, which also allows for the use of thicker targets to increase the overall yield.

Proton beams with energy up to 70 MeV are a reasonably good choice for the production of n-rich nuclei far away from beta-stability. The total fission cross-section for protons on ^{238}U saturates at around 50 MeV and is close to 2 barn, as shown in Figure 3.13 (Ba 71, Me 79, Is 08, Kh 16). It should be noted that with a low power (20–30 MeV) proton beam, a number of n-rich nuclei in the symmetry region have been produced for the first time (e.g., ^{117}Rh with $A/Z = 2.6$ and half-life = 440 ms) and studied at Jyvaskyla using the Ion Guide Isotope Separator On-Line (IGISOL) (Ay 88, Pe 88). Also, a number of exotic n-rich isotopes

Nuclear Reactions for Production of RIBs

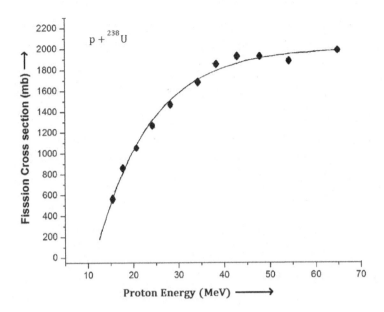

FIGURE 3.13
Total fission cross-section for proton-induced fission of ^{238}U as a function of proton energy.

were produced at the SARA IGISOL facility in Grenoble using a 40 MeV alpha projectile and a ^{238}U target (As 92). Estimations based on empirical formulations show that proton-induced fission could be quite competitive, cross-section wise, for the production of very exotic n-rich nuclei lying on and around the r process nucleo-synthesis path (Kh 16).

At incident proton energy of 70 MeV, quite a thick ^{238}U target of about 4 mm thickness containing 1.9×10^{22} atoms /cm^2 can be used. In the 4 mm thick target the protons loses about 55 MeV and comes out of the target with an energy of about 15 MeV, below which the fission cross-section is small. In view of the quite high average cross-section in this energy range, the number of fissions that can be produced in the target for a proton beam current of 1 mA is expected to be about 6×10^{13}/s. This is indeed a very good fission rate but in practice it is very difficult to design a target capable of handling 55 kW beam power; high power density makes it worse. To reduce the power density, the proton beam can be defocused to a size of about 20 mm and use a uranium carbide target of lower density, but the target design still remains very challenging. At ISAC, TRIUMF, it has been possible to design composite uranium carbide targets bonded to graphite foils (Bri 16) that are capable of absorbing about 30 kW of proton beam power. Thus, proton beam of energy in the range of 50–70 MeV offers a competitive way of producing n-rich fission products and a fission rate of about 3×10^{13}/s could be reached.

3.3.4 Fission Induced by Energetic Neutrons

The conversion of deuterons to neutrons for inducing fission in uranium is a very useful concept offering a number of advantages (No 93). The production of n-rich isotopes in the SPIRAL II project at GANIL is based on this concept, which is schematically shown in Figure 3.14a. The use of high-energy neutrons in the range of 10–40 MeV, instead of protons of comparable energies, seems to be a better option from the viewpoint of target (^{238}U) design since neutrons do not lose energy by interaction with electrons (ionization)

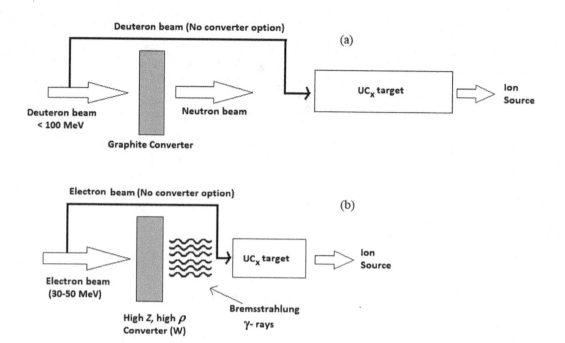

FIGURE 3.14
Schematic presentation of (a) neutron-induced fission using deuteron beam (b) γ-induced fission using electron beams.

and lose energy mainly by elastic collisions. Consequently, with neutrons the target can be much longer (thicker) resulting in a lower power density in the target. Also, greater target thickness helps in the enhancement of yield. Fast neutrons can be produced very efficiently by stopping the deuteron beam in a converter (Be, C, etc.). The SPIRAL II project at GANIL would be using a 40 MeV, 5 mA (200 kW of beam power) deuteron beam, produced using a superconducting linear accelerator. The deuteron beam would be stopped in a thick converter target to produce neutrons up to about 40 MeV, with an average value centered around 14 MeV. The neutrons resulting from reactions in the converter would be forward-focused and the total fission cross-section for these neutrons using a ^{238}U target is much higher compared to low energy (say about 3 MeV) neutrons. Measurements on the production of isotopes also tend to indicate that the neutron option is slightly superior to the energetic proton option for the production of exotic n-rich isotopes.

The converter target should be thick enough to stop the protons resulting from the same break-up reactions that produce the neutrons and would result in a beam of neutrons peaked in the forward direction with a very broad energy distribution centered around 14 MeV. About 10^{14} neutrons/s are produced within a forward cone of about 20 degrees if a carbon converter is used. The beam of neutrons would then produce fissions in a thick ^{238}U target placed downstream in close proximity to the converter target. With 200 kW of beam power the estimated fission rate is about 5×10^{13} /s. The beam power deposited in the uranium target would be only about 1.7 kW for this fission rate and thus the target design is relatively simpler. However, since the beam power is to be deposited mainly in the converter, very efficient cooling of the converter is necessary. For SPIRAL II, the converter envisaged is a rotating carbon wheel of about 6–8 mm thickness, which is enough to stop 40 MeV deuterons and 20 MeV protons. Carbon is chosen as the material owing to its high melting point. Slow rotation of the carbon wheel helps in distributing the beam power over a larger surface area and keeps the temperature of the carbon wheel under about 1850°C (Le 08). However, it is a challenging task to design a converter that can take 200 kW of deuteron beam power at 40 MeV, even if the deuteron beam is de-focused to a spot size of about 20 mm. The actual usable beam power is therefore likely to be less, which would bring down the achievable fission rate proportionally. Alternately, of course, the converter can be avoided and the deuteron beam allowed to directly fall on the UC_x target. In this case, the target would need to handle much more power because of energy loss of the deuteron beam in the target and only a fraction of the available beam power can be used to limit the target temperature to within say 2200 K. However, the converter option gives better control over the target temperature since in this case the target needs external resistive heating to maintain the optimum temperature. In either case, the neutron-induced fission using the break-up of deuteron as envisaged in SPIRAL II represents one of the most efficient ways of producing high intensity exotic n-rich beams in the fission product range. Additionally, the SPIRAL II facility would use the high flux of neutrons produced in the converter to carry out a wide variety of studies (neutrons for science facility) involving neutrons (Led 17).

3.3.5 Fission Induced by Gamma Rays

The photofission of uranium is considered to be a promising and cost-effective route for the production of n-rich RIBs in the fission product range (Di 99, Oga 02). Rough estimations based on simple analytical formulations show that the Bremsstrahlung photons produced from stopping of a 30 MeV, 100 kW electron beam can produce about 3×10^{13} fission/s in a thick ^{238}U target. This is quite a competitive fission rate and many laboratories including the ARIEL RIB facility at TRIUMF are going to utilize this option. The electron-gamma-fission option is shown schematically in Figure 3.14b.

Fission is induced by photons through the excitation of Giant Dipole Resonance (GDR) in $^{235,238}U$. The yield from a ^{238}U target is less by about a factor of two compared with ^{235}U (since the GDR cross-section is higher in the case of ^{235}U) but ^{238}U is mostly preferred as the target material because less regulatory compliance is involved. For ^{238}U, the fission cross-section peaks around 15 MeV of photon energy and the peak cross-section is nearly 160 mb (GA-SPI2/007-A) as shown in Figure 3.15. The GDR peak is quite broad and photons in the energy range of 7–25 MeV contribute to the photofission of ^{238}U, although the major contribution to photofission comes from photons with energies between 10 and 20 MeV. The absorption of gamma rays in the target medium occurs through (γ, f), (γ, n) and $(\gamma, 2n)$ reactions, through $e^+ e^-$ pair production, and through Compton and Raleigh scattering. Although photofission has the dominant contribution to the production of n-rich nuclei, the neutrons produced in (γ, xn) reactions, including those produced in photofission, also induce fission in ^{238}U, contributing to the final fission yield if the target is sufficiently thick. Also, the pair production would lead to photons that can induce further fissions. All these contributions make the fission yield substantial and the mass and charge distribution resemble more that of fission induced by low-energy neutrons than fission induced by medium energy (say 40 MeV) neutrons or protons.

Since mono-energetic sources producing photons in the energy range of 10–20 MeV are not available, electron accelerators are used to accelerate intense (a few mA) beam of electrons in the range of 25–50 MeV and to stop the electron beam to create Bremsstrahlung radiation where the energy of the photons varies from zero energy to the initial energy of the electron beam. The Bremsstrahlung yield falls sharply with photon energy. Thus,

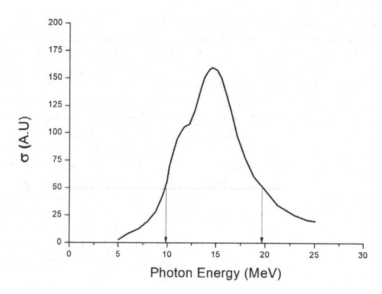

FIGURE 3.15
Cross-section of photon-induced fission of ^{238}U as a function of photon energy.

to produce photons in the most effective GDR energy range for inducing photofission in ^{238}U, an electron beam with energy higher than 20 MeV is needed. Figure 3.16 shows the Bremsstrahlung yield per electron as a function of photon energy when 10, 25 and 45 MeV electrons are stopped in a suitable heavy target. For electron energy of about 50 MeV, the stopping of each electron produces about 20 photons. However, the number of the most useful photons, with energies between 10 and 20 MeV is only about 0.5 to 0.7 per electron.

The choice of optimum electron energy is decided by the photofission yield per electron. As the electron energy increases, say starting from 25 MeV, the number of useful Bremsstrahlung photons increases, resulting in higher fission yield. Calculations show that photofission yield per electron increases rather slowly after 30 MeV and almost saturates beyond 45 MeV (GA-SPI2/007-A) as shown in Figure 3.17. The photofission yield per unit of beam power thus increases very slowly beyond electron energy of 30 MeV. Thus, beam energies lower than 45 MeV, say 30 MeV can be used and the corresponding loss of useful photons can be compensated by increasing the beam current. The total beam power also does not increase allowing the selection of electron beam energy in the range of 30–50 MeV depending upon the budget and other constraints.

The photofission proceeds from relatively low level of excitation of the fissioning compound nucleus (compared to say 30 to 50 MeV proton or neutron-induced fission), which limits the neutron loss from the system, thereby favoring the production of *n*-rich nuclei. Empirical simulations of photofission cross-section (Bh 15) show that at the mean photon energy of 13.7 MeV (which corresponds to Bremsstrahlung end point energy of 29.1 MeV, resulting from the stopping of an electron of about 30 MeV) photofission can produce some of the very *n*-rich nuclei lying around the *r* process nucleo-synthesis path with considerable cross-sections; for example, the expected cross-sections for production of ^{80}Zn (A/Z = 2.66), ^{98}Kr (A/Z = 2.72), ^{110}Zr (A/Z = 2.75) and ^{134}Sn (A/Z = 2.68) are around 2.5 μb, 67 nb, 16 fb and 0.18 μb respectively.

The thermal management in the photofission case is simpler than the other production routes: that is, fission by medium energy protons and indirect fission by neutrons formed

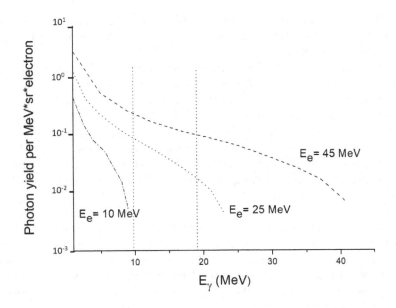

FIGURE 3.16
The Bremsstrahlung photon yield as a function of photon energy for three different electron energies when electrons are stopped in a heavy (W) target. The vertical dotted line shows the photon energy band most effective for producing fission.

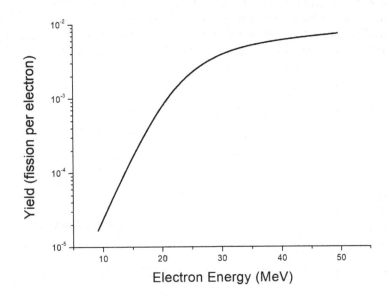

FIGURE 3.17
The photofission yield of ^{238}U as a function of electron energy.

from the break-up of deuterons. This is because an electron beam has a much lower rigidity than either proton or deuteron beams and can easily be scanned over a large target/converter area reducing the power density. Instead of allowing the electron beam to directly hit the target, a converter is often chosen for the production of Bremsstrahlung radiation that can be transported through air to a uranium carbide target placed downstream in close proximity to the converter for production of *n*-rich isotopes through photofission. This way a major fraction of the beam power is handled by the converter that makes thermal management issues for the target much simpler.

The converter material and thickness need to be chosen so that it produces effective γ-rays optimally and absorbs a large fraction of the electron beam power so that the target design becomes simpler. Since the radiated power is proportional to the square of the deceleration (Jackson 98), the converter material should be a high Z and high-density material so that the electron beam rapidly decelerates in the converter within a short distance The converter material should also be refractory so that it can withstand a high temperature. Thus, tungsten is usually considered to be the most suitable converter for the production of Bremsstrahlung photons through the slowing down of a high-power electron beam (liquid lead is often preferred if the beam power exceeds 100 kW). The target should be placed as close as possible to the converter so that it can intercept all the Bremsstrahlung γ-rays that are emitted in a forward cone whose angle depends on the electron beam energy.

The choice of converter thickness is not straightforward. A converter that is too thick would absorb a good fraction of effective photons and would result in a lower number of fissions in the uranium target positioned behind the converter in close proximity. Figure 3.18 shows the estimated number of fissions in a 16 cm³ UC$_x$ target, positioned very close to the target downstream, per second per kW of electron beam power as a function of the thickness of a tungsten converter (Gott 16a). The thickness is expressed as a fraction of the electron range in the converter. The simulation shows results for two electron energies: 35 and 50 MeV. It can be seen that for an optimum fission rate the converter thickness should be between 30 and 60% of the electron range in tungsten. Thus, the optimum converter thickness should be around 3 to 4 mm for tungsten if electron energy is in the range of 35–50 MeV. The converter-to-target distance is quite critical and quite often needs to be decided based on practical considerations related to converter cooling, etc. The distance

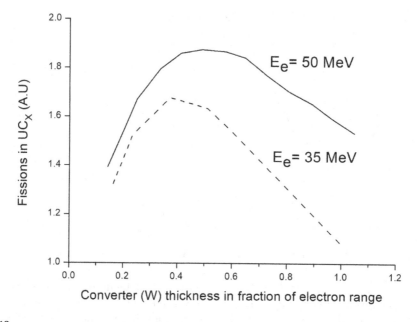

FIGURE 3.18
Fission yield vs converter thickness for electron fission using a W converter for production of γ– photons for two different electron energies.

should however be the minimum possible so that the beam power can be effectively utilized to produce fission in the target.

The target thickness is chosen such that the target can utilize a major fraction of the useful photons. Simple analytical formulations (Be 70) can provide some initial guidance in estimating the converter and the target thickness. However, for realistic estimates the converter thickness, the converter-to-target distance and the target thickness need to be optimised using Monte Carlo-based codes such as FLUKA. In any case, since pair production, which is proportional to Z^2, becomes the dominant process of photon attenuation above 10 MeV, the penetration depth of effective γ-rays is small in a UCx target and is usually in the range of 10–20 mm (Gott 16a). This tends to limit the achievable fission rate per unit beam power to an extent.

In the absence of a converter, the electron beam falls directly on the target and the target itself becomes the Bremsstrahlung converter. In this case the loss of fission-producing photons is minimum and almost an order of magnitude less beam power (electron beam current at the same electron energy) is required to produce the same fission rate. But power dissipated in the target increases, making thermal management of the target more challenging; in particular, the high rate of energy deposition by electrons in front layers of the target facing the electron beam becomes a serious concern. Further, the power deposition would make the temperature profile in the target inhomogeneous which would adversely affect the extraction efficiency of the fission products produced inside the target. However, depending on the beam power, either of the options (with and without converter) can be chosen but the converter option appears to be more suitable at a higher electron beam power.

In the converter option, realistic estimation of photofission yield using detailed calculations using the FLUKA code shows that, as a benchmark number, about 2.5×10^{10} atoms/s of doubly magic n-rich ^{132}Sn, with $A/Z = 2.64$, using 500 kW electron beam at 50 MeV can be expected. The simulation used a liquid lead converter and a 20 g/cm^2 UC$_x$ target (Di 14). The same simulation shows that very n-rich isotopes such as ^{144}Xe and ^{144,146}Cs are also expected to be produced with intensities almost comparable to that of ^{132}Sn.

The photofission route is, however, not very efficient for the production of nuclei with $A > 150$ since the mass distribution is comparatively narrower compared to fission caused by energetic neutrons/protons. Photofission is also not optimum for the production of isotopes in the symmetry region of the fission mass distribution. This however is not a major disadvantage since in this region a majority of the elements is of a refractory nature, which is anyway not suitable for an ISOL-type facility. Photofission remains a very good choice for the production of exotic n-rich nuclei since it can produce reasonably high fission rates at a much lower cost (the cost of the electron accelerator, as well as the cost of shielding, is lower) and the target design, in principle, is comparatively simpler.

3.4 Production of RIBs Using Low-Energy Heavy Ions above the Coulomb Barrier

3.4.1 Fusion–Evaporation Reactions for the Production of Neutron-Deficient Nuclei

In the reaction of a heavy ion projectile with a heavy ion target at an energy just above the Coulomb barrier, the fusion–evaporation reaction has a major share of the total reaction cross-section. This reaction has been extensively used for the production and study of exotic neutron-deficient isotopes close to the proton drip line (refer among others to Th 04). In this reaction, the projectile and the target nuclei fuse together to form a single compound nucleus (CN) and the excited CN decays via evaporation of light nucleons (Bo 36; Bl 52).

For the fusion of two heavy ions, the projectile should have enough kinetic energy (in the classical picture) to overcome the Coulomb barrier. Nuclear forces become operative as the projectile comes close enough to the target nucleus and that leads to the fusion of the projectile and the target nucleus into one composite nucleus. The kinetic energy needed for the projectile to overcome the Coulomb barrier is usually much more than the Q value needed for the formation of the composite nucleus. This excess kinetic energy is randomly distributed among all the nucleons in the composite system through multiple collisions among the nucleons, leaving the combined nuclear system in an excited quasi-stationary state, which is called the CN. The excited CN then releases its excitation energy by emission of gamma rays, evaporation of neutrons, protons and light particles like deuteron, alpha, etc. The excited CN retains no memory of how it was formed (the particular entrance channel leading to its formation) and its decay leads to a final product with a mass number very close to that of the CN, since only a few particles usually evaporate out of the CN. Although the colliding nuclei fuse quite fast, typically in about 10^{-22} s, thermal equilibrium followed by particle evaporation is a slow process. This makes the fusion-evaporation a slow process, having a time scale in the range of 10^{-14} to about 10^{-20} s.

The CN formed out of the fusion of projectile and target nuclei lies almost always on the neutron-deficient (p-rich) side of the beta-stability line, owing to the curvature of the beta-stability line towards the neutron-excess side. After evaporation, the final product nuclei can often reach the proton drip line and beyond. The fusion–evaporation reactions are one of the most optimum (reasonably higher production cross-section, production of less number of unwanted isotopes) reaction routes for the production of exotic neutron-deficient isotopes all over the nuclear (N–Z) chart, especially in the range of $50 \leq Z \leq 82$. A majority of the proton drip line isotopes in this Z range have been produced using fusion-evaporation of heavy ions.

Two heavy ions with energies just above the Coulomb barrier do not necessarily fuse. Depending upon the impact parameter there are various possibilities as shown schematically in Figure 3.19.

Fusion takes place only at low impact parameters. This is because heavy ion collisions bring a lot of angular momentum in the target-projectile di-nuclear system. Depending upon the impact parameter this angular momentum could exceed 100 ℏ. Theoretical calculations, however, show that a rapidly rotating nuclear drop becomes unstable against spontaneous fission beyond a critical angular momentum or spin of l_c and the value of l_c never exceeds 100 units of ℏ (Co 74). Beyond the critical angular momentum l_c the disruptive Coulomb and the centrifugal forces together exceed the attractive nuclear forces, and the projectile cannot be trapped by the target nucleus, preventing complete fusion. The value of l_c depends upon the mass number and the maximum is around $A \approx 130$ for beta-stable nuclei (Figure 3.20). The limit decreases both for lighter and heavier nuclei; lighter because of their small size/moment of inertia, and heavier because of increasing Coulomb energy. Fusion thus takes place for low impact parameter central collisions rather than peripheral collisions that bring in high angular momentum in the projectile-target system (Figure 3.20). The ratio of fusion cross-section to the total reaction cross-section thus decreases with the increase in the beam energy. So, a projectile energy just above the Coulomb barrier is chosen. Also, since for a given Z of the compound nucleus, the Coulomb barrier is highest for

Nuclear Reactions for Production of RIBs

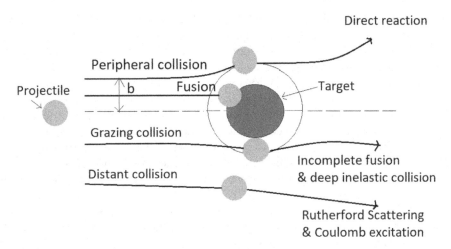

FIGURE 3.19
Different types of low energy heavy ion reactions depending on the impact parameter.

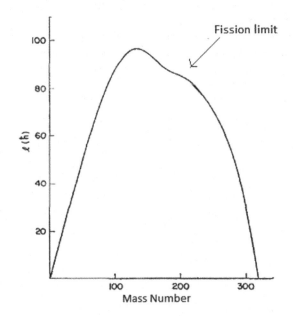

FIGURE 3.20
The angular momentum limit for fusion as a function of mass number.

Z_1 (projectile) = Z_2 (target), asymmetric rather than symmetric projectile-target combinations are generally preferred for fusion reactions, to keep the beam energy relatively lower. However, since the symmetric combinations ($Z_1 \approx Z_2$) lead to the most neutron-deficient compound nuclei, the optimum choice for a given situation needs to be figured out. In fact, symmetric combinations become, quite often, the only choices in many cases, especially for the production of most exotic neutron-deficient isotopes beyond Pb.

The choice of optimum projectile-target combination and the projectile energy are made quite often on a case-by-case basis guided by the particular fusion-evaporation product (the neutron-deficient nuclei) of interest. To make this choice, use is often made of available fusion-evaporation codes like PACE (Gav 80). PACE and similar codes provide quite useful guidelines for the choice of optimum projectile-target combination, beam energy, the useful target thickness and in the overall planning of the experiment for which information regarding the production cross-sections of other nuclei (contaminants) produced in the same reaction is also important to decide on the separation and measurement techniques.

As an example, we have shown in Figure 3.21 the cross-sections, calculated using the PAES code, of a few fusion-evaporation products in the reaction of ^{64}Zn (projectile) and ^{96}Ru (target), as a function of projectile energy. More than the cross-section values, the example is sought to bring out some of the important features of the fusion–evaporation reaction, especially from the viewpoint of an experimentalist. In Figure 3.21, the fusion-evaporation cross-sections of six evaporation channels, viz., *1n, 1p, 2n, np, 2np* and *2pn* leading respectively to six reaction products ^{159}W, ^{159}Ta, ^{158}W, ^{158}Ta, ^{157}Ta and ^{157}Hf are plotted as a function of ^{64}Zn (projectile) energy. It can be seen that the cross-section for any product, say ^{157}Hf, is significant only in a rather narrow range of the projectile energy (roughly from 260 to 310 MeV in the case of ^{157}Hf; the Coulomb barrier (CB) in the laboratory frame being about 256 MeV) and in that range of projectile energy, there are only a few other reaction products. Although, all the possible reaction products are not shown in Figure 3.21 (to avoid it becoming overcrowded with overlapping curves), it can be verified that for any energy range of the projectile which is optimum for the production of a particular product, the number of other reaction products (contaminants) remains low, usually less than ten. This is to be contrasted with the situation in spallation, projectile fragmentation and fission reactions in which more than a few hundred isotopes are produced. Thus, the magnetic separation of the reaction product of interest from the rest is expected to be much

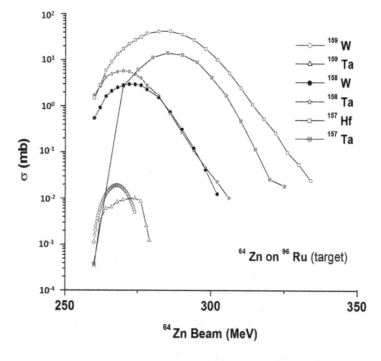

FIGURE 3.21
Fusion reaction cross-sections for different reaction products, calculated using PAES for the ^{64}Zn (projectile) + ^{96}Ru (target) combination as a function of projectile energy.

simpler and more efficient in the case of fusion–evaporation reactions. Also, since the total fusion cross-section is shared by only a few reaction products, relatively high production cross-sections for the individual reaction products can be expected, making it the optimum route for the production of neutron-deficient exotic nuclei.

It is important to note, in the above example, that the cross-sections of predominantly neutron evaporation channels (^{159}W, ^{158}W, ^{157}Ta, etc.) are comparable, or even less when compared to reaction channels involving the evaporation of protons (^{159}Ta, ^{158}Ta, ^{157}Hf, etc.). Usually, a CN decays predominantly through neutron evaporation because protons need to overcome additionally the Coulomb barrier. But near the proton drip line, the situation is different because the binding energy of the last proton decreases sharply while the binding energy of the last neutron increases. Since the ratio of neutron to proton evaporation probability from an excited CN is given roughly by:

$$\frac{P_n}{P_p} \sim \exp\left[-\frac{B_n - B_p^*}{T}\right] \quad (3.4)$$

(Vo 89), the proton evaporation from an excited CN of temperature T (MeV) near the proton drip line might become comparable, often more probable, than the neutron emission. B_p^* in Equation (3.4) is the effective binding energy of proton which is equal to Coulomb barrier + binding energy of proton. For example, the values of B_n and B_p^* for the CN ^{160}W are roughly 12.1 and 12.5 MeV (B_p = 2.17 MeV and CB = 10.3 MeV) respectively, which makes the evaporation probability of a proton and neutron comparable. It might also be noted that the loss of protons from the CN means loss of exoticity (loss of neutron-deficiency in this case) which limits the cross-section of production of proton drip line nuclei. So, for the production of proton drip line nuclei the incident projectile energy needs to be kept just above the Coulomb barrier to keep the excitation energy of the CN (and hence the number of protons evaporating out from the CN) to the minimum possible.

One of the factors that limits the yield in a fusion–evaporation reaction is that the targets used in these reactions are quite thin. The target thicknesses are often just a few times 10^{19} atoms/cm^2, which is about three to four orders of magnitude lower compared to typical target thicknesses used in PF and spallation reactions. This is because low-energy heavy ions have a small range in the target and also the fusion-evaporation cross-section leading to a particular reaction product is optimum only for a narrow range of energy of the incident projectile. However, the beam intensities achievable for low-energy heavy ions are often more than one order of magnitude higher compared to relativistic heavy ions. Also, cross-section wise, the fusion–evaporation reaction, in many cases, is the most optimum route for the production of exotic neutron-deficient species. In such cases, the gain in the yield on account of beam intensity and production cross-section often compensates/over-compensates for the lower target thickness, resulting in a net gain in the production yield. Additionally, the comparative ease of separation of reaction products offers a great advantage. Overall, the fusion–evaporation route remains a very competitive one for the production of neutron-deficient exotic species for a wide range of isotopes on the neutron-deficient side of the beta-stability line.

The efficacy of the fusion–evaporation reaction for the production of nuclei near, on and beyond the proton drip line depends to a large extent on the availability of a proper projectile-target combination among the naturally occurring beta-stable species. Usually for the production of proton drip line nuclei, the most neutron-deficient ones both for the projectile and the target need to be chosen from among the stable isotopes. For example, the choice of ^{64}Zn as the projectile and ^{96}Ru as the target to produce ^{160}W CN satisfies this criterion. It also needs to be ensured that to produce the particular CN, the Coulomb barrier for the projectile-target combination is the minimum so that the CN can be produced with the minimum possible excitation. The availability of optimum projectile-target combinations among the naturally occurring beta-stable isotopes is thus very important, especially if the production of the most neutron-deficient nuclei is intended. However, not many optimum combinations are available. Thus, although the fusion-evaporation reaction is generally quite suitable for the production of neutron-deficient nuclei almost all over the nuclear chart, it is the most optimum reaction route (because optimum combinations are available) for the production of proton drip line nuclei only in the range of $50 \leq Z \leq 82$. For example, most neutron-deficient isotopes of all elements starting from $Z = 52$ (^{106}Te (Sc 81)) to $Z = 82$ (^{180}Pb (To 96)) were produced for the first time in fusion-evaporation reactions using heavy ion reactions, as listed in Table 3.3.

TABLE 3.3

Production of Proton Drip Line Isotopes of $50 < Z < 82$ Using HI Fusion Reactions ((Th 04)

Z	Isotope	Reaction
52	^{106}Te	^{58}Ni(^{58}Ni, $2p4n$)^{110}Xe(α)
53	^{108}I	^{54}Fe(^{58}Ni, $p3n$)
54	^{110}Xe	^{58}Ni(^{58}Ni, $2p4n$)
55	^{112}Cs	^{58}Ni(^{58}Ni, $p3n$)
56	^{114}Ba	^{58}Ni(^{58}Ni, $2n$)
57	^{117}La	^{64}Zn(^{58}Ni, $p4n$)
58	^{121}Ce	^{92}Mo(^{32}S, $3n$)
59	^{121}Pr	^{96}Ru(^{32}S, $p6n$)
60	^{125}Nd	^{92}Mo(^{36}Ar, $3n$)
61	^{128}Pm	^{96}Ru (^{36}Ar, $p3n$)
62	^{129}Sm	^{96}Ru (^{36}Ar, $3n$)
63	^{130}Eu	^{58}Ni(^{78}Kr, $p5n$)
64	^{135}Gd	^{106}Cd(^{32}S, $3n$)
65	^{139}Tb	^{106}Cd(^{36}Ar, $p2n$)
66	^{139}Dy	^{106}Cd(^{36}Ar, $3n$)
67	^{140}Ho	^{92}Mo(^{54}Fe, $p5n$)
69	^{145}Tm	^{92}Mo(^{58}Ni, $p4n$)
70	^{149}Yb	^{112}Sn(^{40}Ca, $3n$)
71	^{150}Lu	^{96}Ru (^{58}Ni, $p3n$)
73	^{155}Ta	^{102}Pd(^{58}Ni, $p4n$)
74	^{158}W	^{106}Cd(^{58}Ni, $2p4n$)
75	^{160}Re	^{106}Cd(^{58}Ni, $p3n$)
76	^{162}Os	^{106}Cd(^{58}Ni, $2n$)
77	^{164}Ir	^{92}Mo(^{78}Kr, $p5n$)
78	^{166}Pt	^{92}Mo(^{78}Kr, $4n$)
79	^{170}Au	^{96}Ru(^{78}Kr, $p3n$)
80	^{172}Hg	^{96}Ru(^{78}Kr, $2n$)
81	^{177}Tl	^{102}Pd(^{78}Kr, $p2n$)
82	^{180}Pb	^{144}Sm(^{40}Ca, $4n$)

However, if a neutron-deficient RIB is used as the projectile, it can very efficiently produce drip line and beyond drip line neutron-deficient (proton-rich) nuclei over a wider Z range. To show the advantage of using a neutron-deficient RIB in combination with a suitable target, a typical case is illustrated in Figure 3.22, where a fusion-evaporation cross-section of ytterbium (Z = 70) isotopes calculated for a Zn (projectile) + Ru (target) system is plotted for the projectiles ^{64}Zn (stable), ^{60}Zn and ^{56}Zn (Ghos 02). The figure shows the gain in calculated cross-sections for very exotic species, such as ^{149,150}Yb, both of which are neutron-deficient isotopes very close to proton drip line, if ^{60}Zn instead of ^{64}Zn is used as the projectile. It should be noted that ^{60}Zn, although a secondary RIB, can be produced with quite a high yield since it is not far from stable ^{64}Zn and the gain in the absolute as well as relative cross-section, leading to a significant improvement in the signal-to-noise ratio, can result in an overall advantage in favor of production using secondary beams. If the more exotic ^{56}Zn is used, ^{146}Yb can be produced, an isotope most probably beyond the proton drip line, with a reasonable cross-section of about 1 mb. Thus, neutron-deficient RIBs, if they can be produced with sufficient intensities, have the potential to produce proton drip line isotopes, which otherwise cannot probably be produced using stable projectile-stable target combinations.

In the Z range of 20 < Z < 50, optimum projectile-target combinations for fusion–evaporation reactions are not available in many cases and it is the Projectile Fragmentation (PF) reaction that can produce exotic neutron-deficient nuclei with optimum cross-sections. In fact, a majority of proton drip line isotopes in this Z range are produced using PF reactions, as has already been mentioned in the section dealing with PF reactions. Beyond Z = 82, the fission channel opens up and to produce the most neutron-deficient isotopes, rather symmetric projectile-target combinations that bring in a lot of excitation energy in the system need to be used. This reduces the production cross-section of exotic neutron-deficient isotopes. Nevertheless, this route remains the most convenient route for the production

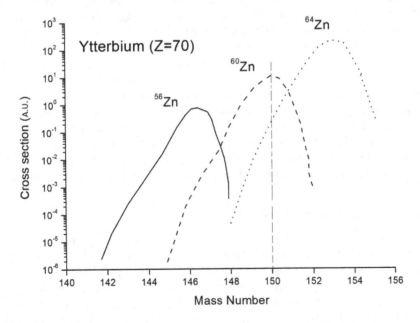

FIGURE 3.22
Simulated fusion cross-sections of ytterbium isotopes as a function of mass number using a ^{96}Ru target and stable (^{64}Zn) and unstable (^{56,60}Zn) projectiles.

of neutron-deficient nuclei and has been used at various laboratories to produce a good number of exotic neutron-deficient isotopes beyond $Z = 82$ (Th 04).

Any discussion of heavy ion fusion reaction remains incomplete without mentioning the use of this reaction in producing super heavy elements (SHE). It is the fusion-evaporation reaction that has been used to synthesize all the trans-actinide elements that have been discovered so far (Mun 18 and references therein). The production cross-sections for these isotopes are small because the formation of compound nuclei and subsequent decay of compound nuclei by evaporation of neutrons (charged particle evaporation is highly suppressed because of a high Coulomb barrier) need to face stiff competition from fission which is by far the dominant decay mode. In this case, the emphasis shifts from producing the most neutron-deficient compound nuclei to compound nuclei with more neutrons. This is because for a given atomic number, an isotope with a higher mass number (lower Z^2/A) has a better chance of survival against fission. In fact, fusion reactions involving n-rich projectiles (such as ^{90,92}Kr, if it can be produced with sufficient intensities) are perhaps the only way to produce isotopes in and around the predicted island of stability near $Z = 114$ and $N = 184$.

The elements up to $Z = 113$ have all been synthesized using either a ^{208}Pb or ^{209}Bi target and the most n-rich stable projectiles such as ^{70}Zn, ^{64,62}Ni and ^{58}Fe as the projectiles. The elements $Z = 114$ and 116 have been produced using actinide targets such as ^{244}Pu, ^{248}Cm and ^{48}Ca as the projectile. When a tightly bound doubly magic nucleus like ^{208}Pb is used, a rather cold compound system with excitation energy in the range of 10–20 MeV is produced. As an example, for the ^{70}Zn + ^{208}Pb projectile-target combination, the Coulomb barrier (CB) is about 257 MeV and reaction Q value is −244 MeV. Thus, if a center of mass projectile energy of 257 MeV to cross the CB is chosen, the excitation energy of the compound system would be CB + Q, that is about 13 MeV. Therefore, evaporation of only one neutron, or two at most, is required for the compound system to shed its excitation energy, but when actinide targets are used, the compound nucleus is created typically with 40–50 MeV excitation energy that requires evaporation of four to five neutrons. Since there will be very strong competition for fission at each stage of neutron evaporation, the survival probability of a heavy compound nucleus is much higher when the fusion leads to comparatively less excitation of the CN, often termed as cold fusion, that is when either a Pb or Bi target is used. But formation probability of the compound nucleus (probability of complete fusion facing strong competition from fission) is also an important consideration, and this prefers more asymmetric systems, which leads to the choice of actinide targets. Thus the elements with $Z = 114$ and 116 are produced using actinide targets and a ^{48}Ca projectile. The production cross-sections for element 112 and beyond have been found to be about 1 pb, that is, an observation of about one event or even less per ten days, given the presently available intensities of the projectiles. But a detection method through a sequence of α decays makes identification possible even with one single atom.

3.4.2 Deep Inelastic Transfer Reactions

As already mentioned, two colliding heavy ions do not necessarily fuse. It has been found that at incident beam energies higher than the Coulomb barrier and for grazing collisions (an impact parameter greater than l_c but less than that for a pure Coulomb interaction), a new type of reaction takes place (Figure 3.19). This reaction cannot be explained by invoking either the direct reactions or the compound nuclear reactions since the reaction products exhibit features of both direct reactions and compound nuclear decay. The reaction is called the Deep Inelastic Collisions (DIC) or Deep Inelastic Transfer (DIT) reactions (Vo 78; Le 78).

In this reaction a di-nuclear system forms for a short time. The di-nucleus rotates almost like a rigid body, because of the high angular momentum brought in by the projectile. The di-nuclear system is far from spherical and the angular momentum brought in is high enough not to allow fusion of the di-nucleus into one nucleus. After a while, about half a rotation, the di-nucleus separates into two fragments. The di-nuclear system remains in contact for about a few times 10^{-22} s during which a good amount of kinetic energy is transformed into internal excitation resulting in the exchange of a number of nucleons between the projectile and the target. The characteristic features of the reactions are: (i) the reaction products that are projectile-like are forward-peaked, having a trajectory in between the compound nucleus trajectory and the grazing angle trajectory; however, the peak of the angular distribution of reaction products that result from larger net nucleon transfers move towards more forward angles; (ii) the kinetic energies of the products are much lower than the initial kinetic energy of the projectile. The kinetic energies of the reaction products are closer to the exit Coulomb barrier, as is found in the case of a compound nucleus decaying through fission; (iii) the material transfer is governed by N/Z equilibration in the di-nuclear system; the mean N/Z of the reaction products increases with the N/Z of the di-nuclear system; (iv) the reaction products have a broad charge and mass distribution since a large number of nucleons are exchanged during the collision.

The DIT reactions have been used for the production of both n-rich and p-rich nuclei. The cross-section for DIT reactions usually substantial for energies exceeding the Coulomb barrier but less than about 10 MeV/u. Since the N/Z ratios of the reaction products have been found close to that of the composite system, a light projectile on a heavy-target is used to produce light (heavy) n-rich projectile-like (target-like) reaction products. Thus, heavy targets like ^{181}Ta, ^{197}Au, ^{232}Th and ^{238}U, and projectiles such as ^{16,18}O, ^{20,22}Ne, ^{34}S, ^{40}Ar, etc., with incident energies typically less than 10 MeV/u for the production of light n-rich nuclei are often chosen.

In Dubna, Volkov and his collaborators were the first to demonstrate the efficacy of DIT reactions for the production of n-rich exotic nuclei. They used ^{22}Ne (7.9 MeV/u) and ^{40}Ar (7.25 MeV/u) projectiles on a ^{232}Th target to produce 29 new n-rich isotopes (Figure 3.23) of elements ranging from carbon to chlorine (Ar 70, 71). Among these was ^{24}O that lies on the neutron drip line. It has been found by Guerreau et al. (Gu 80) that by increasing the energy of an Ar beam to about 8.5 MeV/u, new exotic isotopes with atomic numbers much higher than the projectile (such as ^{54}Ti, ^{56}V, ^{59}Cr, ^{61}Mn, ^{64}Fe, etc.) could be produced. A good number of new n-rich isotopes have also been produced using a ^{56}Fe projectile and a ^{238}U target (Br 80). The DIT reaction cross-sections follow Q_{gg} systematics quite well, where Q_{gg} is the reaction Q value corrected for pairing. The cross-section is given by: $\sigma \sim \exp[Q_{gg} + \Delta Ec - \delta]/T$, where the second term is the change in the Coulomb energy due to redistribution of protons and the third term is the pairing correction term and T is the temperature of a di-nuclear system. Thus, the cross-section of an exotic isotope can more or less reliably be predicted using the Q_{gg} systematics.

In DIT reactions the n-rich isotopes of light elements are often produced with higher cross-sections compared to spallation and often compared to projectile fragmentation reactions. This is because DIT reactions have the advantage that the product's exoticity depends upon the exoticity of the di-nuclear system. Thus, for example, by choosing a target and projectile that are both n-rich the production of n-rich nuclei is enhanced. However, since DIT reactions involve low energy heavy ions, the target thickness is lower by many orders of magnitude (typically three to four) compared to the target thickness used in projectile fragmentation and spallation reactions. Also, projectile fragmentation has the

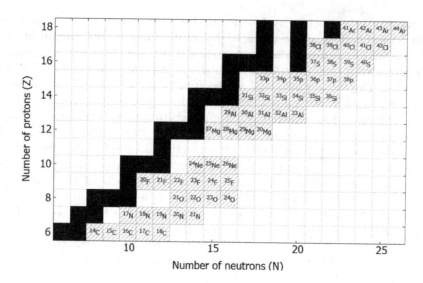

FIGURE 3.23

n-Rich isotopes (hatched squares) produced in Deep Inelastic Transfer reaction using a ^{232}Th target and ^{22}Ne and ^{40}Ar projectiles.

advantage of much narrower forward-peaked angular distribution and ionic charge state distribution. So, from the consideration of overall yield at the focal plane of the separator, projectile fragmentation has almost always been preferred over the DIT reaction. DIT reactions, however, can still have an edge but only in specific cases.

For example, the DIT reaction route is promising for the production of new n-rich isotopes beyond lead. Projectile fragmentation of ^{238}U is not suitable for the production of these isotopes, which in any case cannot produce n-rich isotopes of elements beyond uranium. So, for the production of trans-uranium n-rich species, DIT reactions using targets like ^{238}U, ^{248}Cm and projectiles like ^{18}O, ^{22}Ne, ^{40}Ar, ^{48}Ca, ^{64}Ni, ^{70}Zn, ^{86}Kr, etc., seem to be the most optimum route. For example, an 8.75 MeV/u ^{18}O beam and a ^{238}U target have been used to produce ^{240,241,242}Np and to study the fission of these nuclei (Hi 17). Also, for isotopes not too heavy ($A < 90$), DIT reactions can be used to produce extreme n-rich isotopes with much higher cross-sections than projectile fragmentation reactions. To produce these isotopes, most n-rich combinations of stable projectiles such as ^{18}O, ^{22}Ne, ^{40}Ar, ^{48}Ca, ^{64}Ni, ^{70}Zn, ^{86}Kr, etc. and targets such as ^{64}Ni, ^{238}U, etc. are used (for example please refer to (Pa 18) and the references therein). The projectile energy is chosen typically in the range of 15–30 MeV/u. This energy range is advantageous since it allows forward focusing and efficient collection of projectile-like products and the use of a comparatively higher target thickness. At the same time, this energy range (15–30 MeV/u) allows enough overlap of the projectile and the target surfaces to allow the transfer of many neutrons. It has been estimated that using a 15 MeV/u ^{48}Ca projectile and a 20 mg/cm^2 ^{238}U target, a production rate of about 5×10^3/s and 5×10^5/s of exotic isotopes ^{54}Ca and ^{46}Ar respectively can be expected. In turn, for example, the radioactive ^{46}Ar beam can be allowed to impinge on a ^{238}U target to produce the more exotic ^{51}Ar with a rate of 5 per hour (Pa 18). Thus, the DIT reactions have a good potential for the production of new n-rich isotopes in certain areas of the N–Z chart, especially through using radioactive n-rich projectiles.

4

Targets for RIB Production

4.1 Introduction

The targets for production of Rare Isotope Beams (RIBs) should be thick enough so that RIBs, often very exotic species, are produced with sufficient intensities to allow: (a) identification of new exotic nuclei on or around the drip line; (b) spectroscopic measurements on these nuclei to determine properties such as mass, excited states, decay half-life and decay modes; and (c) study of reactions using RIBs as projectiles to determine reaction cross-sections of astrophysical interest, to study reactions to extract spectroscopic information and to produce new exotic species using nuclear reactions with suitable secondary targets that are difficult to produce with stable ion beams. Whether in the Isotope Separator On Line (ISOL) method or the in-flight method, the use of thicker targets (typical thickness ~1 g/cm^2 to 100 g/cm^2) to enhance the production rate of RIBs poses a difficult technological challenge because of heat deposition in the target by the incident beam. The challenge is much greater in the ISOL method because the target is comparatively thicker, and a large fraction of the beam's power is deposited in the target. The fact that reaction products need to come out of the target through diffusion further complicates target design in the ISOL method. Also, in the ISOL method a large number of targets need to be developed to optimize the production yield and release exotic atoms from the target. This is not the case with the PFS technique since in a projectile fragmentation reaction different beams need to be used to produce different exotic isotopes and not different targets (cross-section almost independent of target). Thus, in the case of PFS-type facilities, the development of a few targets, such as Be, C, etc. is usually sufficient. In fact, the success of an ISOL-type RIB facility depends largely on the ability to develop suitable targets.

The target design considerations in the ISOL technique are more crucial and complicated and will be taken up first, followed by a discussion on targets for projectile fragment separators. Although the beam power that an ISOL target needs to handle at present is a few tens of kW, in future facilities the requirement will increase and targets capable of handling beam power up to a few tens of MW will need to be designed and developed. The development of thick targets is a highly active field of research and many researchers are working on the topic at present. Understandably, there is a large volume of published literature, mainly dealing with experimental studies on ISOL targets. Our discussion here will be limited to the basic considerations for target design, the desirable properties of an ISOL high-power target and different types of targets that have so far been used for RIB production. Interested readers may consult further four recent articles by A. Gottberg (Gott 16b), P. Bricault (Bri 16), T. Stora (Sto 13) and J. P. Ramos (Ram 16), and the references therein, to obtain more complete information on the present status and future directions of studies in this field. For initial developments of targets at ISOLDE, the readers may consult the article by H. L. Ravn (Ra 89).

4.2 High-Power Targets for ISOL Facilities

Targets that qualify for the production of RIBs should have certain thermal, physical and chemical properties. As already mentioned, the reaction products produced in high-energy proton-induced reactions are all stopped in the target, and a large fraction of beam power is also deposited in the target. ISOL targets should therefore be of such materials that can withstand high temperature. The target material is usually placed inside a target container made of a high-temperature material (Ta is the most common material) and the target container is connected to the ion source through a short tube usually made of tantalum. The reaction products produced inside the target reach the surface of the target by diffusion. Once the reaction products reach the surface of the target, the product atoms first undergo desorption from the surface of the target and then a series of absorption–desorption cycles with the inner surface of the target container until they reach the ion source through a small tube connecting the target chamber and the ion source. Collectively, the desorption–absorption process is called the "effusion process," and the combined efficiency of diffusion and effusive transfer to the ion source is often termed as the release efficiency. The "release" should be fast enough to minimize loss due to decay. With the optimum choice of target materials, it has been possible, so far, to achieve a release time fast enough to study exotic nuclei of some elements with half-lives of a few ms (for example ^{14}Be, ^{33}Na, ^{11}Li, etc.). It should be mentioned that the ionization process also involves a delay of several ms and the release time of the same order is what a target designer wants to achieve. However, the targets should be able to sustain, at the same time, damages caused by beam-induced thermal cycles and beam irradiation over a period of several weeks (or more). It is also important to ensure that the beam-induced changes of the physical properties of the target do not adversely affect the release time over this long period of irradiation.

Assuming that a reaction product is formed inside a spherical target grain of radius d, the time needed for the product atom to reach the grain surface by the process of diffusion is given by $t_D \sim d^2/D$, where D is the diffusion coefficient given by $D = D_0 \exp(-Q/RT)$, where Q is the activation energy, R is the gas constant, D_0 is the diffusion pre-exponential factor with the dimension of m^{2}s^{-1} and T is the temperature. Smaller grain size and high temperature help achieve faster diffusion. Thus, quite often an ISOL target is in powder form comprising small grains of a few tens of microns in diameter, stacked or loosely bound together to make the bulk of the target. Also, powdered targets are usually of much lower density compared to the density of the material(s) that the target is made of. This means a lot of open space in between the target grains (porosity) which would aid the effusion process. The ISOL targets are usually operated at temperatures exceeding 1500 K. The high temperature is also necessary for efficient desorption. Ideally, the surface temperature of the target should be higher than the boiling point of the radioactive product atom of interest that is to be released from the target. The desorption delay of the product atoms from the target surface depends on its vapor pressure. Since refractory-type elements have very low vapor pressure even at high target temperatures, the release of isotopes of these elements is very much delayed. It is thus usually difficult to get enough intensity for isotopes of elements with atomic numbers between $Z = 40$ and 46 and also to an extent for elements between $Z = 72$ and 78 in ISOL-type RIB facilities. It should be mentioned that the efficiency of desorption is highest for the alkaline earth and a few other elements and it is much easier to produce these beams in an ISOL-type RIB facility.

Although maintaining targets at higher temperatures is a necessity in most cases, an optimum choice of temperature is very important. First of all, the temperature should be much below the boiling point of the target material. Further, at the operating temperature the vapor pressure of the target material should be below the optimum pressure for the ion source operation. Usually, this pressure is about 10^{-4} torr or even below. Also, the rate of loss of target material by evaporation at the operating temperature should be small enough to ensure that the target only loses a small fraction of its mass during long-beam irradiation of a month or so. At high operating temperatures, it also becomes difficult to maintain the porosity and the grain size constant over the long irradiation time because of sintering. The sintering rate increases with temperature, resulting in the coalescence of smaller grains into bigger grains. Sintering thus often adversely affects the long-time release rate of reaction products.

The primary beam irradiation-induced damage often has some positive effects on the release rate. It creates a lot of defects such as vacancies and interstitial atoms in the crystalline structure of the target that can lead to enhanced diffusion. This effect has a dependence proportional to I^2 or $I^{3/2}$ (I: primary beam intensity) as experimentally observed at the TRIUMF-ISAC facility (Dom 03). However, once again at higher temperatures many of these defects get annealed and a weaker dependence on the beam intensity is generally observed. Thus, to get the advantage of radiation-induced enhanced diffusion the target temperature needs to be kept much lower (typically below 1000 K). This would also reduce the sintering rate considerably and help in maintaining grain size in the range of a few tens of μm over the time period of a typical beam run, which would ensure, in turn, good and uniform release efficiency throughout the experiment. The choice of optimum temperature thus needs to be made considering all the aspects and is to be determined experimentally. It would vary from target to target and would also depend on the isotope of interest in the particular experiment.

The thermal conductivity of the target is another very important consideration for beam power exceeding 5 KW. Poor thermal conductivity would not allow effective dissipation of the beam power and it would be very difficult to limit the rise in the target temperature when the beam power exceeds about, say, 10 kW. Also, poor thermal conductivity results in a large temperature gradient within the target that affects the physical integrity of the target.

The chemical properties of the target material also play a crucial role with respect to extraction/release efficiency of a particular isotope. The isotope of interest should not chemically react with the target material which would prevent its release. Thus, for example, the use of an oxide target should be avoided for the preparation of beams of oxygen isotopes since atomic oxygen is chemically very active. Oxygen isotopes are often extracted from the target by forming a volatile compound with other elements added to the target e.g., in the form of CO or CO_2 in the reaction of the proton beam with a molten LiF target contained in a graphite matrix (as done at Louvain-la-Neuve). It is also important that the target material does not react chemically with the materials around the target. The material for the target container and the transfer line to the ion source is often chosen to be tantalum and it is important that the target material does not react with tantalum, otherwise it would have a detrimental effect on the release (effusion) efficiency. The tantalum transfer tube is usually heated to high temperatures to enhance effusion but sometimes it is kept at a much lower temperature to selectively allow the transfer of more volatile species to the ion source. In fact, transfer line properties are often used to suppress a set of contaminants, often isobars of the desired isotopes that are difficult to separate using a set of di-pole magnets after the ion source.

To summarize, an ideal target is one which (a) allows optimum production cross-section for the isotope of interest; (b) is refractory in nature to allow high temperature operation; (c) allows quick and efficient release for the radioactive atoms; (d) has a good thermal conductivity to allow efficient heat dissipation at large beam power; and (e) does not react chemically with the reaction products and the surrounding materials. No single target can satisfy all these criteria for all possible reaction products simply because the spallation and target fragmentation reaction with high-energy light ions produce a great number of reaction products of a wide range of elements and a wide range of half-lives. In all cases, however, the target needs to be heavier (higher mass number) than the RIB of interest.

4.3 Types of Target Material

Targets that have been used for the production of RIBs using high-energy protons at the ISOLDE facility at CERN and at the ISAC facility at TRIUMF can be grouped into molten materials, thin metal foils of refractory elements, powder targets, often in the form of pellets, high melting point compounds such as oxides and carbides of various elements and some special targets for production of some specific isotopes. A list of targets that have been most commonly used for the production of short-lived isotopes (half-life less than 1 min to ms) is given in Table 4.1.

TABLE 4.1

A List of Targets That Are Commonly Used with High-Energy Proton Beams at ISOL Facilities

Target Material	Type of Target	Limiting Temperature (K)
UC_x	Powder	2373
ThC_2	Powder	2923
Pb & Pb:Bi	Molten	620
Re	Solid/foil	2873
W	Solid/thin foil	3136
Ta	Solid/foil	2923
TaC	Powder	2613
Y_2O_3	Powder	2223
La	Molten	1200
CeS	Powder	2173
Sn	Molten	520
ZrO_2	Powder	2323
TiC	Powder	2323
SiC	Powder	1973
Al_2O_3	Powder	1998
C	Solid	2240
BeO	Powder	2503

Molten targets, such as liquid lead, are characterized by high density (approaching the density of metal) and consequently a high rate of isotope production. However, high density also results in a high rate of energy loss in the target that demands special attention. Also, the release times in molten metals are often quite long, exceeding a few tens of seconds or even minutes. The slow-release time is a common property of all the molten targets that have been tested so far. So, molten targets are most suitable for volatile species and for isotopes that are comparatively longer lived. The high evaporation rates of liquid metals also make it difficult to maintain the desired level of vacuum in the ion source. Moreover, many liquid metals are highly corrosive, requiring special material for their container. Out of the molten targets used so far, those of Pb, La and Sn do not pose any special problem and have been used for the production of RIBs using high-energy proton-induced spallation reactions for optimum production of Hg, Cs and Zn isotopes respectively. The isotopes of product elements in these cases have a higher vapor pressure than the neighboring elements which allow a good degree of chemical selectivity.

Quite often thin (thickness of approximately a few microns) metallic foils of high-melting-point materials such as W, Re, Ta and Ti are used to enhance the production rate of isotopes of certain elements. The metallic foils can be made quite thin, often a few microns, which allows quick diffusion. Also, thin foils have much smaller surface area compared to a powder target that helps in more efficient desorption once the reaction products reach the surface of the crystals. This is especially true for comparatively more refractory-type elements for which the efficiency of release and delay is governed largely by desorption (Bjor 87). The other advantage of metallic foils is good thermal conductivity which makes them well suited for high-power incident beams. To give an example, 2 µm Ta foils were used for the production of very short-lived Li and Be isotopes with half-lives in the range of a few ms. Using this target and surface ionizer, ^{11}Li could be produced with an intensity of 7000 ions/s per µA of 1.4 GeV proton beam. The metallic targets can also sustain radiation damage up to quite high beam powers. It should be mentioned that the thin foil metallic targets of refractory elements can be very pure, and this is an important advantage since for composite/compound targets, certain impurities in the target or of certain elements present in the target, would reach the ion source volume through outgassing. These often adversely affect the efficiency of the ion source. Despite its many advantages, the number of refractory-type metallic targets is limited to only a few and therefore powder targets need to be used, often in the form of carbides and oxides that are refractory in nature. For example, uranium is the best target for the production of n-rich exotic species in the fission product range but pure uranium metal has a melting point of about 1400 K which is not high enough for high-beam power operation and to achieve a short diffusion time less than a few tens of ms. So, uranium carbide, which allows very high temperature operation up to about 2400 K, is used and the release time of this target is fast enough to allow the study of short-lived n-rich isotopes.

Uranium carbide is one of the most preferred targets in ISOL-type RIB facilities using high-energy proton beams, since apart from the neutron richness of this target, the release times of isotopes are also small in this target. This allows efficient production of short-lived n-rich nuclei in the mass range of $70 < A < 160$ in high-energy proton-induced spallation reactions. The target is prepared in pellet form from a mixture of small grains of UO_2 and graphite with an average size of about 10 µm. The target-making procedure, although complex and involved, has been standardized and requires carbo-thermal reduction of UO_2 using a high-temperature vacuum furnace. Finally, UC_x discs are made by casting and then cutting to the final target shape and typically around 200 targets or more, each about 200 µm thick are loaded in a Ta target holder for online operation. For high-power operation, the thermal conductivity of the target needs to be increased and the UC$_x$ target is often bonded on graphite backing using a ceramic casting technique. The target discs are highly porous, having average densities usually in the range of 3–4 g/cm^3 or often even lower, which is only a small fraction of the density of either uranium (19.1 g/cm^3) or uranium carbide (~10.6 g/cm^3). A density of 3 g/cm^3 of UC$_x$ corresponds to a porosity of roughly 70%.

A number of oxides of different elements are refractory in nature and powder targets of many oxides are often used for RIB production. Oxide targets can be in the form of pressed pellets. Usually backing foils that provide good binding between the target and the backing metal are used, such as Ni used as the backing foil for NiO target and Mo is used for uranium and thorium oxide targets. Once again, a ceramic casting technique is used for providing bonding with the metal foils. For oxide targets, the target containers are usually made of rhenium or platinum since tantalum tends to get oxidized. Reaction products of metal tend to get oxidized in an oxide target. This might prevent the release of these products if the oxide molecule is less volatile. Also, sintering poses a problem for the oxide targets that often restricts the operating temperature to comparatively lower values.

Some oxides, such as Al$_2$O$_3$, ZrO$_2$, etc., are also available in fibrous form. The small dimensions of the fibers and their good release properties make them suitable for use as a target for RIB production. For example, using a low density (1.15 g/cm^3) HfO$_2$ fiber target, [17,18]F isotopes were successfully produced at the HRIBF facility, Oak Ridge (Alt 02), which is no longer in operation. Another highly permeable matrix is Reticulated Vitreous Carbon Fiber (RVCF) that could be used as a backing material for deposition of metals and carbides of different target materials. The RVCF is often coated with a thin layer of tungsten to avoid reactions with the reaction products (Alt 98). Both RVCF and fiber targets have good potential for their use as RIB targets but their suitability at high-beam power (exceeding a few kW) has not been extensively studied.

It should be mentioned that for metallic as well as powder targets, the pulsed structure of the driver light ion beam (for example proton beam at ISOLDE) usually offers an advantage: it often leads to an enhancement of the release of the short-lived isotopes. This is because even if the average power is relatively low (about 1 kW for the ISOLDE beam), the instantaneous power in a pulse is higher by several orders of magnitude (depending on the pulse width and time difference between the two pulses) compared to the average power. This leads to defect-induced enhanced diffusion/effusion of reaction products. However, there is a negative side too. The sudden temperature increase following a beam pulse often creates a thermal shock that might affect the integrity of the target and the target container, and also might result in rapid sintering of the target. The temperature thus needs to be controlled and one way to control the temperature is to defocus the proton beam on the target, thereby reducing the power density.

Instead of light ions, ISOL facilities also use high energy heavy ions to produce a wide range of exotic species. In this case the reaction products resulting from the fragmentation of the heavy ion projectile are all stopped in the target. A typical example is the present facility at GANIL where heavy ions of energies up to about 100 MeV/u are allowed to bombard a thick graphite target to produce reaction products from fragmentation reactions. The granular and porous nature of the graphite target allows quick diffusion of reaction products that are stopped in the target. The target was designed for a beam power of about 6 kW. This beam power is, as such, not very high for a thick target using light ions but the fast rate of energy deposition of heavy ions in the target makes the target design complicated. The target used at GANIL consists of a number of 500 μm-thick graphite disks of increasing diameter (to effect better radiation cooling), spaced at a designed distance (heat transfer codes were developed and used for the design) to allow radiation cooling and quick effusion towards the Electron Cyclotron Resonance (ECR) ion source downstream and are held in place with a central rod. The whole target assembly is contained in a water-cooled box. The neutral reaction products are then ionized, mass separated and accelerated using cyclotrons.

4.4 R&D for Future ISOL Targets

The present R&D on ISOL targets has two main goals: improving the power handling capability to the level of a few MW and improving the release properties of reaction products produced in the reaction. Right now, targets that can be operated up to about 50 kW of beam power have been developed. To go to still higher power, the thermal contact between the target disc and target container needs to be improved for efficient heat transfer to the latter and also the heat dissipation capability of the target container to a surrounding heat sink needs to be improved. The target is often designed to be a hollow annular ring where the central portion of the target is cut away. This way high temperature at the target disc center can be avoided. The target container is fitted with fins to increase cooling by radiation. These fins are usually diffusion-bonded to the target container. As the beam power increases to more than about 100 kW, molten targets such as Pb–Bi, NaF–LiF, etc., eutectics need to be used. A circulating loop of liquid metal can pass, after irradiation, through a diffusion chamber where the metal is sprayed like a shower of droplets to reduce the diffusion length for enhanced diffusion (Sto 13). The heat may be removed in a separate chamber in the loop cycle. This technique is potentially capable of resulting in a gain in the production of exotic neutron-deficient isotopes by several orders of magnitude.

Another way to handle high beam power is to use a converter that converts the primary beam (deuteron/electron) into neutron/gamma beams that induce fission in a uranium target. This makes target design less complicated, but the converter design poses a great challenge. For example, the rotating graphite wheel converter to be used for SPIRAL II for radiative heat dissipation has to be very large (almost 2 m in diameter) and components such as the bearings should be able to withstand high temperature operation under a very high radiation environment. In the case of electron beam-based photofission, a disc-type tungsten converter is good only to take an electron beam power of a few kW, primarily due to the high density of power deposition and limited capacity of heat dissipation through edge cooling. A more complicated design involving a conical converter attached to a water-cooled aluminum body has been found to be suitable for handling electron beam power in excess of 30 kW (Gott 17). Essentially, indirect methods (converted beam and not the primary beam irradiating the target) make the target design simpler, largely at the expense of transferring the complicacy to the design of the converter.

The other direction of target R&D is aimed at enhancing the release efficiency from the target. Since diffusion efficiency is roughly inversely proportional to the grain size, reducing the grain size from tens of μm to tens of nm is expected to result in increased release efficiency if the sintering can be arrested. Using low density (~1.4 g/cm^3) UC$_x$ target made from engineered multi-walled carbon nanotubes and engineered nanometric uranium oxide, high release efficiency over long proton irradiation time exceeding 300 hours has been obtained (Gott 16b). This is an order of magnitude improvement over the conventional UC$_x$ target and thus the technique of making targets using engineered nanomaterials has great promise for ISOL targets.

4.5 Target Station in ISOL Method

Target assembly comprises not only the target, target holder and the target chamber but also the ion source and the ion extraction optical elements in an integrated manner. A typical experiment lasts for a week or two, which makes not only the target highly radioactive but the target chamber, the ion source and all the beam line components also become highly radioactive. This is also true for the beam diagnostics components and the beam selection slit just before the target. These locations contain a huge inventory of radioisotopes including gases which need to be contained within the target station. The target chamber therefore needs to be shielded, as well as the entire beam line in the vicinity, with shielding blocks. Also, the target ion source integrated system needs a number of service connections for vacuum, power connections for heating the target, the target chamber and the ion source, cooling of target heat shield, high voltage connections for the ion source and focusing optical elements, etc. All these connections are to be provided from outside the shielding and should have quick connect–disconnect features so that after a beam run (or a breakdown) the entire target ion source assembly can be removed remotely to a hot cell (and finally to a highly shielded radioactive storage area) and then replaced with a new one using cranes or robotic arms. The target module and all these components together are usually called the target station.

For high-energy proton beams, the shielding of prompt secondary radiation comprising neutrons and gamma rays requires thick concrete shielding with thickness exceeding 3 m. Thus, the target module is not only complicated, it is also bulky and the whole operation of removing an irradiated target and replacing it with a new one needs to be planned in a 100% failure-proof manner in order to satisfy the radiation and other kinds of safety guidelines. The target station is thus located in a highly shielded area that can be accessed only by dedicated overhead cranes. As an example of a typical target module, the one used in the ISAC facility at TRIUMF is shown schematically in Figure 4.1.

The target station for low energy fission reactions induced by gamma rays (ARIEL at TRIUMF, for example) and neutrons (SPIRAL II at GANIL) would be quite similar in nature except that these have, additionally, the converter integrated to the target assembly. The integration of a converter gives rise to additional complications arising not only from thermal load but also from geometrical and mechanical constraints.

4.6 Targets for PFS Facilities

As mentioned earlier, in PFS facilities only a few targets (light targets such as Be, C and heavy targets such as Pb) need to be developed, since production cross-sections of projectile fragments are largely independent of the target. Also, the targets in PFS facilities would need to handle comparatively lower beam power. This is because the available beam intensity of heavy ions from high energy heavy ion accelerators is lower than the beam intensity achievable for light ions and also the target in PF reaction typically absorbs about 20% of the beam power. However, considering the future upcoming facilities like FRIB and FAIR and also the existing RI Beam Factory facility at RIKEN and its planned up-gradation, a beam power handling capacity exceeding 50 kW is considered necessary for future projectile fragmentation targets. This level of power is by itself very high and

Targets for RIB Production

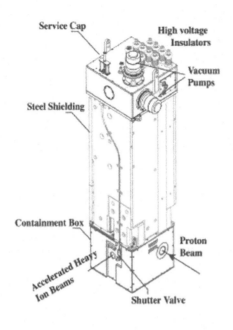

FIGURE 4.1
Schematic diagram of the target module used in ISAC, TRIUMF. (© 2020 TRIUMF. All rights reserved.)

the faster rate of energy loss of heavy ions in the target makes the target design even more challenging. The targets should survive beam irradiation without any significant degradation in their thickness and properties at least for a few weeks to allow an experiment to be completed. It means it should withstand the radiation damage caused by high-energy high-power heavy ion bombardment. Also, the temperature gradient over the target material should be kept small to reduce the thermal stress that could adversely affect the physical integrity of the target.

The beam spot area on the target needs to be restricted typically to about 1 to 2 mm². This means extremely high power densities reaching as high as a few tens of GW/m². Further, at FAIR, the high energy uranium beam would be pulsed with a time period of 1.5 s and pulse width of about 50 ns. This means all the beam particles of ^{238}U containing about 5×10^{11} particles at 1.5 GeV/u would bombard the target within a very short time span of 50 ns. The instantaneous power in this case is extremely high and the target design needs to ensure that the thermal shock produced by this huge instantaneous power does not destroy the physical integrity of the target.

The production target used at the RIKEN RI Beam Factory is a rotating disc-type carbon (graphite) target with the beam heating the periphery of the rotating target as shown in Figure 4.2a (Ku 03). The target is fixed to a water-cooled Al plate connected to the rotating shaft that enters the target chamber under vacuum through a magnetic fluid based vacuum seal. To reduce the heat load on the graphite target, a number of thinner targets instead of a single target can be used. This is known as the multi-slice target concept. The advantage is that the heat load gets distributed to a number of targets. However, the

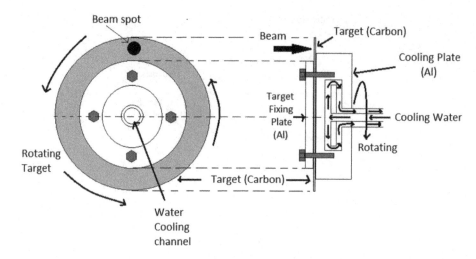

FIGURE 4.2A
Schematic diagram of a rotating disc-type target assembly for high-power heavy ion beam at the RIKEN RI Beam Factory.

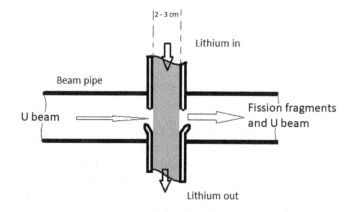

FIGURE 4.2B
Schematic presentation of the concept of a windowless liquid Li target for handling heavy ion beam power ~100 kW.

number hardly exceeds a few, about five, since beam optical requirements restrict the maximum extent of the target in the beam direction to about 50 mm or so. FRIB plans to use such a concept (Pelle 16). The FRIB multi-slice target is designed to be about 30 cm in diameter (beam heating the periphery as in Figure 4.2A) rotated at a high speed of about 5000 rpm. The target is fitted with fins and is radiation-cooled. The diameter and the rotation speed together limit the temperature rise as well as the temperature gradient. The optimum target temperature is high and is about 1800 K so that beam-induced damage is annealed efficiently. The multi-slice target is designed to operate reliably up to a beam power of about 100 kW. For even higher power operation, the use of a windowless (a thick window separating the liquid target would be very difficult to cool in the case of heavy ions) liquid lithium target (Nol 03) is being considered. This is shown schematically in Figure 4.2b.

Targets for RIB Production 117

To handle beam power in excess of 100 kW, the calculated volume flow rate of lithium is about 10 l/s. A permanent magnet-based liquid pump would be used for this purpose. For long-time irradiation the impurities developed in the Li because of beam irradiation need to be filtered out. This adds to the complications in the design since impurity control traps in the loop circuit need to be added.

It should be mentioned that the production target and the entire zone all the way up to beam stoppers stationed after the first di-pole in a PF separator need to be heavily shielded, quite the same way as in the case of light ions. The shielding requirement and obtaining of safety clearance for the operation of high intensity relativistic heavy ions, especially for using ^{238}U as a projectile, involves many issues that include remote handling, contamination of cooling water and long-term contamination of the target chamber and the first di-pole vacuum chamber.

4.7 High-Power Beam Dumps

A beam dump is of course not a production target but is an important component of a RIB facility employing high-power beams. The design of a beam dump constitutes a difficult technical challenge for high-power beams whether to be used for light or of heavy ions and has a lot of features that are common with the target design. In projectile fragment separators involving high-energy heavy ion beams, the heat dissipation and radiation damage issues in beam dumps are quite similar to the one that is encountered in target design. The beam dump is meant to absorb the entire beam energy so as to protect all the downstream optical elements in a separator. The dump needs to absorb the particle energy and the huge thermal load needs to be dissipated efficiently. The beam often creates a huge thermal gradient and also thermal shock waves against which the beam dump material has to be protected. Radiation damage caused by the beam creates an additional technological challenge since beam dumps are required to operate for months. Radioactivity in the beam dump also creates additional issues for handling, replacement, etc.

The beam dump for the Big RIPS fragment separator (Ku 13) in the RIKEN RI Beam Factory facility has been designed to handle all ion beams up to ^{238}U at 345 MeV/u with a beam current of 1 particle μA (82 kW power). The beam dump is water-cooled and uses specially designed swirl tubes as water-cooling channels to enhance the heat transfer coefficient. Screw tubes for enhancing the heat transfer can also be used. The beam dump does not receive the beam normally but at a highly slanted angle of about 6 degrees that effectively increases the beam irradiated area by a factor of ten, thereby reducing the power density by the same factor. Also, the slanting angle allows the use of thinner walls (wall thickness enough to stop beam traveling at an angle of 6 degrees) for the beam dump that helps in faster conduction of heat to the surface cooled by water. The beam dump at Big RIPS has been working reliably since 2007 but has never been tested at its full designed power since the maximum beam intensity reached so far is about one-tenth of the design goal. Unlike the static beam dump at RIKEN, FRIB has developed the prototype for a high-power beam dump (for handling beam power up to 325 kW) that uses a rotating water-filled drum to intercept the unreacted beam (Avil 16). The drum has a diameter of 70 cm and a height of about 8 cm in order to distribute the heat and radiation damage the beam will cause. The beam dump is made of titanium alloy Ti-6Al-4V with a wall thickness of only 0.5 mm in order to limit the power deposited into the wall material. Fabrication of this

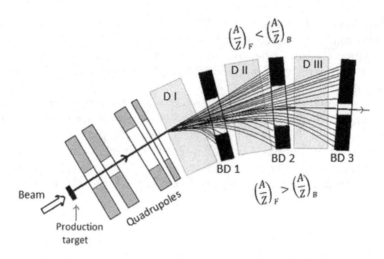

FIGURE 4.3
Distributed beam stoppers (BD1-3) to be used at FAIR, GSI.

large size thin shell-type beam dump constitutes an engineering challenge, the R&D and prototype development for which has been undertaken.

Depending upon the rigidity of the fragment of interest, the primary beam in a fragment separator would get stopped at various positions in the beam line, which include the inner surface of the vacuum chamber walls of the first di-pole magnet and the beam stopper positioned after the di-pole magnet on both sides of the momentum selection slit. To avoid the stopping of the beam inside the vacuum chamber of the di-pole, the di-pole is often broken into three or more pieces of smaller bending angles; for example, in the Super-FRS beam line for the FAIR project, the first di-pole comprises three separate 11-degree di-poles spaced suitably to house the beam stoppers. This is schematically shown in Figure 4.3 (Chat 17).

In an ISOL-type facility, a single beam dump for the unreacted light ion/electron/neutron beam is required, positioned after the target. Here also, the design issues and challenges are similar except that for future facilities beam dumps capable of handling even higher beam power exceeding 100 kW would need to be designed.

5

Ion Sources for RIB Production in ISOL-Type Facilities

5.1 Introduction

An ion source is the first component of any primary accelerator which accelerates ion beams of stable isotopes. In RIB production, it is only in an ISOL-type RIB facility that an ion source is needed for ionization of radioactive atoms produced in the target due to the interaction with the primary beam. Ionization efficiency, emittance, charge state of ions, universality or the capability to ionize atoms of a wide range of elements and the reverse of that, the specificity or the capability to ionize a particular element or a group of elements are the properties that vary from one ion source to the other and characterize the different ion sources. Different ion sources are thus used according to the demand of the application.

The emittance of an ion source is a property that decides the divergence of an ion beam extracted from the ion source and is a characteristic property of an ion source closely related to the process of ionization. A number of parameters, depending upon the particular type of ion source, contribute to the emittance. Some of these parameters are ion temperature and ion velocity distribution, shape of the ion-emitting meniscus, electric field configuration at the ion-extracting zone, variations in temperatures on the plasma surface, presence of axial magnetic field at the extraction region, etc. The emittance results in ion velocity components in the transverse directions X and Y as well as in the direction of ion beam extraction Z. In the transverse directions, it contributes to beam divergence and in the beam direction Z it results in energy or velocity width, which in turn adds to beam divergence when the beam passes through accelerating gaps and magnets. The ratio of V_x/V_z (or P_x/P_z, where P represents momentum) and V_y/V_z (or P_y/P_z) give the divergence angles θ and φ respectively in the X and Y directions and V_z is given by $(2qV_{ext}/m)^{1/2}$, where q is the charge state, V_{ext} is the extraction voltage and m is the mass of the ion. The beam can be considered as a collection of points representing the ions in a six-dimensional phase space $X - P_x$ (θ), $Y - P_y$ (φ), $Z - P_z$ or in independent two-dimensional sub-phase spaces like $X - \theta$, $Y - \varphi$, etc., if there is no coupling between the X, Y and Z motions. The contour of the phase space is usually an ellipse (under a linear restoring force) in two dimensions and the area of the phase space divided by π is defined as the emittance and is denoted by ϵ. The emittances in X (ϵ_x) and Y (ϵ_y) might be equal or different depending upon the extraction orifice geometry of the ion source and other parameters. Liouville's theorem states that under the action of conservative forces the phase space area is conserved. Thus, emittance would not change when the beam moves through drift spaces containing no optical elements or if the beam is focused using focusing elements or passes through a bending (di-pole) magnet. The emittance defined as above, however, decreases (the phase space shrinks as V_z increases) when an ion is accelerated. the normalized emittance $\epsilon_n = \beta.\gamma.\epsilon$, where $\beta c = V_z$ and $\gamma = 1/(1-\beta^2)^{1/2}$ is thus often used instead of emittance ϵ, since ϵ_n remains conserved under acceleration. It is thus obvious that a low-emittance ion source is preferred for acceleration of ion beams in a cyclic accelerator or in a chain of linear accelerators where the acceleration zones are interspaced with drift spaces. The emittance or normalized emittance, however, changes if there are dissipative forces like collision, emission of synchrotron radiation, etc.

Compared to ion sources for β-stable primary beams, the design of ion sources for ionization of unstable radioactive isotopes for RIB production should take into account some additional requirements. *First of all*, the ionization efficiency should be as high as possible since the reaction products are not plentiful. *Secondly*, the reaction products that are fed into the ion source are the ones that come out of the production target by diffusion as neutral atoms. In order to minimize the loss of neutral reaction products in the transportation from the target to the ion source, the ion source needs to be placed in close vicinity to the target. This requires the materials used for the ion source, including the vacuum seals, to be radiation hardened. Apart from this, the ion source, along with the target chamber, has to be assembled in a way that it can be removed from the beam line and replaced by a new one by remote handling after the experiment. This easy-to-connect/disconnect feature adds to the complexity in the design but is necessary since both the target chamber and the ion source become highly active after a beam run that prohibits in-situ handling. Also, the need to separate the ions of interest from hundreds of other reaction products produced in the same reaction, which are almost always produced with orders of magnitude of higher cross-section than the one of interest, puts some additional demands on the ion sources: the ion source should ionize all the neutral reaction product atoms to 1^+ (or a fixed charge state) so that efficient separation of the ion of interest can be achieved using di-pole magnets (multiple charge states would make separation almost impossible); the emittance of the ion source should be small since higher emittance makes magnetic separation more difficult; the ion source should have some selectivity since otherwise clean separation of 1^+ ions of the selected isotope will not often be possible, resulting in a cocktail beam even after magnetic separation. *Lastly*, it needs to be ensured that the time needed for ionization is less than the half-life of the isotope (RIB) of interest. This is very important for production of ion beams of very exotic nuclei that have half-lives in the range of a few tens of ms. It should be mentioned that there is no universal ion source that satisfies all the above criteria and can ionize atoms of all elements that have widely different ionization potentials, varying from about 3.8 eV for francium to about 24.6 eV for helium. different kinds of ion sources are thus used for 1^+ ion production, depending upon the ionization potential and for optimizing other requirements.

In ISOL post-accelerator-type RIB facilities a second ion source for increasing the charge state is often required, which is necessary for post acceleration of low energy RIBs. These ion sources are called charge breeders or charge state boosters. The 1^+ ion beam after mass selection is injected into the second ion source for conversion to a higher charge state, depending upon the mass number of the heavy ion to be accelerated. Typically charge states corresponding to $q/A \geq 1/10$ are required (the higher the better) for efficient and cost-effective acceleration of heavy ions, especially for ions heavier than $A \sim 30$. Some specific experimental studies at low energy, such as Penning trap mass spectrometry, also need ions of higher charge states since the precision of the measurement increases in direct proportion with the charge state of the ion under investigation. The choice of possible charge breeders is dictated by issues such as the efficiency for the production of the particular high charge state required for the ion of interest, the contamination from undesired ions, the width of charge state distribution and the emittance. The charge breeder is stationed far from the target in a low radiation zone so radiation damage of its components and the physical access are not issues, as they are with the first 1+ ion source. It should be mentioned that charge breeders are also required in combined PFS—ISOL type facilities for post-acceleration of exotic fragment ions that are slowed down first in metallic foils and then captured in a helium gas catcher in a 1+ charge state.

5.2 Ion Sources for 1+ Charge State Production

5.2.1 Surface Ion Source

This ion source is the simplest of all ion sources, which is essentially a hollow heated tube of a high work function material like Ta, W, Re, etc. This ion source is suitable for ionizing isotopes of elements with ionization potential up to about 6.5 eV. Alkalis (Group I), alkaline earth (Group II), the elements aluminum, gallium, indium and thallium (Group III), scandium, yttrium and rare earths (Group IIIB) fall in this category. The surface ion source is thus quite versatile in the sense that it can ionize a good number of elements in the periodic table and all the ISOL-type RIB facilities use surface ion sources that are integrated to the target assembly.

In surface ionization, a neutral atom gets ionized through collision with a surface of high work function material. The ionization efficiency is given by the Saha–Langmuir equation although in a cavity ionizer somewhat higher efficiencies are usually obtained (Kir 81). During collision the atom loses an electron and then gets re-evaporated from the heated surface as an ion. In an ISOL facility, the surface ion source is connected to the target through a small transfer tube as shown in Figure 5.1. Both the target and the ion source are in vacuum (and connected to the downstream ion beam line under vacuum). As mentioned already, the surface ion source is very simple in construction, just a tube of high work function material like Ta/W/Re with a diameter of a few mm, connected to the transfer tube. The neutral atoms (reaction products) diffuse through the transfer tube to reach the ionizer tube. The ionizer tube is heated directly or indirectly to temperatures exceeding 2000°C. The transfer tube is also maintained at a high temperature to ensure that the neutral atoms do not stick to the walls of the transfer tube and reach the surface ionizer quickly enough, typically in about 1 ms. The neutral atoms, after reaching the ionizer,

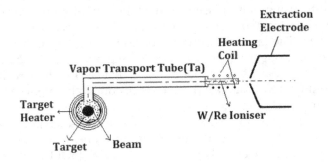

FIGURE 5.1
Schematic diagram of a surface ionization source.

get surface ionized through collisions with the inner walls of the ionizer and the ions are extracted through a small hole of about 1 mm diameter at the end of the ionizer tube. The extraction potential is usually in the range of 20–60 kV and the extraction is done using the extractor electrode assembly. The ionization process is quite selective in the sense that isotopes of elements with ionization potential of greater than 6.5 eV are not ionised. Also, the ionization efficiency is rather insensitive to the other kinds of materials (coming out of the target because of outgassing) present in the ion source as long as those do not react with the material of the surface ionizer. The typical ionization efficiency for the alkalis often exceeds 50%. The surface ion source has been extensively used at CERN-ISOLDE (Ra 89) and at ISAC, TRIUMF (Bri 14).

5.2.2 The Resonant Ionization Laser Ion Source for Metallic Ions

In Resonant Laser Ionization, an electron in the ground state of the atom is excited by lasers step-by-step to various electronic levels of the atom, ultimately leading to ionization. This ionization scheme (Alk 83) allows element selective ionization producing ion beams free from isobaric interference. Isotope selective ionization is also possible in many cases by making use of the isotopic shift of neighboring isotopes and operating the lasers in a narrow band mode. Laser ionization is often used in an ISOL-type RIB facility for all elements whose ionization potential is less than about 10 eV. Several hundreds of isotopes of more than 40 elements have already been ionized using lasers (Fe 12). The great advantage of a laser ion source lies in its element-selective ionization which requires only very moderate resolution, less than about 300 ($A/\Delta A \sim 300$; $\Delta A = 1$), to separate the ion beam of interest from other isotopes of the same element using a di-pole magnet. All other kinds of ion sources ionize all or a good fraction of a large number of reaction products produced in a reaction like spallation and thus the separation of the ion of interest needs separation of isobars, for which a resolving power exceeding a few tens of thousands is necessary. Such a resolving power is difficult to achieve even after multiple separation stages using di-poles, and multiple separation stages also result in a loss of RIB beam intensity.

The number of resonant excitation steps necessary to take a valance electron (in the atomic ground state) to the continuum, an auto-ionization state in the continuum or to a long-lived Rydberg state just below the continuum is usually two or three (Figure 5.2). Often the excited levels are not known, and it is necessary to scan the laser frequencies around some theoretically calculated values. The atoms in a Rydberg state are ionized easily in the presence of a DC electric field/by collision/by radiation in the hot cavity or by the residual laser photons. Maximum ionization efficiency can be obtained when all the two/three color transitions are saturated. This requires lasers of high spectral radiance which is achieved by using pulsed lasers with a high repetition rate to ensure that all the short-lived isotopes are ionized. Usually, in most cases an ionization efficiency exceeding 10% can be achieved. The laser system used for laser ionization is often a set of tunable dye lasers pumped by copper vapor lasers. Following advances in solid state lasers, the recent trend is to replace the copper-vapor pumped dye-laser system by titanium sapphire (Ti:Sa) lasers, all pumped by a single Nd:Yag laser of enough high power (~50 W) and high repetition rate up to 10 kHz.

The energies of the laser photons should add up to the ionization potential; thus, the laser ionization method can only be applied for ionizing isotopes of elements with ionization potential up to about 10 eV. In a novel experiment at CERN, ISOLDE, Resonance Ionization Laser Ion Sources (RILIS) was used to ionize astatine (At) isotopes with ionization potential of around 9.3 eV, produced in a spallation reaction of a 1.4 GeV proton with a ^{238}U target

Ion Sources for RIB Production

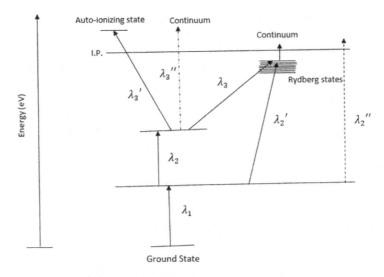

FIGURE 5.2
The basic scheme for laser ionization.

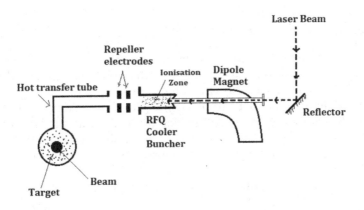

FIGURE 5.3
The schematic arrangement for laser ionization in an RFQ cooler buncher.

(Roth13). Although laser ionization is highly selective, the extracted ion beam is often not so pure if the ionization is carried out in hot cavities. This is because the surface ionization will always take place and ionize all reaction products that are surface ionizable and these act as contaminants to the laser-ionized ions. One way to reduce this contamination is to replace the surface ion source at the outlet of the hot transfer tube with a set of positively biased electrodes and a Radio Frequency Quadrupole (RFQ) ion trap consisting of four segmented electrodes and carry out the laser ionization in the RFQ ion trap (Bl 03). The arrangement is schematically shown in Figure 5.3. The biased electrodes repel the ions (surface ionized by collision of reaction product atoms with the hot inner surface of the target chamber), thus allowing only neutral atoms to enter the RF ion guide and interact with the laser. The RFQ trap (explained in more detail in Chapter 7) cools the laser-ionized ions by collision with helium buffer gas, confines the ions radially and traps them in a narrow localized zone before they are extracted. In addition to purity, the trap also results in a beam of much better quality in terms of transverse and longitudinal emittances.

Another advantage of the laser ion source lies in the fact that all the critical components of the ion source, the lasers as well as the optical elements, can be housed in a clean air-conditioned laboratory situated far away from the radiation area. The laser beams are usually injected into the transfer tube through a zero-degree port of the first di-pole magnet as shown in Figure 5.3. However, laser beam alignment with the ion/atomic beam could still be problematic since the target ion source area is not easily accessible. As already mentioned, for elements with high ionization potential (e.g., He, B, C, N, O, F, Ne, Kr, Xe, etc.) a laser ion source is not suitable. Plasma ion sources need to be used to ionize isotopes of such elements.

5.2.3 Forced Electron Beam Arc Discharge (FEBIAD) Ion Source

The elements that are not surface ionizable, that is elements with ionization potentials greater than about 6.5 eV, the preferred ion source is often one of the arc discharge-type plasma ion sources. In these sources the plasma is created by generating fast electrons (usually electron energy of approximately a few hundred eV) produced in a heated cathode. The arc discharge-type ion sources most often used are the Hollow Cathode Ion Source (Si 65), and the Forced Electron Beam Arc Discharge (FEBIAD) Ion Source (Ki 76). Among these, the FEBIAD ion source (improved compact versions) is used both at ISOLDE, CERN and at ISAC, TRIUMF (Ra 89, Bri 14). The ion source is schematically shown in Figure 5.4. The reaction product atoms enter the anode chamber of the ion source through a central hole in the cathode which is usually a tantalum disc with an opening at the center. The cathode, the anode chamber and the transfer line from the target are all heated using the same current source. Electrons are extracted from the cathode using a grid and magnetic field generated by a co-axial solenoid which helps in collimating the electron beam. The anode and the discharge chamber are usually made of molybdenum and the anode is insulated from the cathode using boron nitride (BN), alumina or AlN insulators. With electron impact energies in the range of a few 100 eV, all elements with high ionization potential including helium and neon can be potentially ionized to 1$^+$ charge state in this ion source.

It should be noted that the FEBIAD ion source has certain advantages over other kinds of plasma ion sources. The FEBIAD can work at lower arc support gas pressure down to about 10^{-5} torr and the ion source efficiency is not critically dependent on support gas pressure. This is because the primary electrons produced in the hot cathode are extracted by means of a grid. Also, it can work efficiently in the pressure range from 10^{-3} to 10^{-5} torr and the pressure fluctuation does not lead to instability in the operation. Further, the ion

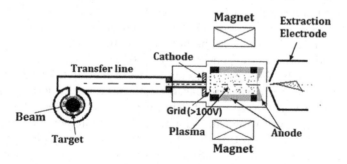

FIGURE 5.4
Schematic diagram showing the FEBIAD ion source.

source has a longer lifetime (of about three weeks) since the cathode is not a filament but an indirectly heated disc made usually of tantalum. The ionization efficiency of a FEBIAD-type ion source is largely independent of pressure inside the ion source (in the range of 10^{-3}–10^{-5} torr) and for many elements exceeds 20%.

A FEBIAD ion source is often used to ionize isotopes of metals, elements that are reactive or condensable with a tendency to stick to the walls, and elements with high ionization potentials such as helium, boron, carbon, nitrogen, fluorine, neon and argon isotopes. The ionization efficiency for helium is usually less than 1% and for neon about 1.5% but efficiency increases for elements heavier than Ne. For example, argon and xenon can be ionized with almost 20% and 50% efficiency respectively. The failures of insulators, degassing from the insulator and radiation damage of the components of the ion source including permanent magnets (if permanent magnets are used for the solenoid for compactness) are some of the problems faced with this ion source.

5.3 Electron Cyclotron Resonance (ECR) Ion Source

ECR ion sources (Gell 76) provide the best ionization efficiency for noble gases and elements with high ionization potential (He, C, N, O, F, Ne, Ar, Kr, Xe, etc.). In an ECR Ion Source (ECIRS) plasma is formed inside a plasma chamber, maintained at low pressure (usually in the range of 10^{-4}–10^{-8} torr), by stochastic heating of electrons using electromagnetic (RF/microwave) fields that are injected into the plasma chamber using a waveguide. For resonance absorption of microwave power, a magnetic field is superimposed with the electromagnetic field such that the frequency of the electromagnetic field is equal to the electron gyrofrequency in the magnetic field B given by $f_{RF} = f = eB/2\pi m$, where e is the elementary charge and m is the mass of the electron. If this condition is satisfied, electrons in circular motion would resonantly gain energy from the external electromagnetic field (from the electric field component perpendicular to the B field) and would ionize the low pressure gas-forming plasma. In order that the electrons gain enough energy required to ionize all elements, the electron should cross the ECR zone many times. This is ensured by confining the electrons axially by using two solenoids that create an axial mirror field and radially by permanent magnets that create a sextupolar (or octupolar) confining magnetic field (Figure 5.5). The electrons will have a helical motion in the magnetic field and would undergo multiple reflections from the magnetic mirrors during which they gain energy (very much in the same way that ions gain energy in a cyclotron while crossing the Dees) producing plasma in a limited egg-shaped zone defined by the shape of the mirror fields, as shown in Figure 5.5. In a typical ECRIS that uses solenoid coils for axial confinement and permanent magnets for radial confinement, the electron energy has a broad distribution from a few tens of eV to a few hundred keV. The high energy electrons ionize the gas by knocking off electrons from outer shells. The electrons that have low velocity or become slow after inelastic collisions are less affected by the confining magnetic fields. A good number of these electrons leave the plasma along with positive ions, thereby conserving the plasma neutrality.

One of the greatest advantages of an ECRIS over other plasma type ion sources lies in the fact that the ECRIS has a very long lifetime since it has no cathodes. The other advantage is that ECRIS can efficiently produce ions both with charge state of 1+ as well as with very high charge states, depending upon the microwave frequency used. For the production of

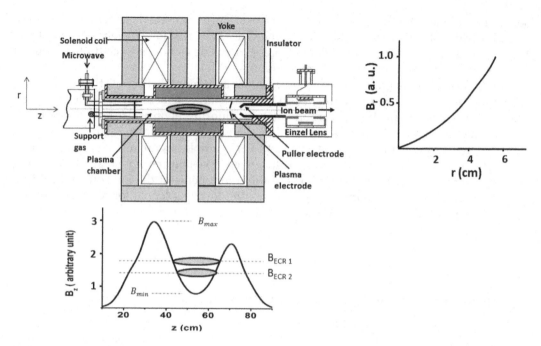

FIGURE 5.5
The schematic of an ECR Ion Source with its magnetic field configurations.

higher ionic charge states, higher microwave frequency and higher magnetic mirror fields are needed. Generally, the microwave frequency used in an ECRIS for the production of singly charged ions varies from 2.45 GHz to 6 GHz. Higher frequencies up to 56 GHz are used/have been proposed for the production of ions with high charge states.

5.3.1 ECIRS for 1⁺ Charge State

Very intense beams of protons and light ions with currents exceeding 100 mA have been produced using an ECRIS operating at 2.45 GHz. For noble gases, the 2.45 GHz ECRIS developed at GANIL for singly charged ions reported ionization efficiencies as high as 35% and 72% for Ne and Ar respectively (Ler 04). For the production of singly charged ions ($q = 1^+$), it is most common to use the commercially available 2.45 GHz microwave source that requires a magnetic field (B_{ECR}) of 0.0875 Tesla for resonance. The support gas pressure inside the ECR is usually maintained in the range of 10^{-4} to 10^{-6} torr.

In an ISOL-type RIB facility, the online ECRIS placed close to the target should be able to produce 1⁺ charge states of radioactive isotopes with high efficiency. The plasma chamber pressure would be high, about 10^{-4} to 10^{-5} torr due to gas load from the heated target and transfer line. It is just convenient to use low frequency (2 to 6 GHz) ECRIS. The high radiation field around the target creates problems for the permanent magnets used for confining the ECRIS plasma radially, which become demagnetized due to damage induced by neutrons in a matter of a few days. For this reason, ECRISs that use coils for axial as well as radial confinement have been developed (Hu 05; Lab 08). In an all-coil ECRIS, such as the mono-charged ion source for the TRIUMF and ISAC complex (MISTIC) the axial and radial mirror fields necessary to attain good confinement at low frequencies (typically between 2 to 6 GHz; MISTIC operates at 5.6 GHz) can be produced by using a number of coils surrounded by iron in an appropriate geometry. The use of electromagnets also allows the strengths and shapes of the confining fields to be adjusted to achieve a better ionization condition. One of the disadvantages of the ECIRS is its relatively high emittance compared to other 1+ ion sources used in ISOL-type RIB facilities. This makes the separation of the isotope of interest downstream more challenging.

5.3.2 ECIRS for High Charge State Production

The capability of an ECRIS to produce high ionic charge states has revolutionized the acceleration of stable heavy ions. The ECRIS has made possible injection of high charge state ($q/A = 1/2$ for lighter ions to about 1/7 for uranium) into an accelerator, making acceleration to higher energies much simpler and cheaper. For example, in a cyclotron the final energy of an ion after acceleration is given by $K\, q^2/A$, where K is a constant ($K \sim B^2 R^2$, B: magnetic field and R: extraction radius). Thus, the final energy increases as a square of the ionic charge state. For linear accelerators the final energy is proportional to q. Thus, in all kinds of accelerators, injection of low energy ions with high charge states gives a unique advantage and for this reason the ECRIS is now used to inject ion beams of high charge states into cyclic (cyclotron, synchrotron, etc.) as well as linear (RFQ linac for example) accelerators. The only other option is to accelerate low charge states to some energy initially and use charge stripper foils to increase the charge state. But the stripper produces a number of charge states out of which one needs to be selected and accelerated in an accelerator downstream. Usually, a number of such (stripper followed by selection and acceleration) stages are needed to reach high energies of 100 MeV/u or higher. The stripper option is costlier, requiring more accelerators and space. Also, since at each stage only one charge state is usually accelerated after the stripper, a lot of beam is lost. Injection of high charge state ions helps to avoid charge strippers or at least reduces the number of stripper stages.

High charge states of heavy ions are produced in the ECRIS mainly through the process of successive (stepwise) ionization and partly by Auger transitions. Successive ionization requires high electron density (n in number per cc), high electron velocity (v in cm s^{-1}) and longer ion confinement time (τ in s). It has been estimated that for obtaining very high charge states the product of these three factors $nv\tau$ should exceed at least about 10^{19} cm^{-2} (Gell 96). The maximum electron density is proportional to the square of the microwave frequency assuming that the electron density is equal to the cut-off electron density in plasma. The electron confinement time increases with the mirror ratio (B_{max}/B_{min}) of the confining magnetic fields and the longer the average electron confinement time the longer would be the average ion confinement time. In ECRISs designed for high charge state production, the operating frequency is usually chosen in the range of 14–28 GHz, although it is more common to use microwave frequencies of 14 GHz (Hi 00) and 18 GHz (Naka 96). The mirror ratio of the confining magnetic fields is usually kept around three. It should be noted that the mirror confinement principally affects the confinement time of the hot electrons, having energy greater than, say, 10 keV. These electrons can be kept confined (by proper choice of mirror ratios) in the plasma for times exceeding a few tens or even a few hundred milliseconds. The mirror fields have less influence on the cold electrons, arising out of the ionization process, that leave the plasma in about a few hundred microseconds. Thus, to obtain longer average electron confinement time, the number of hot electrons should exceed the number of cold electrons by orders of magnitude and the ECR needs to be operated at higher microwave frequencies. The microwave power also needs to be simultaneously increased to the extent that it does not make the plasma unstable. The density of hot electrons can also be increased substantially by two-frequency heating (Xie 97).

For example, if in an ECRIS, two microwave frequencies are used, say 10 and 14 GHz instead of only 14 GHz, the charge state distribution shifts significantly to higher charge states. This happens because two frequencies would form two well-separated ECR surfaces and electrons would be heated four times in their travel from one mirror point to the other instead of two times when a single frequency is used. The density of hot electrons would thus increase, resulting in an enhanced production of high charge states.

Although the ratio of hot to cold electrons is needed to exceed unity by orders of magnitude, the cold electrons play an important role in high charge state production. This is because the ion confinement time is inversely proportional to the mean ion velocity which is strongly governed by the plasma potential; lower plasma potential would mean longer ion confinement time of highly charged ions. It has been observed (Naka 91) that coating the plasma chamber surface with Al_2O_3 leads to a significant enhancement in the intensity of highly charged heavy ions. The enhancement is because the aluminum oxide has a high secondary electron emission coefficient under electron impact and coating of the plasma surface by Al_2O_3 results in the supply of a large number of secondary cold electrons to the ECR plasma. This helps to reduce the plasma potential, thereby increasing the ion confinement time. Additionally, lowering of the plasma potential also makes the plasma more stable.

An ECRIS designed to operate at frequencies greater than 18 GHz needs superconducting magnets since it requires magnetic fields exceeding 2 Tesla. Even for 18 GHz operating frequency for which B_{ECR} = 0.6428 Tesla, the mirror field (B_{max}) required is 2 Tesla or more. A number of ECR sources operating at 18 GHz in fact use superconducting solenoids for generating the axial field. As the frequency is increased further to 28 and 56 GHz, the design complications increase because of strong magnetic forces, intense Bremsstrahlung radiation and heat removal issues from the plasma chamber walls.

For production of high charge states, it is also important to minimize the loss of charge states due to recombination. The predominant process of recombination in ECR is through charge exchange collisions of the multiply charged ions with the neutral residual gas atoms. The gas pressure in the ECR plasma chamber therefore has to be low to minimize the loss of charge states due to charge exchange. In fact, the ratio of background neutral atom density (n_0) to the electron density (n) is an important factor that decides the population of a particular high charge state for a given ion (Gol 86). To produce a charge state q of a given atom of mass number A, the neutral atom density should be less than a value given by the expression: $n_0/n \leq 7 \times 10^3 \, \epsilon \, T_e^{-3/2} A^{1/2}/q$, where T_e is electron energy and ϵ is the total number of electrons in the outer shell. Since the electron density in ECR hardly exceeds $10^{11-12}/cm^3$, the neutral atom density should be less than about 10^7 atoms/cm^3. This requires the plasma chamber vacuum to be maintained at the level of 10^{-8} torr for obtaining high charge states of heavier ions.

5.3.3 ECRIS as Charge Breeder

The idea to use an ECRIS as a charge breeder was pioneered by the Grenoble group (Dom 98, Lam 02). To use an ECRIS as a charge breeder (often referred to as the 1^+ to n^+ scheme) for boosting the charge state of 1^+ (singly charged) ion beam of unstable isotopes, the mass separated singly charged ion beam of the desired isotope needs to be slowed down (decelerated) from a few tens of keV (extraction voltage of the first 1^+ ion source) to a few tens of eV (typically in between 20 and 50 eV) before injection into the charge breeder ECRIS. This "soft landing" into ECR plasma is necessary so that the singly charged ions can be trapped efficiently in the plasma for charge state boosting (to $q = n^+$). The deceleration of singly charged ions can be achieved gradually (Ban 00) or in a single step (Lam 98). The gradual deceleration scheme, shown schematically in Figure 5.6, has some potential

Ion Sources for RIB Production

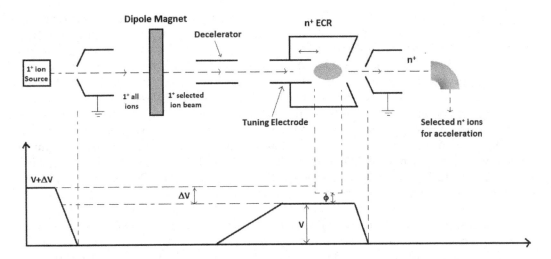

FIGURE 5.6
ECR charge breeder concept schematically shown with its potential scheme.

advantages. It minimizes the parasitic loss of multiply charged ions from the 1+ ion injection side, which otherwise accounts for the loss of almost 50% of the ions. The stepwise deceleration along with an additional provision to adjust axially the position of the tuning electrode with respect to ECR plasma by a few mm, allows a much better focusing of the low energy decelerated beam at the position of the ECR plasma. This is especially important for obtaining good ionization efficiencies for isotopes of condensable elements which if not focused properly into the ECR plasma would hit the plasma chamber walls and be lost. Charge breeders designed for dual frequency operation offer some advantage in terms of providing larger plasma zone for 1+ ion to soft land, apart from helping in higher charge state production. It should be noted that in many cases the accelerator downstream of the charge breeder accepts only fixed velocity ions, which necessitates the scaling of the extraction potential of the charge breeder (and hence the potential of the 1+ ion source) according to q/A of the ion beam extracted from the charge breeder.

Charge breeding in an ECRIS is quite efficient; for the selected charge state efficiencies in the range of 5–10% are usually achievable for a large number of elements (Wen 08, Lam 06). Although an ECRIS produces wide charge state distribution, the peak of the charge distribution can be shifted to higher charge states by increasing the frequency and the confinement time. The charge breeding time (closely related to confinement time), defined as the time interval in which the extracted intensity of the ion of the selected charge state increases from 10 to 90%, has been experimentally determined in a number of cases by injecting the 1+ ions into the ECR charge breeder in a pulse mode. The required breeding time for optimum production of high charge states is often in the range of a few hundred milliseconds to a few seconds. For example, the measured breeding time for $^{132}Xe^{29+}$ in an ANL ECR charge breeder is about 1.3 s (Von 16). The breeding time of approximately a few seconds is fine if the radioactive isotope has a half-life longer than this time but not so for very exotic species with half-lives of approximately a few tens of ms. In such cases a compromise is required between the charge state achievable and the number of radioactive ions lost due to decay. However, confinement time of a few tens of ms usually allows optimum production of charge states that are still high enough (say $q/A > 1/10$) to allow efficient post-acceleration to a certain stage followed by a charge stripper to increase the charge state.

In spite of so many advantages, an ECRIS has one serious limitation: the extracted beam from the ECRIS not only contains a number of charge states of the isotope of interest but it also contains a huge background from other contaminant ions, which is usually orders of magnitude higher in intensity than the ion of interest. The contaminant ions cover a wide range of A/q making magnetic separation not effective enough. Although relatively clean A/q zones where the contribution from contaminant ions is less can still be found, it is usually not enough if the intensity of the selected ions is not high enough (>10^6 pps). A detailed study on the contaminants and their sources has been carried out (Ame 14, Von 16). One of the major sources of contamination is from materials present in the aluminum alloy of which the plasma chamber is made. These contaminants come into the plasma due to plasma sputtering of the inner surface of the plasma chamber. By coating the plasma chamber with high purity (99.9995%) aluminum these contaminants could be brought down substantially. Another major source of contamination comes from the presence of micron and sub-micron loose particulates on the vacuum chamber surfaces of the plasma chamber and the beam line. These can be reduced significantly by pre-treating the plasma chamber by sand blasting followed by high-pressure rinsing. The degassing from the "O" rings used to maintain the plasma chamber vacuum constitutes another major source of contamination. This can be reduced significantly by replacing Viton "O" rings by metallic "O" rings which have much lower vapor pressure. The support gas also gives rise to contaminations and the leak rate can be optimized to reduce the contamination without compromising the efficiency. Magnetic analysis of the extracted n^+ ion beam shows that optimization leads to A/q zones with much reduced contamination. At the present state of development, after taking all these precautions and choosing the relatively clean A/q zones, an ECRIS charge breeder can be used for production (post-acceleration) of RIBs if the beam intensity is ~10^4 pps. The developmental efforts in an ECR-based charge breeder are continuing and are aimed at further developing and adopting "clean techniques" to construct the ECR so that overall background is reduced further and broader relatively clean A/q zones are available for injection of n^+ ions into the post-accelerator downstream. The use of a high-resolution mass separator before injection into the post-accelerator to further clean up the ion beam is also being considered. The goal is to push the present intensity limit of ~10^4 pps down by an order of magnitude or more so that charge breeding in ECR can be used for very exotic isotopes that are produced with very small intensities. The degree of contamination can be further reduced by accelerating the cocktail beam since an accelerator itself acts as a filter. Ultimately, after acceleration to a certain energy (usually greater than 1 MeV/u) a charge stripper and magnetic separation can be used to select a particular charge state of the isotope of interest to obtain a beam without appreciable contamination. In the process, however, it is necessary to sacrifice a good fraction of the beam intensity caused by charge stripping that produces not one but a number of higher (than the one before the stripper) charge states after the stripper. Charge stripping, however, is almost unavoidable if acceleration of RIBs to very high energies is needed.

5.4 The EBIS: For High Charge State Production and as Charge Breeder

In an Electron Beam Ion Source (EBIS) (Don 69; 81; Lev 81), the ionization is carried out by electron impact using a high density and high current mono-energetic electron beam. The ionization is carried out in a stepwise manner and for production of highly charged ions they need to be trapped for a sufficiently long time to allow the production of multiple charge state ions (MCI). The ion source is thus often referred to as EBIT (Electron Beam Ion Trap). Thus, for MCI production, the EBIS is always operated in the pulse mode.

The basic principle of working of an EBIS for high charge state production is shown in Figure 5.7 (Zs 14). An electron beam with energy usually in the range of 10–200 keV and current in the range about 100 mA to 10 A is injected into the ion source. To increase the charge state, high electron density that cannot be delivered by electron guns is needed. Thus, strong magnetic fields are used to compress the beam so that an electron beam density in the range of 200–10,000 A/cm^2 could be achieved in the ionization zone. At comparatively lower electron energy, say up to ~30 keV and lower electron density, the required magnetic field can be supplied by NdFeB permanent magnets but for higher electron energy and density the use of superconducting solenoids to produce axial magnetic field up to ~6 T is needed. The electron beam is ultimately decelerated and collected in a collector. The 1$^+$ ion beam is injected axially into the source and in the ionization zone high charge states are produced by electron impact in a stagewise manner. The negative space charge of the electrons confines the ions radially while the axial confinement of ions is provided by applying DC voltages to a number (greater than three) of drift tubes/cylindrical electrodes, where the outer ones are kept at higher potentials, forming an electrostatic mirror trap. By switching off the axial trap potential of the outer electrode periodically, the n^+ ions can be extracted and then magnetically separated (Figure 5.7). The ionization/breeding time can therefore be controlled by changing the switch on-off time of the potential of the outer electrode. To minimize loss of highly charged ions by recombination, a vacuum ~10^{-10} torr is usually maintained in the ion source.

As shown in Figure 5.7, for the EBIS to be used as a charge breeder, the 1$^+$ ion beam needs to be cooled in a RFQ cooler and buncher before injecting into the EBIS, and thus the EBIS when used as a charge breeder operates in the pulse mode. The 1$^+$ beam accumulated in the RFQ trap enters the EBIS ionization region by lowering the potential barrier (of D3 in Figure 5.7) for about 10–20 μs. The barrier then kept high for a time varying from a few ms to a few hundred ms during which the 1$^+$ ion is converted into n^+. The barrier

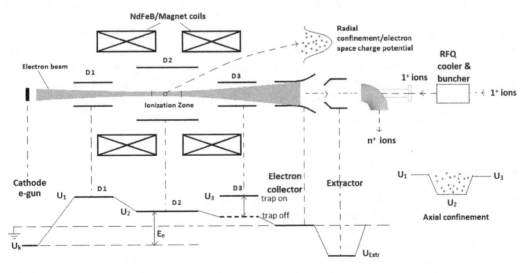

FIGURE 5.7
The schematic of an EBIS charge breeder.

height (once again of D3) is then lowered for about 10–20 μs during which the n^+ ions are extracted. In the pulsed mode of operation, the 1^+ injection energy (potential) and the n^+ extraction potentials are independent of each other, unlike in the ECRIS charge breeder but pulsed n^+ beam extraction gives rise to high instantaneous count rates in the ion detectors downstream.

In EBIS, the final charge state that can be produced depends upon the product of electron beam density and the confinement time. Compared to an ECR charge breeder, an EBIS can produce higher charge states with less confinement time, if the electron density exceeds a few hundred A/cm². For example, the breeding time for Cs 28+ as measured in EBIS at ANL was 28 ms. For this measurement a solenoidal field of 5 T and electron beam density of 385 A/cm² were used and the breeding efficiency was found to be about ~20% (Von 16). Future EBISs plan to use much higher current density, exceeding 10^4 A/cm². In such a source the breeding time for high charge states is expected to come down by orders of magnitude; for example, the expected breeding time for Sn 40+ charge state is only about 3 ms for current density of ~10^5 A/cm2 (Wen 08).

The charge breeding efficiency in an EBIS depends, in addition to the EBIS parameters (electron density, etc.), on how well the 1^+ beam overlaps with the electron beam. The beam optics for 1^+ beam injection needs to ensure that the envelope of the 1^+ beam has a spiral motion in the solenoid field smaller than the electron beam diameter in the ionization zone. If that is ensured, higher charge breeding efficiencies are obtained in EBIS compared to that in an ECRIS. Charge breeding efficiencies have been measured at the REX-EBIS at CERN for a wide range of elements and for a number of charge states ($2.5 < A/q < 5$). The measured efficiencies lie in the range of 3–20 % (Wen 10). For example, the measured efficiencies for ^{224}Ra 51+, ^{133}Cs 34+, ^{65}Cu 19+ and ^{23}Na 9+ are about 4%, 10%, 20% and 23% respectively. These efficiencies are higher than those obtained in an ECRIS charge breeder. Given the fact that REX-EBIS is only a rather low electron current (200–300 mA) and low electron energy (20 keV) source where current density does not exceed 500 A/cm², much higher efficiencies are expected from EBIS operating with higher current densities and stronger solenoidal fields.

Compared to ECRIS, the EBIS charge breeder has many fewer problems from the background contamination. This is because the vacuum inside the EBIS is better by at least two orders of magnitude compared to the ECRIS and also because in an EBIS the ion chamber's inner walls are not subject to ion sputtering. However, although the background is much less, the selection of proper A/q regions with less background contamination is still needed. Selecting such A/q regions, an almost contaminant-free high charge state ion beam of the selected isotope can be obtained (Wen 10). The only disadvantage of an EBIS charge breeder, compared to an ECRIS charge breeder, lies in its pulsed operation that requires changes in the accelerator operation.

5.5 Positioning the First Ion Source away from The Target (the HeJRT Technique)

As already mentioned, the requirement to position the first 1^+ ion source close to the target creates a lot of practical difficulties. It is possible, however, to position the ion source away from the target by using the helium jet transport technique (Macf 69). In this technique, the reaction products coming out of the production target by diffusion or by recoil momentum are transported by helium gas to a distant (~10 m or more) low background ion source chamber using differential pumping through a capillary (diameter ~1 mm) connecting the target chamber and the ion source chamber. The helium gas is loaded with macromolecules or aerosols and the reaction products get attached to the aerosols in the target chamber. The aerosols loaded with radioactive atoms are then transported by the helium gas to the ion source chamber (Molt 80). The aerosols break down inside the ion source releasing the radioactive atoms which are then ionized in the ion source. The transport time of the reaction products from the target chamber to the ion source chamber is usually a few hundred ms that restricts the use of this technique to the production and study of exotic isotopes with a half-life greater than about 300 ms. Subject to this limit, the technique offers, in principle, a lot of advantages.

The transport efficiency for transport from the target chamber to the ionization chamber is usually more than 50% (could be even close to 100%) and is largely independent of the chemical nature of the isotope; the helium cooling of the target would allow use of higher primary beam power for the production of RIBs; decoupling of target from the ion source and the first magnetic separator after the ion source makes things much simpler mechanically and from the viewpoint of radioactive handling and storage; the ion source life would not be dictated by radiation damage since the ion source and the first separator are placed in a distant low radiation zone. The pressure inside the ion source can be kept better than 10^{-3} torr by using a skimmer positioned upstream of the ion source in the ion source chamber (Figure 5.8). The skimmer takes off most of the helium load coming out of the capillary as a jet which is eventually pumped out by a mechanical pump but allows a few orders of magnitude heavier (than helium) reaction products loaded aerosols/clusters to pass through and enter the ion source. The skimmer efficiency is usually more than 50%, resulting in an overall efficiency exceeding 25% for the transfer of activities from the target to the ion source.

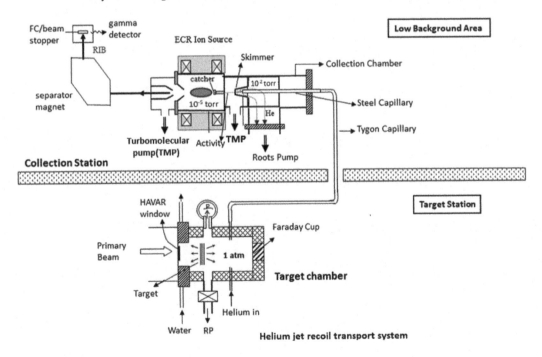

FIGURE 5.8
Schematic diagram showing a helium jet transport system coupled to an ECR Ion Source.

The gas load, however, restricts use of an ECRIS as the first ion source since an ECRIS requires a better vacuum (10^{-5} torr) for operation. To reduce the gas load and improve the vacuum level inside the ECRIS a second skimmer with pumping in between can be used, which of course brings down the efficiency roughly by another factor of two. But even then, it has been observed experimentally that the activity loaded clusters/aerosols do not get captured in the ECRIS plasma. This problem can be overcome by using a porous catcher (W coated reticulated vitreous carbon fiber, RVCF) inside the ECR chamber (Figure 5.8), positioned close to the plasma region (Naik 13). The porous catcher stops the clusters and at the same time acts as the second skimmer with practically no loss in efficiency. The catcher, being porous and heated by the ECR plasma (additional electrical heating can be provided for elements that are less volatile), releases the radioactive atoms which are then ionized in the ECR plasma. The technique has already been used to produce a few RIBs and a combined efficiency of 15–20% could be obtained for skimmer, diffusion through porous catcher, ionization and extraction of 1+ beam (Naik 13).

6
Accelerators for RIB Production and Post-Acceleration

6.1 Introduction

Accelerators are an integral part of any RIB facility. Several types of particle accelerators and associated new technologies have been developed over the years, ushering progress in nuclear and particle physics research with particle beams of different types and increasing energies. The energy scale probed by the accelerators over the years depicted in a so-called "Livingston plot" is shown in Figure 6.1.

Driver and the Post-Accelerator

The history of RIB production using an accelerator dates back to 1951, when Otto Kofoed-Hansen and Karl Ove Nielsen at the Niels Bohr Institute in Copenhagen bombarded a uranium oxide powder target with deuteron from a cyclotron to produce neutron-rich krypton isotopes (Cor 04). This was then transported via gas-flow technique, energized using high voltage DC and mass separated. It was, however, in the 1970s that exotic nuclei studies picked up momentum because of the realization that these nuclei might possess unexpectedly different properties compared with nuclei that are beta-stable. The need to produce these nuclei in sufficient numbers using different kinds of primary beams was also strongly felt. It was clear that the *driver or the primary accelerator* delivering the primary beam of stable isotopes on the target for production of these nuclei should be capable of delivering high intensity beams at high relativistic energies so that the low production cross-section of these exotic isotopes (for very exotic species the cross-section is often as low as a few pb or even less) are compensated by higher beam intensity and higher useable target thickness at higher beam energy.

In any driver accelerator, a common problem encountered for high current acceleration is that arising due to the space charge that adversely affects the beam dynamics of the accelerating beam. Space charge is essentially the electrostatic Coulomb forces between the charge particles in the rest frame of the ions that create a self-field which acts on the ions very much like a distributed lens resulting in radial (longitudinal also, in case of a pulsed beam) defocusing. Since the ions are also moving, an attractive magnetic force (like the force between two parallel conductors carrying current in the same direction) is produced that tends to cancel the electrostatic repulsive force. The cancelation becomes more and more effective as the ion velocity increases and the full cancelation occurs when ion velocity reaches the velocity of light. The space charge thus affects the beam dynamics more severely at low ion velocity. However, there is an indirect space charge effect too that does not cancel out at high energies. The presence of vacuum pipes and beam

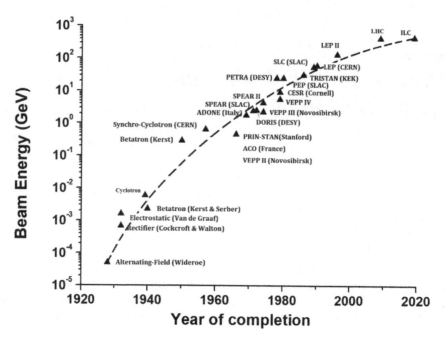

FIGURE 6.1
"Livingston Plot" showing development of accelerators over the years.

chambers through which the beam passes give rise to this indirect space charge effect due to image charges and currents. This indirect effect does not get canceled at high energies and becomes important at very high relativistic energies.

The RIB facilities employing the ISOL technique for the production of RIBs, where the reaction products are stopped in the target, need also accelerators to accelerate RIBs to the desired energy. These accelerators are termed "post-accelerators." It should be emphasized that RIB facilities employing the PFS technique do not need post-accelerators since the RIB, after fragmentation, moves almost with the beam velocity. Unlike the primary or the driver accelerator, the space charge creates no problem in post-accelerators since it is not the beam intensity but the lack of it which is of major concern in the case of RIBs. Since RIBs are produced in minuscule quantities, it is important that the post-accelerator accelerates the beam with minimum loss, that is with maximum efficiency. Also, since the RIB production process involves the production of a large number of products, it is an advantage if the accelerator itself has the capability of mass selection, like in a cyclotron. The combination of drivers and post-accelerators that are being used in some of the vibrant ISOL-type RIB facilities are listed in Table 6.1 (Lin 04). It may be noted that all the driver accelerators in Table 6.1 are cyclic accelerators, whereas for post-accelerators both cyclic and linear accelerators have been used. This does not mean that a linear accelerator is an inferior choice for the driver accelerator. In fact it is not so. Actually, most of these facilities were built by suitably modifying/upgrading the existing accelerators, and although the combinations have worked well to produce excellent physics, they are not necessarily the most optimum ones if production of a wide variety of RIBs with maximum possible intensities at widely different energies is desired.

It is an advantage if the post-acceleration of RIBs is done in stages so that RIBs of widely different energies varying from a few keV/u to a few hundred MeV/u could be

TABLE 6.1

Driver and Post-Accelerator for ISOL-Type RIB Facilities

Facility	Driver	Post-Accelerator	Comments
Louvain-la-Neuve Belgium	Cyclotron p, 30 MeV, 200 μA	Cyclotrons K = 110, 44	1989
HRIBF Oak Ridge, USA	Cyclotron p, d, α, 50–100 MeV, 10–20 μA	25 MV tandem	1997–2012
ISAC: TRIUMF Vancouver, Canada	Cyclotron p, 500 MeV, 100 μA	Linac up to 1.5 MeV/u (ISAC I) Linac up to 6.5 MeV/u (ISAC II)	2000 (ISAC I) 2008 (ISAC II)
SPIRAL: GANIL Caen, France	2 cyclotrons heavy ions up to 95 A MeV, 6 kW (SPIRAL I) SC linear accelerator LINAG deuterons up to 40 MeV heavy ions up to 15 A MeV (SPIRAL II)	Cyclotron CIME K = 265, 2–25 A MeV	2001 (SPIRAL I) 2008 (SPIRAL II)
REX ISOLDE: CERN Genève, Switzerland	PS booster p, 1.4 GeV, 2 μA	Linac 0.8–3.1 MeV/u Linac up to 5 MeV/u (upgrade)	2001 2008 (upgrade)
EXCYT Catania, Italy	K = 800 cyclotron heavy ions	15 MV tandem 0.2–8 MeV/u	2004

TABLE 6.2

Required RIB Energy and Intensity for Conducting Different Measurements

Measurement	Energy	Intensity (pps)
Mass and moments	keV–MeV/u	10^{1-3}
Decay	keV–MeV/u	10^{1-4}
Cross section and momentum	Tens of MeV/u	10^{1-4}
Coulomb excitation and break-up	Tens of MeV/u	10^{2-4}
Elastic and inelastic	All energies	10^{4-5}
Quasi Free Knockout	80–300 MeV/u	10^{4-6}
Transfer Reaction	1–50 MeV/u	10^{5-6}
Fission–Fusion	<30 MeV/u	10^{5-6}

delivered to the users' target for different types of experiments. Thus, for example, both in the ISAC and Rex-ISOLDE facilities, the post accelerators comprise a number of linear accelerator modules where the acceleration to the highest energies is achieved by stage-wise acceleration. To decide on the optimum choice of the driver and post-accelerators an idea of the energy and intensity of RI Beams required for carrying out different kind of studies is required. This is given in Table 6.2 (Le 13) for studies of interest in nuclear physics and nuclear astrophysics. Usually much higher intensities are required for material science studies, but the isotopes used are generally less exotic which can be produced with orders of magnitude in higher cross-sections. The intensities mentioned in Table 6.2 give an idea of the minimum intensity required for the different types of measurements. However, these numbers are mostly indicative and are likely to come down further with innovations in experimental techniques which, over the last three decades have brought the requirement on minimum intensities down by orders of magnitude, resulting in the numbers mentioned in Table 6.2.

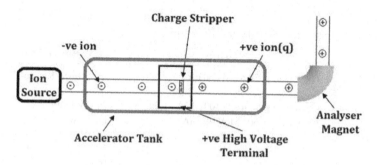

FIGURE 6.2
Layout of a typical DC Tandem Accelerator.

6.2 DC Accelerators for RIB Production

A DC/Tandem accelerator, shown schematically in Figure 6.2, is the simplest and a comparatively low-cost accelerator. For operating a DC accelerator in Tandem mode a negative ion is accelerated first, followed by charge stripping in a carbon foil that turns the negative ion into a positive ion, which is then accelerated further. Negative ions can be produced in the ion source for elements with positive electron affinity. Thus, for noble gases and for elements such as Mg, Sr, Ba, Sc, Zn, Cd and Hg, it is not possible to produce negative ions as electron affinity is negative. Thus, elemental dependence of negative ion source limits the range of beam that can be accelerated using a Tandem DC accelerator. Moreover, the maximum energy that can be achieved by a DC accelerator is given by $V(q+1)$, V being the terminal DC voltage and q being the charge state of the beam after the charge stripper that strips the negative ion to produce a positive ion of charge state q, which is then accelerated by the same potential to produce the final energy. The terminal voltage is limited by breakdown and thus can hardly exceed 20 MV. So, for a heavy ion with $A > 50$, it is not possible to exceed the energy necessary to overcome the Coulomb barrier. Because of these limitations, DC accelerators are not considered optimum for RIB production.

6.3 Cyclic Accelerators for RIB Production

The cyclic accelerators that are used/planned in RIB facilities as the driver accelerator are Azimuthally Varying Field (AVF) cyclotrons of single magnet and sector type and the synchrotron, as shown schematically in Figure 6.3.

In cyclotrons the ion orbit grows in size radially (spirally) as the ion is accelerated whereas in synchrotrons the ion radius is fixed, and the ion revolves in the same orbit many times while it gets accelerated. In both types, acceleration is achieved by applying radio frequency (RF) fields and much higher energy as well as beam intensity compared to DC accelerators can be obtained. We discuss first the cyclotron and then the synchrotron.

6.3.1 Cyclotrons

The cyclotron is perhaps the most popular accelerator because it is compact (small footprint), cost-effective and at the same time has the capacity to accelerate high current (a few

Driver and Post-Accelerator for RIB

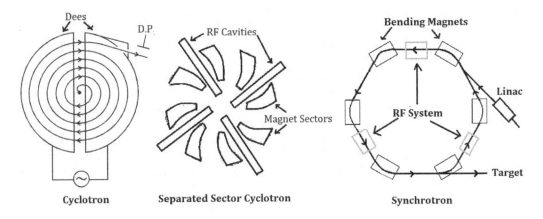

FIGURE 6.3
Schematics of three classes of cyclic accelerators.

particle mA) beams of light and heavy ion beams to relativistic energies. In a cyclotron, the time period of an ion's orbit under magnetic field should be in resonance with that of the oscillating voltage applied on the electrodes providing energy gain in each turn. The isochronism condition and final energy are given by:

$$\omega = qB/m, \; E = \left(q^2 B^2 R^2\right)/2m = e^2 B^2 R^2 / 2m_p * \frac{q^2}{A} = K\frac{q^2}{A}, \; r = mv/qB \quad (6.1)$$

In Equation (6.1), ω, q, m and A are the angular frequency, charge, mass and mass number of the ion respectively; B, r and R are respectively the magnetic field, radius of an orbit and extraction radius of cyclotron; m_p is the mass of proton, e is the electronic charge and K represents the bending limit of the cyclotron in MeV. Usually, the cyclotrons operate at comparatively lower frequencies in the range of 6–60 MHz, the higher limit being imposed, among other factors, by the maximum magnetic field that can be produced. The synchronism condition demands a constant magnetic field independent of radius as long as mass remains constant. However, for RIB production accelerating the beam to high energies is required. The ion in a cyclotron designed for the purpose would attain, after a number of revolutions, an energy for which relativistic increase in mass would become important. The synchronism condition in such a case would demand an increase in the magnetic field with a radius so as to keep the time period of revolution constant (ions can be accelerated to roughly 10% of their rest mass if the magnetic field is kept constant). The required increase in the average magnetic field (known as the isochronous field since it restores the isochronism) with radius is given by:

$$<B(r)> = \gamma B_0 = \left(1 + \frac{E}{E_0}\right); \quad (6.2)$$

Where B_0 is the magnetic field at the center of the cyclotron; E: kinetic energy of the ion and E_0 is the rest mass energy. For heavier ions with higher rest mass energy the relativistic correction required is obviously less compared to lighter ions. In any case, the required increase in the magnetic field to restore isochronism would lead to axial defocusing pushing ions away from the median plane and most of the ions would hit the pole faces instead of getting accelerated.

The solution to this problem was provided in an AVF cyclotron where the azimuthal variation of the magnetic field is created by adding hills on the magnet faces that create hills and valleys as shown in Figure 6.4 (left). This idea was proposed by Thomas and thus the radial hill and valley sectors are named after him. The gap between the pole pieces is less in the hills; so hills are regions of stronger magnetic fields compared to valley regions in which the gaps between the pole pieces are larger. An ion entering a valley after crossing a hill would experience a lower magnetic field. The ion path would be flatter in this region compared to its path in the hill region. Thus, the orbit of the ion would not be a circular one and would oscillate around an imaginary perfect circle that the ion would have traced had the magnetic field been uniform with the value , which is the average of the hill and valley fields at that radius (the actual ion orbit would be much different since the ion would also gain energy at specific places called dees). Because of its scalloped orbit, the ion gets a radial velocity component v_r which is maximum at every hill–valley boundary (Figure 6.4). The radial velocity component together with the B_θ component of the magnetic field arising due to the fringing field at the hill–valley boundaries would produce an axial focusing force pushing ions moving away from the median plane towards the median plane. It should be noted that the signs of radial velocity component and B_θ in the entry and exit of a sector are opposite, keeping the direction of axial force always towards the median plane. Thus Thomas's prescription could solve the problem of axial focusing and the magnetic field can be increased with radius by gradually increasing the angular extent of hill regions at the expense of valley regions. However, increasing the average magnetic field in an isochronous manner as given by Equation (6.2) by increasing the azimuthal width of the hill sector works perfectly only for a single ion. In cyclotrons designed for multi-particle acceleration, a number of circular trim coils are additionally used along the radius on the hill sectors and by varying the trim coil currents and polarities the isochronous condition for different ions can be met.

The hills are often twisted into a spiral form (Figure 6.4, right) which results in an additional axial focusing, discovered by Kerst and Laslett, due to the azimuthal velocity of the ion. For the spiral sector, B_r is non-zero away from the median plane and therefore it

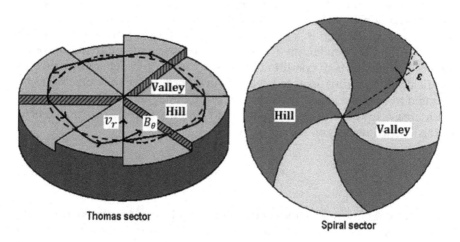

FIGURE 6.4
Four-sector Thomas magnet and three-sector spiral shape pole piece.

changes sign at the entry and exit of a sector. Focusing is thus achieved by alternate focusing and de-focusing. The Thomas and Spiral tunes are expressed as:

$$\Delta v_{z\,Thomas}^2 = F^2 = \frac{(B_H - <B>)(<B> - B_V)}{<B>^2}$$

$$\Delta v_{z\,Spiral}^2 = 2F^2 (\tan \varepsilon)^2 \tag{6.3}$$

In this expression, F is flutter factor; B_H, B_V and $<B>$ denote hill, valley and average field respectively. ε, the spiral angle, is defined as the angle between the radius vector and the tangent to the spiral as shown in Figure 6.4.

The net axial force thus constitutes a de-focusing force owing to an increase in the average magnetic field, with radius along with the focusing component induced by the flutter factor and the spiral angle:

$$v_z^2 = -\beta^2 \gamma^2 + F^2 (1 + 2\tan^2 \varepsilon) \tag{6.4}$$

With an increase in kinetic energy, the magnitude of the defocusing (first) term would increase and at a particular high energy might become equal to the focusing (second) term inducing axial instability. This is called the focusing limit of a cyclotron. This restricts the achievable kinetic energy using an AVF cyclotron which is more significant for ions with a lower rest mass. However, the focusing limit can be stretched in energy (and hence the maximum achievable energy) by increasing the flutter as well as the spiral angle.

The maximum limit in the flutter can be reached by making cyclotrons only out of sector magnets where the valley field B_v is essentially zero. Such a cyclotron, a variant of the AVF cyclotron, is referred to as a separated sector cyclotron (second diagram in Figure 6.3) in which the valley is absent, and the flutter factor approaches unity compared to about 0.6 in a normal single magnet AVF cyclotron with hills and valleys. The valleys or the space between the pole pieces are utilized mostly for placing the RF resonator cavities. This allows smaller vertical gaps between the sector magnets and results in much better axial focusing. Also, the multiple RF cavities in between the sector dipole magnets can achieve higher energy gain per unit turn that results in a reduction of the total path length inside the cyclotron, making the acceleration more efficient. More importantly, it leads to better turn separation at the extraction radius that allows highly efficient extraction with extraction efficiency approaching almost 100% as in the case of the PSI cyclotron. Further, it can be shown that the intensity limit in a cyclotron scales as a cube of the voltage applied to the RF cavities in between the sector magnets. Thus, separated sector cyclotrons allow acceleration of a much higher intensity beam compared to a single magnet AVF cyclotron. For example, the PSI-separated sector cyclotron can deliver 2 mA proton beams at 590 MeV, the highest intensity achieved in a cyclotron so far.

In a separated cyclotron, the beam injection is done radially into an orbit of higher radius. To do so the rigidity of the beam (product of beam velocity and m/q) should match the rigidity (product of magnetic field and the radius of the orbit) of that orbit. The ion beam therefore needs to be pre-accelerated using another accelerator to the energy required for rigidity matching before injection. The radial injection helps in avoiding the beam losses that are significant in the case of axial injection of low velocity ions from an ion source in a single magnet AVF cyclotron, especially when the ion current is high enough for space charge to affect the beam dynamics. So for high current acceleration, the pre-injector is generally chosen to be a linear accelerator (discussed in the next section) where space charge forces at low velocity can be more efficiently mitigated.

Construction of separated sector cyclotrons with large diameter is more simplified and cost-effective compared to single magnet cyclotrons. Extraction can be achieved by placing thin electrostatic electrodes providing a kick to ions in the last orbit. The other option is to use stripping foil whereby the accelerated beam changes the charge state and thus deviates from the usual orbit in the magnetic field making extraction simple and efficient. At relativistic energies, turn separation decreases with an increase in energy which becomes a bottleneck for going to even higher energies using a cyclotron/separated sector cyclotron.

Acceleration of high intensity beams of light ions, especially of protons to high energy exceeding 500 MeV is achievable in room temperature cyclotrons using a manageable magnet size, since q/A for a proton is unity and (E (MeV) ~ K (MeV) × q^2/A. The TRIUMF cyclotron is one example of a room temperature separated sector AVF cyclotron (with six sectors), where protons are accelerated to 500 MeV and the extracted beam current is quite high (about 100 µA) (Dut 03). PSI has been providing the most intense proton beam (2 mA) in the world at 590 MeV using two separated/ring cyclotrons as shown schematically in Figure 6.5. In the PSI facility, a DC-Cockroft machine delivers 870 keV proton beam to the first $K = 72$ separated sector cyclotron (injector 2). A 72 MeV proton beam is then injected into the second separated sector or ring cyclotron, with eight sectors, for further acceleration to 590 MeV (Sch 06). It should be mentioned that cyclotrons are operated in continuous wave (CW) mode and this divides the beam charge per second (or the beam intensity/current) into a large number of pulses given by the RF frequency. This helps in keeping the average charge per pulse to moderate values from the space charge point of view and allows acceleration of intense beams.

In principle, higher and higher energy (subject to focusing limit) can be achieved in a cyclotron by increasing the radius of the magnet but in practice it becomes prohibitively expensive (cost ~R^3) and also magnets become very heavy (often exceeding several

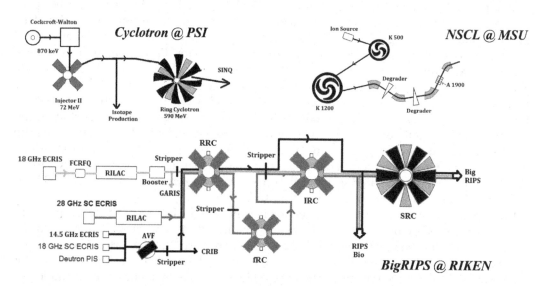

FIGURE 6.5
Schematic layout of multistage accelerators at PSI, RIKEN and NSCL using Separated Sector Cyclotrons.

thousand tonnes). Due to advances in superconducting magnet technology, it is now more common to use superconducting coils for the main magnet to produce an average magnetic field that is about three times the room temperature limit of less than 2 Tesla. In principle, therefore, for the same radius, the energy can be increased by almost an order of magnitude using superconducting coil magnets. The $K = 800$ superconducting AVF cyclotron at INFN-LNS, delivering heavy ion beams up to 80 MeV/u and beam current up to 1 pμA, is one example where a single AVF cyclotron is used as the driver accelerator for the RIB facility EXCYT.

Even with superconducting magnets, the limitations in the magnet size limit the maximum K value and hence the final energy to which an ion can be accelerated. SRC, the K 2600 superconducting separated sector cyclotron built at RIKEN (Yan 07) is presently the largest cyclotron operating in the world and the weight of the magnet sectors in SRC totals to about 8500 tons. Due to the limitation in the magnet size, acceleration of heavy ion beams with $A > 100$ to energies exceeding about 100 MeV/u is a difficult task in a single AVF cyclotron. The energy limit decreases further as the ion gets heavier. This is because A/Z ratio decreases for stable isotopes as A increases and fully stripped charge states ($q = Z$) for heavy ions are impossible to achieve in an ion source. Taking the case of uranium, as an example, it is possible to produce and inject, using a modern Electron Cyclotron Resonance (ECR) ion source, uranium ions with charge state of 34+ (compared to 92+ for the fully stripped ion) only with an intensity/current of about 1 pμA or a bit higher. Assuming a $K1000$ cyclotron, the maximum energy to which the uranium beam of $q/A = 1/7$ can be accelerated is only about 4.76 GeV or 20 MeV/u. This energy should be contrasted with the need to accelerate the uranium beam to at least a few hundred MeV/u in a PFS-type RIB facility.

The final energy of heavy ions can be increased, however, by increasing the charge state of the ions by using a charge stripper. For this it is necessary to pre-accelerate a comparatively lower charge state ion (that allows injection of a comparatively higher intensity ion beam from an ion source) in the first accelerator, use a stripper to increase the charge state of the beam accelerated in the first accelerator and then inject the ion of high charge state into a cyclotron for acceleration to the final energy. The pre-accelerator could be another cyclotron or a linear accelerator, but it is important to pre-accelerate the ions to an energy high enough to get a most probable charge state after the stripper that is much higher than what an ECR ion source (or an EBIS) can produce with comparable intensities. This scheme has been used for heavy ion acceleration at RIKEN and at NSCL in which a number of cyclotrons are coupled with charge strippers in between to increase the final energy of the heavy ions.

RIKEN has been operating since 1986 the K 540 MeV room temperature ring cyclotron (RRC) delivering protons to 210 MeV, heavy ions such as C, O, Ne ions up to 135 MeV/u, Ar to 95 MeV/u and Bi to 15 MeV/u. There are two injectors for RRC: variable frequency linear accelerator (RILAC) providing heavy ion beam with energy 6 MeV/u and K 70 MeV AVF delivering protons of 14 MeV/u and Ca ions up to 5.6 MeV/u. As an energy augmentation program for the RI Beam Factory project, three more cyclotrons were added to RRC: the four-sector K 570 fixed-frequency ring cyclotron (fRC), the four-sector K 980 MeV intermediate-stage ring cyclotron (IRC) and the six-sector K 2600 MeV superconducting ring cyclotron (SRC).

With this unique chain of cyclotrons (shown schematically in Figure 6.5) and charge strippers placed after the RRC (at about 11 MeV/u for uranium) and fRC (at about 51 MeV/u for uranium) a charge state of 86+ could be injected in the SRC to achieve the final accelerated beam energy of 345 MeV/u for uranium. Light ions can be accelerated to 440 MeV/u using this chain of cyclotrons. The design goal has been to achieve heavy ion beam intensity of 1 pμA for uranium at 345 MeV/u (Yan 07, Oku 12).

At NSCL, two superconducting cyclotrons, K 500 and K 1200, are coupled (Figure 6.5) to accelerate light heavy ions up to A ~ 100 to final energies exceeding 100 MeV/u (for example 132 MeV/u for ^{76}Ge beam). The injection to K 500 is done using an ECR ion source that injects relatively low charge state (and higher current) heavy ions axially into the K 500 cyclotron that accelerates the ion beam to a few tens of MeV/u (17 MeV/u typically). The extracted beam from K 500 is then injected radially into a matching orbit in the K 1200 cyclotron that corresponds to the equilibrium charge state of the ions achieved after the beam passes through a stripper foil placed inside K 1200. The increase in the charge state by using the stripper before injection to the orbit of K 1200 resulted in extracted ion beams of much higher intensity and higher final energy compared to stand-alone operation of K 1200 using axial injection from an ECR ion source.

Cyclotrons have many advantages: they can accelerate high intensity beams of light and heavy ions and are an excellent mass separator; they have a small footprint, require less shielding since thick iron for the pole pieces and the return yoke provide an effective shielding, and for a given beam power (product of beam energy and beam current) is cost-effective. But it has limitations if the acceleration of intense ion beams to very high energies is desired. The limitation comes from the practical size of the magnet that can be used, the focusing limit and for light ions the requirement to increase magnetic field for isochronism. In the case of heavy ions, apart from the limitation in the magnet size the need to use multiple stripping stages and the small turn separation at very high energies leads to loss of beam intensity. For example, it is difficult to accelerate protons with energy exceeding about 2 GeV and very heavy ions like uranium to energies exceeding about 400 MeV/u using cyclotrons. To surpass this energy limit, synchrotrons, the other kind of cyclic ion accelerator, need to be used.

6.3.2 Synchrotrons

In a synchrotron the energy limit can be surpassed but at the cost of having a pulsed beam in place of a CW beam from cyclotrons. Synchrotrons like the ring cyclotron need an injector accelerator for operation and acceleration. In a synchrotron, the orbital radius remains fixed while the magnetic fields in the focusing and bending magnets are ramped with time in accordance with energy gain of the accelerating ion. Thus, a bending magnet extends over a limited annular section (roughly the extent of the beam pipe) saving in size and cost. Extraction thus gets rid of turn separation problems as encountered in the cyclotrons at relativistic energies. Early synchrotrons, like the cyclotron, relied on weak focusing by shaping bending magnets (outwardly wedge-shaped pole pieces) with field index $0 < n < 1$, where $n = -\frac{r}{B}\frac{\partial B_z}{\partial r}$. It resulted in a beam with a large cross-section requiring magnets with a large bore. With the advent of alternate gradient or strong focusing (AGS), the bending field is provided by a series of wedge-shaped magnets having alternate positive and negative field index. The combined effect of focusing and de-focusing in the horizontal and vertical direction results in net beam confinement. This ensured a stronger focusing leading to a substantial reduction in the beam size, which made possible the development of the next generation of proton synchrotrons at CERN and BNL. Modern synchrotrons are designed with separate magnets for bending and focusing with the intermediate RF cavities for energizing the beam. Strong focusing made it possible to enhance the intensity of beams at higher energies. The frequency of the RF cavity (f_{rf}) can be in multiples of the particle revolution frequency (f_0). The harmonic number h where $f_{rf} = hf_o$ would result in acceleration of h possible bunches. The typical revolution frequencies of the charged particle in synchrotrons are in the range of few tens of MHz. At such low frequencies the usual resonating cavity structure would be too large. Filling the cavity with magnetic material reduces the size of the cavity. In most of the synchrotrons, thus ferrite-loaded (Ni-Zn being the standard ferrite material) cavities are used. Such a cavity can be easily tuned for a range of frequencies within a short time, making it suitable for ramped operation in a synchrotron (Kli 11).

Longitudinal stability of the ion bunch getting accelerated is ensured by operating the RF cavity within the stable synchronous phase region. The selection of a stable synchronous phase depends on the energy of the accelerating particle. In the non-relativistic regime, particles with velocities lower than the synchronous particle would take more time to circulate and thus arrive late in the accelerating cavity compared to the synchronous particle. Thus, the synchronous phase needs to be chosen on the rising part of the accelerating field to ensure longitudinal stability. When the particle energy becomes relativistic the circulating time would only depend on the length of circulating orbit. For a given magnetic field, particles with lower energy would traverse a shorter path length reaching the cavity earlier in time. In such cases, a stable phase lies on the falling part of the accelerating field. There is another component which affects the revolution frequency of the particle. In the presence of a bending magnet, a change in momentum drives the particle to revolve in a different orbit. This would induce a change in revolution frequency. This is denoted by a term called "momentum compaction" factor α. Net change in revolution frequency due to change in momentum denoted by η is given by:

$$\eta = (1/\gamma^2) - \alpha; \eta = (p/dp)(df/f); \alpha = (p/dp)(dR/R) \qquad (6.5)$$

In the above Equation (6.5), p is the momentum while f and R denotes the revolution frequency and radius of synchrotron respectively. Usually, the RF phase (Figure 6.6) of the cavity should accordingly be changed once the particle crosses the particular energy limit ($\eta = 0$), usually called the transition energy given by $1/\sqrt{\alpha}$. This is not true for electron synchrotrons since electrons are injected above the transition energy. Linacs injecting

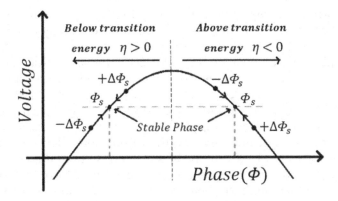

FIGURE 6.6
Synchronous phase in a synchrotron below and above the transition energy (arrows in the figure show that the phase difference with Φ s would decrease in subsequent cycles providing longitudinal stability).

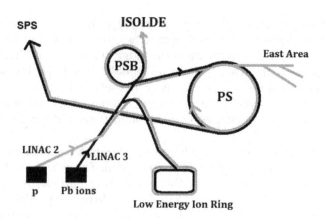

FIGURE 6.7
Schematic layout of the ISOLDE facility based on Proton Synchrotron Booster at CERN.

protons/heavy-ions into synchrotrons operate at the low repetition frequency imposed by the rise time of the synchrotron magnets.

The ISOLDE facility at CERN uses synchrotrons (PS and PSB) as the driver accelerator as shown in Figure 6.7. PSB, a strong focusing synchrotron, delivers a pulsed proton beam with 3.1×10^{13} protons per pulse with spacing of 1.2 s at 1.4 GeV. It receives proton beams from linear accelerator LINAC 2 at 50 MeV. Using the proton beam and suitable thick targets, the ISOLDE facility produces beams of exotic isotopes through fission, target fragmentation and spallation route (Bor 16).

Accelerating heavy ion beams using a synchrotron in the energy and intensity range ideal for production of RIBs through projectile fragmentation requires special design efforts. Although a synchrotron delivers a pulsed beam, compared to the CW operation of cyclotrons, the higher energy of heavy ions allows the use of thicker targets, enhancing the yields of RIBs produced in fragmentation reactions. In order to have intensity in the order of 10^{12} particles per pulse low charge states of heavy ion beams should be chosen. This would induce lower space charge tune suppression (lower de-focusing force in radial as well as longitudinal directions due to space charge) and therefore efficient acceleration through the synchrotron. Most of the present-day synchrotrons (AGS at BNL, PS at CERN, Nuclotron at JINR, SIS 18 at GSI) that are now used to accelerate heavy ion beams to GeV/u produce output beam intensity in the range of 10^7 to 10^9. The main reason for not achieving enough intensity is because these synchrotrons were initially designed for protons. Synchrotrons designed for protons typically operate at a vacuum level of 10^{-9} to 10^{-10} mbar, which does not allow injection of ions with lower charge states since the lifetime of the lower charge state is limited by charge stripping by the residual gas. It is thus necessary to inject ions with very high charge state (such as 73+ for uranium) into the synchrotron, which limits the beam current (Spi 14). The existing heavy ion synchrotrons thus suffer from a lack of suitable high intensity injectors delivering heavy ion beams and thus deliver heavy ion beams of moderate intensity (less than 1 pnA) only. The other option could be to accumulate and stack beams (operating as storage ring) from the injector before being accelerated and extracted from the synchrotron. This utilizes the high acceptance of synchrotrons, but at the cost of high emittance of the extracted beam from the synchrotron.

The synchrotron SIS 100 for the FAIR facility at GSI would be the first heavy ion synchrotron designed for high-intensity heavy ion beams for nuclear physics studies (Spi 08), where a vacuum level of the order of 10^{-11} to 10^{-12} mbar would be maintained. The acceleration

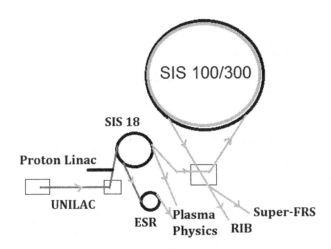

FIGURE 6.8
Layout of accelerator facility at GSI. (GSI/FAIR.© all rights reserved.)

scheme for FAIR is schematically shown in Figure 6.8. A Heavy Ion Linear accelerator called the Universal Linear Accelerator (UNILAC) is the first accelerator that accelerates heavy ion beams to almost 20% of the speed of light before injecting into the existing SIS 18 synchrotron. SIS 18 would then act as the injector delivering beams of lower charge state U^{28+} (instead of U^{73+}) which will be finally accelerated in SIS 100 to energy of 2 GeV/u with intensity of 5×10^{11}. It should be mentioned that injection of the relatively low charge state of 28+ reduces the problem associated with the space charge.

6.4 Linear Accelerators for RIB Production

Linear accelerators, as mentioned in the preceding section, are already in use in many of the RIB facilities as the front-end accelerator accelerating beams to energies suitable for injection into cyclic accelerators. Linear accelerators can be used to accelerate light as well as heavy ions all the way to the highest energy required for RIB production, ADSS, pion and neutrino production, and for cancer therapy that needs acceleration of carbon beams to a few hundred MeV/u. FRIB, the upcoming RIB facility at MSU (Wei 12, Yor 09, Ost 19), for example, will be accelerating high-intensity (~10^{13} pps) uranium and other heavy ion beams to 200 MeV/u for RIB production using only linear accelerators.

A typical acceleration scheme for heavy ions using linear accelerators usually involves both room temperature and superconducting linacs, and also linacs of different RF structures at different energies. Readers interested in learning in depth about linear accelerators and the various kinds of linac structures that are optimum for light and heavy ion acceleration at different energy regimes can refer to the book *RF Linear Accelerators* by Thomas P. Wangler, which covers practically all aspects of linear accelerators (Wang 08).

The basic scheme of a heavy ion, as well as light ion, primary accelerator using only linear accelerators is shown in Figure 6.9. Light ion acceleration does not require a charge stripper. The same scheme can be used for acceleration of unstable heavy ions (RIBs) in an ISOL-type facility by replacing the ion source in Figure 6.9 with an integrated target

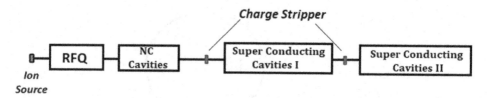

FIGURE 6.9
Schematic layout of stage-wise acceleration of heavy ions to high energy using different kinds of linear accelerators: an RFQ, normal-conducting linacs and superconducting linacs. Charge strippers are not needed for acceleration of light ions. Also, the position and number of charge strippers shown in the figure are only indicative and vary from facility to facility.

ion source for +1 charge state production of RI atoms produced in the target, followed by a low-energy beam line section comprising of magnetic separators and a second ion source to increase the charge state (charge breeder).

The ion source delivers high current, high charge state heavy ions using a state of the art ECRIS which is then accelerated, first in a Radio Frequency Quadrupole (RFQ) linac and then in a series of normal temperature and superconducting linacs. The high repetition rate/CW mode of acceleration) added to the ease with which beams can be efficiently injected, accelerated (simpler beam dynamics, less harmful resonances compared to cyclic accelerators) and finally extracted from linacs without much loss, gives it an edge over its circular counterpart as far as achieving the highest possible intensity is concerned. Since beams pass through the linac once, each accelerating structure at a different energy regime can be optimized to be most efficient for a particular particle velocity or velocity range, allowing the best possible acceleration efficiency for both light and heavy ions over the entire energy range. It should be mentioned, however, that at very high energy with ion velocity approaching the velocity of light and not changing with energy, acceleration in a synchrotron tends to be more efficient. This is because, compared to a linac which is a single pass machine, a synchrotron is a multi-pass machine. However, for acceleration of intense protons up to, say, 1–2 GeV/u and for heavy ions up to 10 GeV/u or even higher (depending upon how heavy the heavy ion is) linacs are more suitable than any other accelerator, especially when the required beam intensity is high. Thus, for RIB production and for other uses such as ADSS, production of pions, neutrinos, etc., the recent trend is to use linacs all the way, starting from keV energies up to the highest energy (usually in the range of a few hundred MeV/u to about 2 GeV/u). The price to be paid for the only linear accelerator option is the distributed footprint that increases the overall cost of the accelerator which in part is due to the enhanced cost for the shielding.

Historically, the linear accelerator was developed to overcome the limitations of the DC accelerator to accelerate beams to higher energies because of issues related to insulation which limited the maximum DC voltage that could be used. In 1928 Wideroe showed that ions can be accelerated by using a single drift tube flanked on both sides by two grounded drift tubes. Wideroe used a frequency of 1 MHz for the oscillator and applied a voltage of 25 kV to the middle drift tube to accelerate potassium ions to energy of 50 keV. For acceleration, the flight time of the ion in between the two gaps has to be equal to the time during which the oscillator phase changes by 180 degrees. The ion beam stays inside the drift tube, an equipotential zone, while the polarity of the drift tube is opposite and thus experiences no decelerating force. The length of the drift tube should be such that the ion traverses a length $L = \beta\lambda/2$, where $\beta = v/c$, v being the velocity of the ion and λ is the wavelength of the RF oscillator, from one gap to the next. Sloan and Lawrence used Wideroe's

Driver and Post-Accelerator for RIB

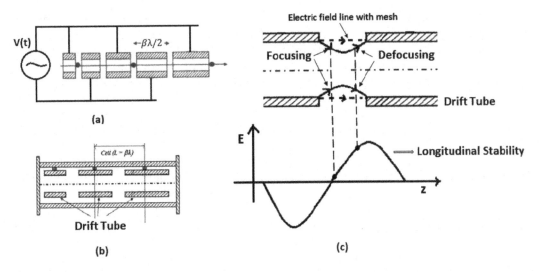

FIGURE 6.10
(a) Wideroe Linac, (b) Schematic of Alvarez Linac showing cell and drift tube, and (c) RF defocusing associated with longitudinal stability.

idea in 1931 to accelerate mercury ions to energy of 1.26 MeV at a beam current of 1 µA in a 30-gap accelerating system and using a 10 MHz alternating voltage source (Slo 31). However, to accelerate ions to higher energies, a good number of drift tubes are needed and to keep the flight time of the ion the same, as the ion gets accelerated in the gaps, the successive drift tubes should be of increasing lengths Figure 6.10a. At a low frequency of 1–10 MHz, the drift tube lengths become too long to be useful as an accelerating structure, especially for protons and light ions (for a given energy the velocity of light ions is more compared to heavier ions). Using higher frequency is obviously a solution (for reducing drift-tube lengths) but the drift tubes remain an equipotential zone only for lengths much shorter than the wavelength of the RF. At higher frequencies, the dimension of the drift tube plus the supporting stem would become comparable with the RF wavelength and the drift tubes would start behaving as antennas, resulting in high RF power losses. Thus the Wideroe structure (Wideroe DTL) is not suitable at higher frequencies exceeding a few tens of MHz.

The first proton drift tube linac (DTL), suitable for high frequency excitation, was built by Alvarez during the late 1940s. The linac was longitudinally closed at both ends by conducting metallic walls (having central holes big enough for a beam to pass through but not RF) that keep the RF fields confined within the cavity. Thus, the problem of radiation was solved, allowing use of a high frequency RF. Alvarez used a structure resonating at 202.5 MHz (the resonant frequency is determined by the diameter of the cylindrical cavity and the diameter and the distance between the drift tubes). The technological developments required to power Radar systems during World War II had made high-power high-frequency oscillators available and Alvarez made use of that. The resonant mode excited at 202.5 MHz is such that it has a longitudinal field along the axis required for acceleration. The metallic drift tubes shield the ions during the decelerating phase to ensure that the ions continue to accelerate only while crossing a gap between two consecutive drift tubes. The paper published in 1955 (Alv 55) described the linac that was 40 feet (~12 metres) in length and consisting of 46 copper-lined drift tubes suspended by stems. The linac, schematically shown in Figure 6.10a, could accelerate protons to 31.5 MeV with a current of 0.25

μA. The Alvarez structure is often termed as a "0" mode linac since the phase difference (or phase shift) between the two consecutive drift tubes is 0 or 360 degrees. In this mode there is no net current flowing through the stems holding the drift tubes. The ion flight time between any two gaps should be equal to the RF time period for synchronism and thus the distance between two consecutive drift tube gaps in an Alvarez linac is given by $L = \beta\lambda$, which should increase (proportionally) as β increases, to keep the flight time of the ion constant. It should be noted that in the "0" mode, there is neither any net charge on the drift tubes nor any net current flowing through the stems from which the drift tubes are suspended.

As a resonant structure, the Alvarez linac can be considered as comprising of cells of length $\beta\lambda$ as shown in Figure 6.10b, where each cell can be thought of as an independent cavity or a basic resonating unit. To form a cavity, the insertion of parallel conducting walls perpendicular to the axis at the middle of each drift tube needs to be imagined. In the "0" mode (TM_{010}) the presence or the absence of conducting walls would not affect the electric field direction but conducting walls, if present, would dissipate power due to eddy current loss, so its absence (as in the Alvarez linac) increases the power efficiency. In the lumped circuit representation, the drift tubes act as a capacitor ($C \sim \epsilon_0 \pi d^2/4g$; g is the gap between the drift tubes and d is its diameter) and the current path connecting the two consecutive drift tubes via the two stems and the outer wall act as the inductance ($L \sim \mu_0 \beta\lambda \ln(D/d)/2\pi$; D is the diameter of the linac tank). Since the walls are removed, there is a strong electric (capacitive) coupling between the drift tubes (cells) and the entire linac can therefore be powered by a single RF source.

It should be mentioned that instead of a multi-cell structure an alternative is to use a number of independent resonating cavities where each cavity is powered by separate RF sources that are phased independently to ensure synchronism, and consequently acceleration along the length. Such cavities, called independently phased cavities, offer a lot of flexibility in operation and are easier to fabricate but the cost and complexity of the RF system increases. The independent cavity option is often preferred for superconducting linacs accelerating heavy ions of different charge-to-mass ratios.

The Alvarez and the Independent cavities are both examples of standing wave structures. It should be mentioned that it is also possible to accelerate particles using traveling waves co-propagating with the ion beam if a resonant mode is excited which has an axial electric field. However, for practical implementation, if a cylindrical waveguide as the linac cavity is considered, it is found that in such a waveguide the phase velocity (not the group velocity) of the traveling waves exceeds the velocity of light making the phase synchronism needed for particle acceleration impossible. To bring down the phase velocity to the velocity of light or even lower, obstacles need to be created in the waveguide by loading it with discs. Thus, a cylindrical waveguide consisting of a periodic array of conducting discs with axial holes big enough to allow the beam, as well as the electromagnetic wave to travel can be used to accelerate electrons and ion beams where the particles need to match the phase velocity. In such a traveling-wave (TW) structure, the RF power needs to be fed into the structure using an input coupler at the beginning, exciting the first cavity and extracting out of the structure by using an output coupler at the end that dumps the remaining RF power in a matched load. It is comparatively simpler to use TW acceleration for electrons that come close to the velocity of light even at an energy of a few hundred keV requiring no change of phase velocity along the structure. But for low velocity (β) acceleration of light and heavy ions (proton to uranium), the standing wave (SW) mode of acceleration is largely preferred. An SW pattern in any cavity develops as a result of multiple reflections of the electromagnetic waves at the two conducting end walls of the cavity and usually takes more than microseconds to build up to its final amplitude simultaneously in all the cavities in a structure. SW structures thus fill the cavity "in time" (as opposed to TWs filling the cavity "in space" cell after cell) and confine the electromagnetic field within the volume enclosed by the cavity. An SW resonant linac cavity can resonate only at certain discrete frequencies with discrete phase changes from cell to cell and allows the cell lengths to be matched to the ion velocity. The modes are named as per the phase difference between two adjacent cells. Thus, if the cell lengths are L = $\beta\lambda$, $\beta\lambda/2$ and $\beta\lambda/4$, the corresponding modes are termed 0, $\pi/2$, and $\pi/4$ respectively. The field profiles for any structure can be calculated by solving Maxwell equations (often numerically) with proper boundary conditions. For acceleration of ions, an axial electric field is needed. The simplest solution is to excite a Transverse Magnetic (TM) mode in the cavity at the designed frequency, which provides the axial electric field. This is done in Alvarez and other types of multi-cell linacs. However, the Transverse Electric (TE) mode of excitation in which the electric field is perpendicular to the axis is used quite often for acceleration by creating an axial electric field by some intelligent trick, for example by modulation of vanes as in RFQ and by connecting drift tubes alternately to the opposite side of the resonator as in the case of IH linac. It is also possible to create an axial field for acceleration using the TEM mode of excitation (in a coaxial-type cavity) which often offers the only practical solution when the required operation frequency is low (~5 to 50 MHz), as in the case of cyclotrons. This is because the resonance frequency for TM and TE mode cavities scale inversely with the transverse dimension of these cavities, making these cavities impractically large at these low frequencies. The resonant frequency for a TEM mode cavity, on the other hand, is determined by the length and not by the transverse dimension. If one end of the cavity (coaxial line) is short and the other end is terminated by a capacitance, a length $\sim\lambda/4$ is needed for resonance. Apart from cyclotrons, TEM mode cavities are also used in superconducting linear accelerators (Quarter Wave Resonator (QWR), Half Wave Resonator (HWR), etc.) for acceleration of low β heavy ions with operating frequency in the range of 100–200 MHz.

It was mentioned while discussing synchrotrons that for longitudinal stability the ions should be accelerated during the rising phase of the RF voltage for $\beta < 1$. This makes phases of individual ions oscillating around the synchronous phase as the ions are accelerated in the successive gaps and thereby keep the beam bunched around the chosen synchronous phase. But this also leads to transverse defocusing (Figure 6.10c) at each of the gaps. It happens since an ion takes a finite time to cross the gap, during which the RF phase changes significantly. While entering the gap the ion experiences, apart from the accelerating force component, a transverse focusing force. However, towards the end of the gap the ion experiences a transverse de-focusing force. Since the RF voltage has increased in the meantime, the de-focusing force would be greater, resulting in a net overall defocusing. The transverse focusing force, being proportional to inverse of γ^2, decreases as the ion velocity increases and vanishes as the velocity of ions become equal to the velocity of light. To digress a little, it should be mentioned that the other effect of the finite gap crossing time is that the ion is only accelerated by a fraction of the peak RF voltage while crossing a gap. This fraction is called the transit time factor (T) which approaches close to 90% or even more if the gap is small so that ion takes a time, which is only a very small fraction of the RF period, to cross the gap. Of course, the gap cannot be arbitrarily small and has to be wide enough to avoid voltage breakdown or sparking between the drift tubes.

Apart from the RF defocusing force, an ion beam from any ion source has finite emittance because the beam will grow in size as it moves through linac unless there is a counter focusing force. Further, if the beam current is high, transverse de-focusing due to space charge becomes a serious issue, especially at low energies. Focusing elements are thus needed to keep the beam size under control and this is achieved by placing magnetic focusing elements like quadrupoles inside the drift tubes at regular intervals. The magnetic focusing, however, is velocity-dependent and not efficient at low velocities. DC pre-accelerators, such as a Cockroft–Walton-type DC generator are thus used, to raise the energy of the ion beam to a few hundred keV so that magnetic focusing becomes effective. The ion sources and DC accelerators deliver DC or continuous (uniformly stretched, as it were, over the entire 360 degrees of the RF phase) ion beams while linac can only accelerate a small fraction of the beam that arrives at the gaps at the right time. Thus, it is necessary to compress the DC beam into bunches of small time-width that are equally spaced in time with a time period that is equal to or an integer multiple of linac time period. This is achieved by an RF buncher placed before the linac and by synchronizing the buncher RF source with that of the linac. However, bunching increases the space charge force requiring more effective transverse focusing. The combination of Cockroft–Walton and linac proved to be a reasonably good option but only at a moderate beam current regime. Thus, for high current acceleration, there has always been a need for an accelerating structure that provides velocity-independent focusing that counteracts the space charge forces in addition to acceleration, so that it can accelerate a high-intensity ion beam at low energy. An added qualification of the new structure would be to have the capability to bunch low energy ion beams.

6.4.1 Radio Frequency Quadrupole (RFQ) Linac

The new structure was invented in 1970 by Russian scientists (Kap 70) and is called a Radio Frequency Quadrupole linac or simply RFQ. An RFQ is essentially a transverse (x and y) focusing (TE mode) structure, as shown in Figure 6.11, which can bunch as well as accelerate ion beams along the z-direction. The four rods assembled in quadrupole configuration provide velocity independent transverse focusing all along the structure. The perturbation

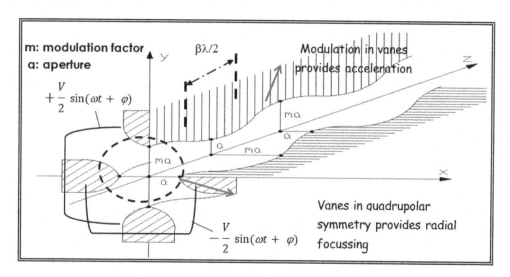

FIGURE 6.11
RFQ structure with four vanes arranged in quadrupolar symmetry facilitating focusing of the beam. X vanes and Y vanes are at opposite potential. Modulation of vanes along beam axis (Z) provides acceleration. In the figure: *a*: minimum distance of a vane from the axis, *ma*: maximum distance from the axis and *m*: modulation parameter.

Driver and Post-Accelerator for RIB

in terms of sinusoidal modulation of the vanes results in an RF field that has a longitudinal component, providing acceleration. The sinusoids of vertical and horizontal vanes are displaced by half a period ($\beta\lambda/2$) ensuring synchronism between the RF and particle required for acceleration. A modulation period consists of two cells and $\beta\lambda/2$ is called the cell length over which the RF should change its phase by 180 degrees.

Due to acceleration the β increases from one cell to the next and hence the cell length increases continuously along the structure. The acceleration takes place continuously and the RFQ has, so to say, "no drift tubes but only gaps," resulting in a constant but somewhat lower transit time factor of $\pi/4$. The modulation factor m can either be greater than or equal to 1; $m = 1$ means no modulation, in which case the RFQ becomes a focusing element only (Figure 6.11). The RFQ can efficiently provide the required focusing effect against the space charge since it is equivalent to a large number of X and Y electrical quadrupole pairs; the number being equal to half the total number of cells in the RFQ. The bunching of a continuous low-energy beam from an ion source is achieved adiabatically or slowly in a RFQ by increasing m as a function of the cell number, slowly from $m = 1$ at the beginning of the RFQ to the final value (usually around 2) in the "accelerating" section after which the modulation is usually kept fixed. So, initially while the DC beam is getting bunched the acceleration is only nominal, but once the beam is bunched it is accelerated at a constant rate in the accelerating section (if m is kept fixed). Thus, at the exit of the RFQ a beam that is accelerated, bunched and focussed is obtained, which is ideally suited for injection into a conventional linac.

In an RFQ, a part (χV) of the inter-vane voltage (V) is used for focusing (A_{01} is the focusing parameter) while the other part ($VA_{10}I_0(ka)$) is used for acceleration, where A_{10} is the acceleration parameter and I_0 is the zeroth-order Bessel function. The expressions in Equation (6.6) show how these parameters depend on the modulation parameter m.

$$\chi = \left[\frac{I_0(ka) + I_0(mka)}{m^2 I_0(ka) + I_0(mka)}\right]; A_{01} = \chi/a^2$$

$$A_{10} = \frac{m^2 - 1}{m^2 I_0(ka) + I_0(mka)} \tag{6.6}$$

From the relations, it is evident that by increasing "m," more acceleration can be acquired but only at the expense of focusing, while decreasing "a" enhances focusing. The energy gain per unit cell for a peak vane voltage of V_p is given by $qA_{10}Vp\cos[\Phi s]\frac{\pi}{4}$. The peak voltage V_p is limited by frequency-dependent maximum allowable surface field on the vane tip, given by the Kilpatric limit: f (MHz) = $1.64\ E^2.\exp(-8.5/E)$, where E is the maximum allowed surface field in MV/m beyond which sparking would become frequent. If a good vacuum and good surface condition of the vanes are maintained, it is usually possible to exceed the limit imposed by the Kilpatrick limit by about a factor of 1.5. For pulsed beam (not CW) the limit can further be pushed to about 2 times of Kilpatrick limit.

An RFQ is usually divided into four sections, namely Radial Matching Section (RMS), Shaper (SH), Gentle Buncher (GB) and Accelerator section (AC), as shown in Figure 6.12. The RMS takes care of the matching of the time-independent DC beam from the ion source with the time-dependent RFQ structure. This is accomplished, following Crandall (Cran 83) by moving away from the axis the first few cells at the entry of the vanes without any modulation and gradually narrowing down to their nominal distance. The focusing strength also increases from 0 to its nominal value. The bunching first takes place at a stable phase

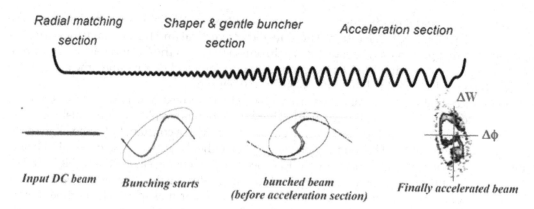

FIGURE 6.12
Longitudinal phase space dynamics along the length of an RFQ.

$\Phi s = -90°$ and the bunch is not accelerated, considering Cos[Φs] variation of electric field. The synchronous phase is then gradually shifted from $-90°$ (zero electric field, bunching phase) to $-30°$ (acceleration phase). The particles ahead of the reference particle are decelerated while the particles behind are accelerated and catch up and fall back into the bunch about the synchronous particle. Gradually, the phase is increased from $-90°$ to $-30°$ with the separatrix (which separates the longitudinally stable from the unstable motions) starting from no enclosed area (accepts the entire DC beam of phase width 360 degrees) to the bunch picking up energy resulting in a separatrix with finite energy and phase widths. A separatrix surrounding the bunch grows in height (energy difference from the synchronous particle) and collapses in length (phase difference from the synchronous particle) during the bunching process. After several phase oscillations, the majority of the injected beam is concentrated near the center of the separatrix (Figure 6.12). The ions within the separatrix are accelerated properly to the final energy at the exit of the RFQ with an energy width largely determined by the bunching process.

The choice of operating frequency in an RFQ depends on the input velocity β of the beam. If the velocity is small, a lower operating frequency needs to be chosen so that the shortest cell length (before acceleration) $\beta\lambda/2$ can be machined. The machinability restriction usually allows for protons a choice of frequency in the range 200–400 MHz, while for heavy ions the choice lies in the range 30–100 MHz. This is because β at typical extraction voltage of a few tens of kV is much higher for protons as compared to heavy-ions. Moreover, for RFQs designed to accelerate very high intensity protons, use of a higher frequency or higher repetition rate allows distribution of the charge particles to bunches (for a given average current), thereby reducing the effect of space charge. So, for high current acceleration higher frequencies are preferred. The resonating structure of an RFQ can broadly be classified into vane- and rod-type structures, as shown in Figure 6.13. In the vane type, cavity walls (equivalent to inductance) and the four vanes form the resonating structure, while for the rod type the four rods (capacitance) and the supporting posts (inductance) form the basic resonating structure with the cavity playing a rather insignificant role. Thus, at lower frequency a rod-type structure is preferred, since the cavity size can be kept small compared to those in the vane type. For a frequency above 200 MHz, vane-type structures (Sta 90) are preferred over four-rod-type structures that are used in the range of 30–100 MHz (Sch 92). Another structure, which is intermediate between the rod type and vane type, is the four-vane with windows (Del 92). This structure is optimum in the range of 100–200 MHz.

Driver and Post-Accelerator for RIB

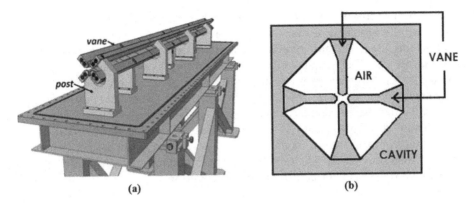

FIGURE 6.13
(a) Rod-type RFQ with post and vane; (b) Cross-section of a vane-type RFQ.

A single RFQ can also be designed to operate over a range of frequencies. The unique example of such a variable-frequency RFQ is the one operating at RIKEN, Japan (Kam 99), which injects heavy ions to the heavy ion linac, called a RILAC. The RIKEN RFQ is based on a folded co-axial structure and works over the frequency range of 17.4–39 MHz. The RFQ can accelerate heavy ions of A/q ratios ranging from 5.3 to 26.4. The variation in frequency is achieved by moving a shorting rod inside the cavity, yet the RFQ is power-efficient requiring maximum power of 30 kW at 39 MHz. Since an RFQ has so much flexibility and the same structure can provide acceleration, focusing and bunching, it is now used in almost all the accelerator facilities as the first accelerator accelerating often intense beams from the ion source. It is important to note that an RFQ is not a power-efficient accelerator since a good fraction of the RF power is spent in bunching and focusing. Also, since the energy gain/cell in an RFQ is independent of cell length ($\beta\lambda/2$), the RFQ becomes less and less efficient in terms of acceleration gradient per unit length as the energy (and the cell length) increases. Thus, an RFQ is used to accelerate ion beams only up to an energy that is suitable for injection into a conventional linac.

6.4.2 Acceleration to High Energies: Room Temperature Linacs

In a purely linear accelerator-based facility, ion beams are accelerated to the highest design energy using a series of linear accelerators. However, a linac's structure changes as the energy of the ion beam increases. As the ion gains in energy the linac can be operated at a higher frequency, which offers many advantages. It reduces the cell length, thereby allowing a compact structure and better accelerating gradient (energy gain per unit length). Also, at higher frequency the voltage allowed by the sparking limit is higher, allowing a higher acceleration gradient. Further, at higher frequency the resonant cavity would be smaller in diameter/size, which reduces the total surface area of the inner surface of the cavity over which the RF current flows. At a higher frequency, therefore, the RF surface losses thus become less, making the acceleration more power-efficient. It should be mentioned that RF power loss usually does not refer to the amount of RF power utilized by the beam (product of beam current and the energy gain) for its acceleration, which is its fundamental purpose. Instead, it refers to the power loss due to the finite conductivity of copper, of which the inner surfaces of resonating cavities operating at room temperature are made. Due to finite conductivity, the RF current does penetrate slightly inside the conducting surface and the extent is expressed in terms of skin depth which is inversely proportional to the square root of frequency and conductivity. Because of the surface losses, RF power needs to be supplied to sustain the standing wave pattern in the resonant cavity even in absence of the ion beam. The power efficiency is usually expressed by two figures of merit: the quality factor Q, which is the ratio of stored energy to the energy dissipated per RF cycle divided by 2π; and the shunt resistance Rs of the cavity given by $V_p^2/2P$, where V_p is the peak voltage and P is the power necessary to compensate for the power dissipation in the cavity walls in order to sustain the voltage V_p. To be power-efficient, a resonant cavity should be designed for maximum Q and R_s. A third figure of merit called the effective shunt impedance per unit length is also often used. This is given by ZT^2, where $Z = R_s/L$ and T is the transit time factor. The effective shunt impedance per unit length gives a measure of how well the RF power is utilized for acceleration.

A number of linac structures were invented after Alvarez and some of them are a variant of the drift tube linac. Each of these structures resulted from the effort to maximize the shunt resistance and the transit time factor, which in turn ensures high accelerating gradient (Rat 05). For details of these cavity structures, interested readers may refer to Wangler (Wang 08). The first accelerator chosen after the ion source is in recent times almost always an RFQ, since it offers a lot of advantages for acceleration of high current beams at low energy (<0.5 MeV/u for heavy ions; <3 MeV/u for protons). The next accelerator for protons is usually an Alvarez-type drift tube linac which is an open structure with quite high effective shunt impedance (~30 to 40 MΩ/m) and allows placement of quadrupoles inside the drift tubes. The Alvarez structure is optimum in the frequency range of 100–400 MHz and for acceleration of protons up to about 100 MeV. At higher β, the structure becomes inefficient, since as the length of the drift tube increases the inductance increases and to keep the frequency constant it is necessary to decrease the capacitance by increasing the gap which spoils the transit time factor. Also, reducing capacitance by increasing the gap becomes ineffective at higher β because of a change in the electric field pattern in the gap as the drift tube length increases. There are some variants of DTL structures like a cell-coupled DTL (CCDTL) and a bridge-coupled DTL (BCDTL) that can operate at higher frequencies than the Alvarez structure and with higher effective shunt impedance. The SNS facility at Oak Ridge, USA, for example, uses an Alvarez linac operating at 402.5 MHz to accelerate protons from 2.5 to 87 MeV (Ilg 03). For proton acceleration above 100 MeV, π-mode structures that are magnetically coupled through slots cut on the walls (where the magnetic field is high), are quite optimum. The Linac 4 at CERN, for example, is a π-mode structure operating at 352.5 MHz that accelerates protons from 100 to 160 MeV (Ger 09). The linac comprises short cavities with seven cells in each. The cell length inside one cavity is fixed since β changes slowly at these high energies and the phase slippage is small, but the cell length changes from one cavity to the next.

It should be mentioned that long linac structures (powered by the same RF source to reduce the cost) comprising of many cells (say, >30) operating in the 0 or π mode are extremely sensitive to mechanical errors and vibrations. This is because in these structures there are resonant modes (normal modes) with frequencies lying close to the desired mode. The mechanical errors in cell dimensions would result in an electric field modified by electric fields of these perturbing modes. Also, because of finite Q value the chosen mode in a resonant cavity would itself have a natural width given by $\Delta\omega = \omega/Q$. If this width is larger than the frequency difference with the nearest adjacent mode, the modes will overlap distorting the electric field. It can be shown that the number of allowed adjacent modes is equal to the number of cells; the number of cells in a linac operating in 0 or π mode thus needs to be limited. The solution is to use $\pi/2$ mode for long structures, known as coupled cavity linacs (CCL), in which the frequencies of adjacent modes are quite far apart from the frequency of the designed mode. In $\pi/2$ mode all the cavities/cells are not excited and for better accelerator efficiency the unexcited cells can thus be removed from the beam axis and used as coupling cells. That is what is done in a Side Coupled Linac (SCL), a variant of CCL, shown schematically in Figure 6.14. A CCL works efficiently in the frequency domain from 700 MHz to about 3 GHz and can be used to accelerate protons to very high energies (β = 0.5 to 1). For example, the LANSE linac facility at Los Alamos uses a 735 m long-SCL operating at 805 MHz to accelerate protons from 100 to 800 MeV and the CCL at Spallation Neutron Source (SNS) operating at the same frequency can be used to accelerate protons from 87 to 186 MeV.

For acceleration of heavy ions (also protons) at low energies, H-type linac structures (Rat 05) that operate in TE mode offer considerable advantage and are often chosen to accelerate low-energy ions after the RFQ. This is because these structures are quite compact and have much higher shunt impedance (~200 to 400 MΩ/m) compared to other structures (~30 to 40 MΩ/m). In TE mode excitation, compact drift tubes with less capacitance can be used and due to the π mode of operation the transit time factor is higher. Further, the magnetic field is less near the cavity walls which results in lower surface current and consequently in less RF losses at the walls. As such, TE mode does not have any field in the axial direction required for acceleration but the longitudinal field can be induced by connecting alternate drift tubes to opposite sides of the resonating cavity as shown in Figure 6.14. The resulting structure, ideal for operation at lower frequencies from about 70 to 200 MHz, is known as the inter-digital H mode structure or simply the IH linac. There is another structure called the cross-bar H-mode structure or the CH linac in which the axial field is created by holding alternate drift tubes by a pair of stems at right angles. The CH linac structure is shown in Figure 6.14. This structure is optimum at a higher frequency (300 to 700 MHz; shunt impedance ~50 to 100 MΩ/m) and when the protons/other ions are already accelerated to about one-fifth the velocity of light.

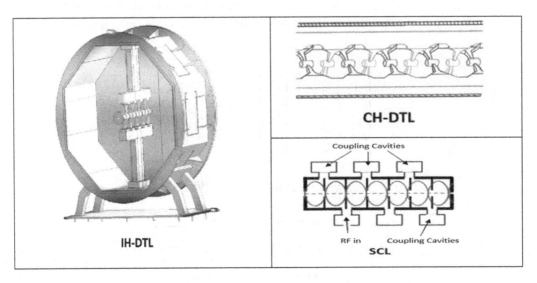

FIGURE 6.14
The IH, CH and SCL (CCL) linac structures.

6.4.3 Acceleration to High Energies: Superconducting Linacs

Above 100 MHz, the cavity dimensions become small enough for use of niobium-based superconducting cavities. Superconducting cavities often turn out to be a better choice in terms of achievable acceleration gradient, especially for CW operation. Since the surface resistance of copper at room temperature is 10^5 times that of niobium at 4 K, the Q value in superconducting cavities is of the order of 10^5 times greater. In superconducting cavities, the surface losses due to the RF current is very small but not zero since there are always some normal-conducting unpaired electrons. The RF power is thus mainly utilized to impart energy to the beam. The normal-conducting cavities are designed to have higher shunt impedance and a more concentrated electric field profile in the accelerating gaps. This dictates that the cavities have a reduced aperture of the drift tubes. But the design considerations are different for superconducting cavities. Instead of aiming to get maximum shunt resistance, which is already very high, an attempt is made to optimize cavity shapes to optimize (minimize) two ratios: ratio of peak surface electric field and accelerating field (E_{pk}/E_{acc}) and the ratio of peak magnetic field and accelerating field (B_{pk}/E_{acc}). These result in cavity shapes that allow a larger aperture through which the beam can pass without any loss. The peak magnetic field B_{pk} should be well below the critical field, above which a superconducting niobium cavity can quench and undergoes transition to its normal state. Normally it is kept below 0.1 Tesla in superconducting linacs. A high surface electric field would result in field emission of electrons and these electrons absorb RF power to contribute to energy loss. This increases the load on the cryogenic system that delivers liquid helium at 2 or 4 K. In principle, compared to superconducting linacs, a higher acceleration gradient can be achieved in room-temperature linacs but high surface losses make the cooling of copper cavities (by low conductivity water) extremely difficult. So, a higher acceleration gradient in normal-conducting linacs can only be achieved for a pulsed beam with lower average power.

Superconducting RF cavity structures are grouped in three classes depending on the velocity of ions: low, intermediate and high β. The structures mostly used are quarter wave, half wave, spoke cavities and elliptical cavities (Figure 6.15). While the first three are TE mode cavities, the elliptical cavity operates in TM mode. The operation regime

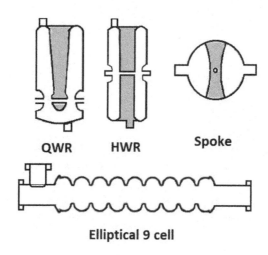

FIGURE 6.15
Superconducting Cavity structures.

for the above cavities in terms of frequency range and velocity of ions are listed in Table 6.3 (Kel 12).

QWRs have been used as heavy ion linac boosters for decades in the low β regime. They have a co-axial transmission line of length $\lambda/4$ shorted at one end where the accelerating gaps are located. The structure has inherent asymmetry in an electric field, which produces a vertical steering of the beam. The steering can be compensated using different techniques such as off-axis injection of the beam and introducing a tilt angle to the beam ports and drift tube (Fac 11). QWRs are constructed mostly as a double-gap structure; increasing the number of gaps reduces the velocity acceptance, or in other words the transit time factor drops sharply for small deviations from the designed beta.

An HWR can be thought of as two QWRs facing each other and is a ($\lambda/2$) resonating transmission lines short at both ends. The symmetry of structure cancels the steering inherent in QWRs and can be used for $0.1 < \beta < 0.5$. Detailed discussion of SC RF structures and design can be found in reference (Pad 14).

For $\beta = 0.5$, multi-gap spoke cavities are used. A spoke cavity is basically a half-wavelength resonant transmission line (spoke) operating in TE mode. The cell-to-cell coupling is magnetic in nature through large openings. The use of a spoke resonator can be further extended with no clear-cut transition from spoke to elliptical cavity. The spoke cavity haa an outer diameter of 0.5λ and are thus smaller than an elliptical cavity.

Elliptical ($\beta = 1$) cavities operating in π-mode are used for electrons, positrons or for protons at the relativistic energies. Acceleration can be achieved for $0.5 < \beta < 0.9$ by compressing the dimensions of the elliptical resonator cavity. SNS, as an example, accelerates protons from 200 to 600 MeV using $\beta = 0.6$ elliptical cavity (Kim 08).

The transition energy from a normal to superconducting structure needs to be carefully considered. While for a CW beam the transition to SC cavity can take place at lower energy, a pulsed beam the transition occurs at higher energies. This is because in the case of the pulsed beam, the duty factor for RF and cryogenic systems almost doubles, requiring more wall-plug power (Pod 13). This is a consequence of high Q of the superconducting cavities which implies a high stored energy that needs to be stored in the cavity before the arrival of the beam. Due to the finite filling and decay time of the RF field in the cavity, the RF pulse length has to be much longer than the beam pulse length leading to a waste of energy while the beam is not there. The wasted RF power needs to be removed by cryogenic cooling, requiring more wall power. Figure 6.16 shows the transition energy from normal to superconducting structures at different facilities classified in terms of duty factor.

Acceleration of electrons in a linac is comparatively simple since electron velocity reaches about 80% of the velocity of light at the energy it is injected from electron guns. So, after a little bit of acceleration, say to about 2 MeV, the cell length does not need to increase for electrons. This is in sharp contrast to ion acceleration where the cell length needs to increase by a factor exceeding 100, requiring different accelerating structures operating at

TABLE 6.3

Parameters of Operation for Different Superconducting Cavities

Cavity Structure	Particle Velocity (β = v/c)	Frequency (MHz)
Quarter Wave (QWR)	$0.001 < \beta < 0.2$	$30 < f < 160$
Half Wave (HWR)	$0.1 < \beta < 0.5$	$70 < f < 350$
Spoke	$0.2 < \beta < 0.7$	$230 < f < 780$
Elliptical	$0.5 < \beta < 0.9$ and $\beta = 1$	$170 < f < 3000$

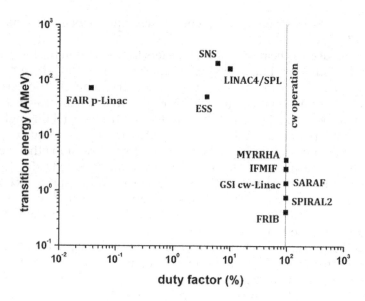

FIGURE 6.16
Transition energy from NC to SC structure for different duty factors (Pod 13).

different frequencies and different standing-wave modes. One great advantage of accelerating electrons to the highest energy using a linac is the absence of synchrotron radiation that limits the highest energy in a cyclic accelerator because of power loss. For the production of neutron-rich RIBs, electron accelerators with electron energy in the range of 30–70 MeV are used where the photons (γ) produced by stopping the electrons are used to induce fission in actinide targets. The nine-cell Tesla cavities developed for linear colliders are most often used for electron acceleration. These cavities are also suitable for the acceleration of protons at high energy. As mentioned earlier, electrons with $\beta \sim 1$ can be accelerated in TW cavities. The typical operating frequency of normal temperature TW cavities lies in the range of 2–12 GHz.

6.5 Beam Acceleration and Charge Stripper

Charge stripping is unavoidable for the acceleration of heavy ions to high energies. Charge stripping results in a higher charge state of the ion, thereby reducing the total acceleration voltage required to attain a given final energy in the ratio of the charge states. Higher charge states also make the ion beams less rigid, making the design of bending magnets (also accelerator main magnet for cyclic accelerators) simpler. Thus, charge stripping reduces the cost of acceleration greatly and brings it down to a viable level. Charge stripping becomes necessary for heavy ions since it has so far been impossible for ECR and EBIS to produce fully or close to fully stripped ions with charge state q approaching atomic number Z. Further, extracted current from any ion source decreases sharply for high charge states. As, for example, for uranium it has only been possible so far in a 28 GHz ECR ion source at RIKEN to produce charge state of 35+ with an intensity of about 10^{13} pps (Higu 12). Since 35+ is smaller by more than a factor of 2.6 compared to 92+ there is huge scope to reduce the total acceleration voltage by using charge strippers.

The charge stripping, however, has a major demerit: it inevitably leads to loss of beam. At each stripping stage roughly 80% of the beam intensity is lost. Increasing the number of strippers at different stages of acceleration would therefore mean serious compromise in the beam intensity of the driver accelerator. The loss of intensity happens because the stripping process produces a number of higher charge states distributed around an equilibrium charge state. The post-stripper maximum fraction of a single charge state is typically about 20%. Since accelerators usually accept a single charge state for acceleration, the rest of the ions with charge states different to the chosen one would not be accelerated. Producing the highest possible charge state after stripping with a significant fraction is one of the most important requirements and both the equilibrium charge and the charge state distribution depend upon the material and on the ion velocity before the stripper, but not on its charge state. Usually, low Z element strippers like thin foils of C and Be and liquid Li produce the highest charge state. Gases like He and N_2 are also used as strippers but they produce lower final charge states compared to metals, since metallic strippers have higher density. The stripper should be just thick enough for charge equilibration. This is because higher thickness would increase the beam straggling. Beam straggling results in an increase of both the longitudinal and transverse emittance and might lead to acceleration losses if it exceeds the limit dictated by the acceptance of the downstream accelerator. Any non-uniformity in the stripper's thickness or gradual reduction of thickness due to beam irradiation would have the same detrimental effect. Further, from an operational point of view the lifetime of the stripper under heavy ion irradiation is one of the most important considerations. Usually, carbon is used as the stripper since it can withstand high temperatures and very thin self-supporting carbon foils can easily be made. Also, carbon produces the highest charge states after stripping compared to gas strippers. However, the lifetime of carbon foils under heavy ion irradiation is often short (Oku 14) and becomes a serious issue at higher intensity and as the ion becomes heavier (dE/dx increases, increasing the total power dissipation in the stripper foil). Thus, gas strippers are often used for longevity although gas strippers lead to lower charge states after the stripper. Extensive R&D is being carried out for development of strippers, which includes gas strippers as H, N and He gas strippers, the liquid Li stripper and the rotating Be disc stripper (Oku 14).

Detailed calculations are available for charge distribution depending on beam energy and species for C foil as stripper (Shim 92). Using a large set of experimental data, a fitting formula, called the Schiwietz formula, has been derived to calculate the mean equilibrium charge state of heavy ion beams at different energies for different strippers (Schi 01). The mean equilibrium charge state for uranium beam using C stripper is shown in Figure 6.17. The figure also shows the charge strippers that are used/planned for uranium beam acceleration at different energies for the FAIR project at GSI, the RIKEN RI Beam Factory and the FRIB project at MSU. In FAIR, it is planned to use a single N_2 charge stripper at 1.4 MeV/u to increase the charge state from 4+ to 28+ for further acceleration of uranium beam to 1.5 GeV/u using synchrotrons. The goal is to achieve a beam intensity exceeding 10^{11} pps. For pulsed operation, the space charge forces tended to become unmanageable unless a comparatively lower charge state of 28+ was chosen. It should be noted that at 1.4 MeV/u, the equilibrium charge state in the case of carbon stripper is much higher (~40+). The RI Beam Factory (RIBF) facility at RIKEN uses two charge stripping stages for acceleration of uranium, xenon and other very heavy ion beams. The uranium beam from a 28 GHz ECR ion source with a charge state 35+ is accelerated first in linear accelerators comprising of a RFQ and the RIKEN heavy ion linac (RILAC), and then in RIKEN Ring Cyclotron (RRC) to reach an energy of 11 MeV/u (Oku 14). At this stage the beam is charge stripped to 64+ (using helium stripper) or 71+ using carbon stripper. The beam is then accelerated to

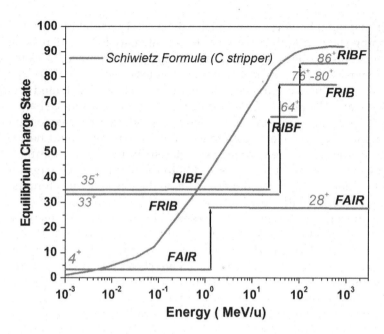

FIGURE 6.17
Equilibrium charge state of U beam (with C foil stripper) at different MeV/u using the Schiwietz formula. Stripping stages at different facilities are also shown.

51 MeV/u using fixed frequency Ring Cyclotron (fRC). After fRC, the second charge stripper (C foil/rotating Be disc) is placed to increase the charge state to 86+. The beam is then accelerated to 127 MeV/u in an Intermediate Stage Ring Cyclotron (IRC) and then finally in a K 2600 Superconducting Ring Cyclotron (SRC) to its final energy of 345 MeV/u. The stripping efficiency for the first stripper has been found to be about 18% and that for the second one 28%, resulting in an overall efficiency of 5%. Of course, there will be some loss of beam intensity in each of the six accelerators (four cyclotrons and two linear accelerators). The total efficiency of acceleration is, at present, about 0.2% for uranium, which means a target beam intensity of 2.10^{10} pps at 345 MeV/u on the production target.

FRIB plans to use only linear accelerators (Wei 12, Yor 09). To increase the intensity two charge states 33 and 34+ are injected from ECR into an RFQ capable of accelerating both the charge states. Dual charge acceleration is possible since linear accelerators (the RFQ in this case) being single-pass machines have wider momentum acceptance and as the charge states extracted are quite high (33 and 34+), the momentum difference is about 1.5% in this case. The dual charge state beams after the RFQ are then to be accelerated in a series of independently phased heavy ion superconducting linacs (QWRs) from 0.5 to 17 MeV/u. At 17 MeV/u, a liquid Li stripper will be used to increase the charge state to 78+ on average and using a series of superconducting HWRs the beam will be accelerated to 200 MeV/u. In HWRs, after the stripper, four charge states from 76 to 80+ will be simultaneously accelerated. This requires HWRs to be fine-tuned (Ost 19) for acceptance of a momentum width of ~2.6% (±1.3% with respect to the mean momentum for charge state of 78+). This can be achieved by sacrificing a bit of the acceleration gradient. Using multi-charge acceleration, as high as 80% of the initial uranium beam can be accelerated in principle in FRIB, delivering beam intensity (projected) of 10^{13} pps at 200 MeV/u.

Charge stripping is more effective for cost reduction if stripping can be introduced at lower energy. Also, while accelerating short-lived RIBs in an ISOL-type RIB facility, the short half-life of the isotopes limits the charge breeding time and hence the maximum charge state that can be extracted from the charge breeder. So, in cases of RIB acceleration a stripper at lower energy needs to be introduced. However, at lower energy the charge state distribution is much wider, and the mean charge state after the stripper is also much lower. So, linacs need to have much wider momentum acceptance to accelerate multiple charge states at lower energy. Simulation shows that in general it is possible to accelerate a number of charge states if the stripper is introduced at low energy—for example at 0.4 MeV/u in the ISAC II facility, after which acceleration is carried out in superconducting QWRs up to about 6.5 MeV/u. However, intensity can be gained at the cost of sacrificing to a considerable extent the acceleration gradient and the longitudinal emittance which increases by a good few times (Lax 02). As such, the scheme can work up to a few MeV/u and is not suitable for acceleration to high energies.

It is shown that Asymmetrical Alternate Phase focusing (A-APF) realized in a series of independently phased superconducting resonators (QWRs) has the potential to handle multi-charge acceleration at lower energy much more effectively (Dec 13). In this structure, the synchronous phase is gradually varied from the positive phase to the negative phase over a period of accelerating gaps. This would provide a net focusing effect both in the longitudinal and transverse directions. Such a structure can be quite compact with no or less need for additional transverse focusing elements. Phase variation can be realized by changing the drift length between the gaps of a single cavity or by changing the phase of independently phased cavities with one or two gaps. Study (simulation) of A-APF structure realized in a sequence of 36 independently phased superconducting quarter wave resonators (QWRs) has been shown to accelerate almost 81% of input uranium beam at 1.3 MeV/u before the foil stripper to energy of 6.2 MeV/u. A-APF structures have comparatively larger transverse and longitudinal acceptance. This in turn ensured almost ten charge states from 34+ to 43+ to be accelerated simultaneously with the phase of resonators tuned for 34+ (Dec 14). Further, the normalized transverse and longitudinal emittance remains within acceptable limits compared to that of a tuned charge state of 34+. It should be mentioned that accepting and accelerating multiple charge states for uranium would result in minimal radioactive contamination of accelerating volume, apart from enhancement in intensity of driver beam.

It should be mentioned that Alternate Phase Focusing (APF and not A-APF where the sign of the synchronous phase is reversed at each gap) has been realized in a normal-conducting IH-DTL linac for muon acceleration by changing the drift length between the gaps (Ota 16). The accelerator consists of 16 gaps with no transverse focusing elements, accelerating muons to $\beta = 0.28$. The output beam satisfies the experimental requirement of having a low emittance beam.

Using a stripper at low energy would also require a second stripper stage at higher energy, say at around 17 MeV/u as in FRIB. But this stripper would not lead to any significant loss in the intensity because of favorable charge distribution requiring narrower momentum acceptance of A-APF compared to that at low energy. Further, stripping at lower energy allows the extraction of a much lower charge state beam from ECR (say 10+) with an order of magnitude higher intensity (compared to 34+). Thus an A-APF structure or A-APF mode of phase tuning has the potential to deliver heavy ion beams (uranium and other heavy ions) with an order of magnitude higher intensity at the highest energy. Further, as has already been mentioned, for acceleration of short-lived RIBs where getting a high charge state from the charge breeder is rather difficult, the A-APF structure offers a unique way of acceleration with a minimum loss in intensity.

6.6 Post-Accelerators for Acceleration of RIBs in ISOL Facilities

Acceleration of low intensity RIBs is required only in ISOL-type and PFS-ISOL combined-type facilities. The accelerators remain the same, but the design emphasis changes. First of all, because of low intensity the space charge does not create any problem in RIB acceleration. Secondly, the charge state of RIBs injected into the first accelerator at low energy is lower compared to a stable ion case since the breeding time in a charge breeder is limited by the half-lives of the short-lived RIBs. Thirdly since short-lived RIBs are produced with very small intensity, it is of utmost importance to minimize the acceleration loss.

Elemental dependence of a negative ion source restricts the applicability of Tandem DC as the post accelerator for RIB facilities. Although HRIBF used a 25 MV Tandem as the RIB post accelerator (Mei 96) before being finally closed in 2012, DC acceleration is not considered a suitable option in any versatile RIB facility. Cyclotrons, being compact, can be used as a post-accelerator. Cyclotrons operated at different frequencies and harmonics can satisfy the required energy variability of accelerated exotic species. Further, a cyclotron is an excellent mass separator. This can compensate for a low resolution separator (in order not to lose the intensity of the precious RIB of interest) usually designed after the high emittance source such as an ECR ion source. The revolution frequency in a cyclotron is dependent on q_0/m_0, the charge to mass ratio of the ion. Thus, ions with a different charge-to-mass ratio would arrive at a different phase of RF field and would be lost before reaching the final extraction radius. The mass resolution of cyclotron is given by:

$$R = \frac{\Delta(m/q)}{(m_0/q_0)} = \frac{1}{2\pi hN} \tag{6.7}$$

Where h is the harmonic number and N the number of turns required to achieve final energy. Increasing the harmonic number would reduce the final energy of the extracted beam, needing a lower gap voltage, which in turn means poor transmission and extraction efficiency. Similarly, increasing N also demands a lower voltage for particular extraction energy. This again would reduce the extraction efficiency due to a decrease in turn separation. Cyclotrons as post-accelerators are used at Louvain-la-Neuve and GANIL. Cyclone $K = 110$ is the post-accelerator for the ISOL-based RIB facility at Louvain-la-Neuve, Belgium, accelerating RIBs from an ECR ion source to energies starting from 600 keV/u to several MeV/u. Another option is to accelerate RIBs using CYCLONE 44 to an energy of about 200 keV/u and provide isobar separation with $\Delta m/m \sim 10^{-4}$ (Gae 02). The CIME cyclotron ($K = 256$) at GANIL is used as the post-accelerator for accelerating RIBs extracted from the ECR ion source up to an energy of 25 MeV/u for light heavy ions. The mass separator preceding the cyclotron has a mass resolution of $\Delta m/m \sim 10^{-3}$. The cyclotron can separate isobars with a resolution of 5×10^{-4}. Additional separation can be achieved by using a stripper foil but that would lead to a loss of RIB intensity. By tagging beam particles with the RF signal of the cyclotron, isobars can be separated based on the timing of the events. This would increase resolution to 2×10^{-5} (Vill 01).

However, in a cyclotron the acceleration efficiency is inferior compared to linacs because of injection and extraction losses. Further, final attainable energy in a cyclotron is limited by the high charge state of the RIB that can be injected. The second problem can be overcome by injecting a high charge state beam into a separated sector cyclotron after initial acceleration in a linac and using a charge stripper. The charge stripper would induce loss of beam intensity but acceleration efficiency in a separated sector cyclotron is close to 100%, compared to about 25% in a normal cyclotron with axial injection at low velocity, compensating the intensity loss due to stripping. So, cyclotrons remain a good choice for acceleration of RIBs and are very competitive if used in combination with a linac pre-injector.

Linear accelerators on the other hand can provide more efficient transmission of precious RIBs to higher energies. But linear accelerator facilities would require carefully designed mass separators involving an RFQ cooler to reduce emittance; and higher resolution is achieved at the cost of loss in the beam intensity. ISAC I and II at TRIUMF-RIB are linear accelerator facilities used for accelerating RIBs to Coulomb barrier energies (Lax 14). The post-acceleration scheme starts with 35.4 MHz RFQ with a post and split ring structure which is a variant of a rod-type structure. It accelerates beams of $A/q \leq 30$ from 2 keV/u to150 keV/u. After the RFQ a carbon stripper is used to increase the charge state ($A/q \leq 6$), and subsequently the beam is accelerated to any energy up to 1.5 MeV/u using five room-temperature IH accelerating cavities resonating at 106.1 MHz. Further acceleration of the RIB is achieved in a series of SC-QWRs, operating at 106.1 MHz (Phase I) and 144.44 MHz (Phase II). The SC QWRs add a maximum accelerating potential of 40 MV. All the linacs operate in CW mode which can preserve RIB intensity. The post-acceleration of RIBs at ISOLDE also involves an array of linear accelerators (Kad 17). As has already been mentioned, acceleration of RIBs to the highest energy using linacs would also involve more than one stripping stage but the stripping losses can be minimized using an A-APF mode of acceleration.

7

Experimental Techniques

7.1 Introduction

This chapter deals with the experimental techniques used for studying the properties of exotic nuclei. The nuclear and atomic techniques used for measurements of the properties of exotic nuclei will be discussed. The experimental techniques that have been used are numerous and often experiment-specific; and all the techniques of experimental nuclear physics that are employed for studying β-stable nuclei and nuclei close to stability are also used in the case of exotic nuclei. However, experiments on exotic nuclei and using ion beams of exotic nuclei (RIBs) involve special challenges in production, separation and measurement. The production aspects have been discussed in Chapters 3, 4, 5 and 6. The separation of exotic nuclei, which constitutes the first and a significant part of the planning of an experiment and which has an important bearing on the planning of the rest of the experiment, will be discussed in some detail in this chapter. In the measurement part we confine our discussions mainly to the basic techniques used for determining the ground state properties of exotic nuclei: mass, ground state spin and static moments, size and matter distribution, and β-decay half-lives. Among the techniques for studying excited states, we mainly discuss the Coulomb excitation and dissociation methods. A few techniques for measurements of reaction cross-sections of importance to nucleo-synthesis are also discussed. The experimental efforts initiated for the study of atomic Electric Dipole Moment (EDM) using RIBs are briefly discussed as the last topic in this chapter.

7.2 Separation of Isotopes in ISOL- and PFS-Type RIB Facilities

Separation of the isotope of interest from the rest of the reaction products, whether in an ISOL-type facility or in a PFS-type facility, is always achieved by using di-pole magnets, which are basically momentum analyzers that bend the ion beams of different momenta in different curved paths and the isotope of interest is selected using a slit (or a number of slits after each di-pole magnet if more than one di-pole magnet is used to achieve the desired separation) that stops all other isotopes except the one of interest, which it allows to pass through for identification and subsequent measurements of its properties. In an ISOL-type RIB facility, radioactive atoms produced at the target are first converted into ions in an ion source and the extracted ions of different isotopes, having the same charge state and the kinetic energy, are then separated using one or multipole di-pole magnets that separate the ions according to their momentums (masses). The undesired reaction product ions can be neighboring isotopes (same Z, different A) or isobars (different Z, same A) or isotones (different Z, same N). The separation of isotopes (and also isotones) in ISOL systems requires only moderate resolving powers; say for separating $A = 100$ and 101, a resolving power exceeding 100 is required. However, the separation of isobars is by no means an easy task since it requires high resolving power in the range of 10,000 to about 100,000. Just to give an example, the mass difference between the two isobars ^{132}Sn and ^{132}Sb is about 3 MeV ($Q_{\beta-}$). The required resolving power to separate the two isobars would therefore be roughly M (^{132}Sn)/$Q_{\beta-}$ ~ 39,000. This is a very high resolving power which is extremely difficult to achieve without heavily sacrificing the beam intensity of ^{132}Sn since a very small opening for the mass selection slit placed at the position of the focus after the di-pole magnet will be required. As the exotic nuclei are produced in limited intensity, loosing major fraction in the slit would be detrimental for conducting meaningful measurement of their properties. The requirement on resolving power for isobar separation, however, varies from lighter to heavier nuclei and with the exoticity of the isotope of interest. It is lower by a few factors for lighter nuclei and for more exotic species since for lighter nuclei the mass is less and for more exotic nuclei the mass difference (Q_β) increases; usually for lighter exotic nuclei a resolving power of ~10,000 is needed.

However, the requirement on resolving power calculated in this way is only valid if the isotopes/isobars to be separated are produced with equal or comparable intensities/cross-sections. This is often not the case for exotic nuclei that are produced with much lower cross-sections compared to neighboring isotopes and isobars and this puts a further demand on the resolving power. The effect of an unequal rate of production on the required resolving power is schematically illustrated in Figure 7.1. Here spatial distribution of two neighboring isotopes, after they are separated using a di-pole magnet, are shown at the focal plane (position of the isotope selecting slit) after the di-pole magnet. Two cases are considered: (i) when the isotope of interest and its neighboring isotope both have same intensity at the focal plane and (ii) when the isotope of interest is produced with three orders of magnitude lesser intensity than its neighbor from which it is to be separated. In the example, the spatial distribution of both the isotopes is taken to be Gaussian, having the same FWHM (2.35 σ) at the slit position, and it is assumed that the di-pole magnet can effect a separation of 6 σ

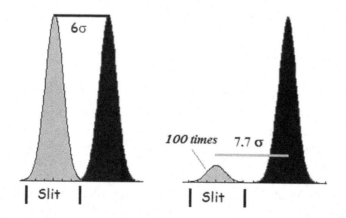

FIGURE 7.1
Spatial separation at selection-slit position necessary for achieving the same degree of separation between the isotope of interest and its neighboring isotope when: both are produced with the same intensity (left); and the neighboring isotope is produced with 1000 times relative intensity (right). In the figure to the right, the intensity of the isotope of interest is amplified 100 times to make it visible.

Experimental Techniques

between the two peaks. The slit after the magnet is placed to select the isotope of interest while rejecting the contaminant. A slit width of ±3 σ will allow more than 99% particles of the isotope of interest, but will also at the same time allow, in the first case, a small fraction of about 0.13% of the neighboring isotope (contaminant). This may be assumed to be tolerable in a particular experiment. However, in the second case where the isotope of interest is produced with 1/1000th of the intensity compared to the neighboring isotope, for every 100 particles of the isotope of interest, 135 contaminant particles will also pass through the slit which is clearly unacceptable. To achieve the same level of purity in the separation as in the first case, the separation between the two peaks in the second case has to be about 7.7 σ. So, for separating nuclei of interest, in the second case a much larger separation (~1.3 times) is needed between the peaks of the two nuclei at the focal plane of the di-pole magnet. The separation or the distance between two peaks is expressed in beam optics by a quantity called dispersion (D) which is defined as the distance between two peaks (in the dispersive plane at the image position) differing in momentum by 100%. In beam optics, the momentum resolving power R (which is twice the mass-resolving power for the same kinetic energy of all ions), is defined, in terms of dispersion as:

$$R = (D)/\text{image size; (image size = magnification} \times \text{object size)} \quad (7.1)$$

As already mentioned, the separations in PFS-type facilities are also achieved using di-pole magnets. However, the ion optical considerations become different because of the different initial conditions of the ion beams. In this case, because of high energy, a large fraction of ions is almost fully stripped ($A/q \sim A/Z$; which means ions of different elements will have different charge states) and have the same velocity but they have a large initial momentum spread because of the reaction mechanism and the finite target thickness. The large spread in the momentum and different charge states for isotopes of different elements makes the separation more challenging in the PFS technique and to achieve adequate selection at least two di-poles with a profiled degrader placed between the first and the second di-pole is required. The first di-pole selects fragments as per the A/Z (A/q) ratio for fully stripped ($q = Z$) fragments. However, because of the momentum width of the projectile fragments and also the presence of multiple charge states, the separation in the first di-pole is hardly sufficient and the first di-pole allows many different nuclei after the momentum selection slit. The degrader placed after the slit introduces a Z and A dependent energy loss in the fragments and thus the second di-pole can effect a much cleaner separation. If the fragment energies are less than a few 100 MeV/u, the combined effect of two di-poles with the degrader in between results in a separation of fragments where isotones are more difficult to separate than the isobars, unlike in an ISOL facility where isobaric separation is extremely difficult. In the sections to follow, the separation in ISOL and PFS techniques are discussed in somewhat more detail.

7.3 Isotope Separation in ISOL-Type RIB Facilities

As already mentioned, the separation of the reaction products is achieved using di-pole magnet(s) placed downstream of the ion source. The ion source and the extractor decide the initial conditions of the ion beam (as given by the emittance). The beam from the ion source contains a cocktail of different reaction products from which the ions of the isotope of interest need to be separated in the ion optical system comprising of the di-pole and quadrupole magnets for focusing, etc. The initial ion conditions (such as ion beam size, divergence, energy spread) are very important for designing the ion optical system and usually more than one di-pole magnet is necessary to achieve just enough resolving power. The initial conditions also often restrict the resolving power that is achievable and necessitate the use of devices to improve the ion beam quality (to reduce emittance, for example) to achieve the desired resolving power.

The dispersion of a di-pole magnet is given by $D = \rho(1-Cos\varphi)$, where ρ is the radius of curvature and φ is the bending angle of the di-pole (Figure 7.2). Maximum dispersion is thus obtained for a bending angle of 180° for a large radius of curvature. However, it is necessary to ensure that the beam, while traversing a long path in the magnet, does not blow up to hit the pole pieces. In the absence of any vertical focusing force, the ion beam from the ion source, having an initial emittance, will blow up and eventually hit the pole pieces above and below (say, in the vertical plane). Moreover, even the ions of interest of a particular mass-to-charge ratio would have some initial energy width and the large dispersion would lead to an increase in beam size in the dispersive plane (horizontal plane) within the magnet and the beam might hit the side walls of the yoke. Thus, a magnet with a large pole gap and large pole width is needed. Such big magnets are difficult to machine accurately and inaccuracies in machining limit the attainable resolving power. This is because the imperfections in the magnet would increase the aberrations in the image and would ultimately lead to an increase in the image size reducing the resolving power. All these factors can be better addressed in magnets with shorter path lengths $(l = \rho\varphi)$.

It is important to note that the separation distance between two beams of different momentum or rigidity at the focus after a di-pole depends not only on the dispersion (($D = \rho(1-Cos\varphi)$)) as given by the expression, but also on the beam optics of the entire separator stage from the source to the slit. This is because any ion optical system for mass separation comprises of, in addition to a di-pole, a number of quadrupoles for transverse focusing (in X and Y, assuming Z to be the direction of the ion beam). Quadrupoles before a di-pole magnet take care of divergence of the ion beam from the ion source and are used to ensure the vertical focusing necessary to keep the beam size in Y small enough so that the beam passes through the di-pole magnet without hitting its poles. However, if the magnetic fields of the quadrupoles, for example, are adjusted

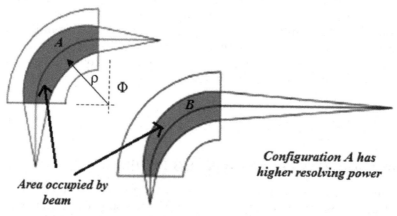

FIGURE 7.2
Dependence of resolving power on the area occupied by beam inside the sector magnet.

Experimental Techniques

such that the beam crosses in the dispersion plane (perpendicular to the magnetic field direction) exactly half-way inside the di-pole, the dispersion produced by the first half of the di-pole magnet gets canceled by the dispersion of the second half, resulting in an achromatic transport. Thus, proper settings of the quadrupoles are necessary to achieve the desired dispersion. Quadrupoles after the di-pole magnet are usually tuned to ensure proper image formation at the position of the mass selection slit. The tuning of the quadrupole–di-pole–quadrupole combination decides the final dispersion at the position of the slit. The combination may also be used to nullify the dispersion resulting in an achromatic transport. Some knowledge of beam optics is thus required to use the different beam optical elements (quadrupoles, di-poles, etc.) optimally to achieve the design goal. For an introduction to beam optics, readers may refer to the book by H. Wollnik (Woll 87).

In the absence of aberrations if proper object–image relation (x position of the image is independent of divergence of the ion beam in the X–Z plan) exists between the source (ion source exit) and the mass selection slit, the resolving power $R = D/(M_x \times 2x_0)$ becomes independent of the bending angle φ (Yav15). In this case, the ratio of dispersion to magnification (M_x) becomes proportional only to the area occupied by the beam (S) in the dispersive plane inside the di-pole magnet (see Figure 7.2). So, in order to enhance the resolving power it is necessary to ensure that the beam is spread out in the dispersive plane inside the magnet. This can be achieved by properly tuning the quadrupoles placed before the di-pole magnet. A decrease in bending angle φ would push the object and image further from the entry and exit of the di-pole/sector magnet. The width of the beam inside the magnet would increase in this case while the path length ($l = \rho\varphi$) of the beam inside the di-pole decreases, keeping the area unchanged.

Once the ratio of D/M_x is optimized, the resolving power may further be increased by reducing the object size. This usually means reducing the aperture of the exit orifice of the ion source. However, for exotic nuclei this will result in cutting down the beam intensity which beyond an extent would be limited by the need to make meaningful measurements over the background. Thus, the reduction of the source size to achieve higher resolution is only possible for stable nuclei or nuclei close to the stability which can be produced with enough intensity, but for exotic nuclei this option offers only a limited advantage.

It should be mentioned that di-pole magnets can be designed to impart focusing in a transverse direction as a second-order effect apart from bending. This is done by providing cut angles at the entry or exit or at both edges of the yoke as shown in Figure 7.3. This creates an angle between the fringing magnetic field and the beam (Figure 7.3) and thus the beam would experience a focusing force in the Y direction. Focusing in the Y direction helps to keep the beam well within the pole gap, especially if the

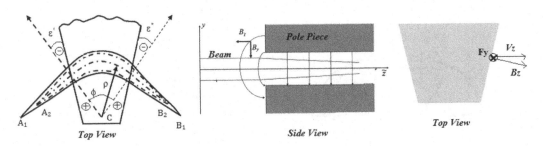

FIGURE 7.3
Transverse focusing owing to cut angle of the di-pole.

magnetic length is long, and also helps in achieving a much higher resolving power. Distortion in the fringing field (flux varies with the distance from the axis) limits the cut angle to 30°.

Considerations discussed so far assume a perfectly uniform magnetic field that is stable. But in actual practice it is hard to get a perfectly uniform magnetic field over the region occupied by the beam and that limits the ultimate resolving power since the mass resolving power ($M/\Delta M$) is proportional to the achievable magnetic field uniformity ($B/\Delta B$). This also means that the stability of the high current power supply creating the magnetic field should be better than the designed resolving power; say better than 10 ppm if a mass resolving power of 10^5 is aimed at. Assuming perfect stability of the magnet power supply, magnetic field simulation using 3D codes is needed to fix up the dimensions of the magnet poles and the yoke to have the desired magnetic field uniformity over the region occupied by the beam. Usually, to achieve a field uniformity of, say, better than about 1 in 10^4, it is necessary to use a pole width that exceeds the beam size by about one to two pole gaps (g), depending upon the shaping of the pole profile at the two edges of the poles. The pole gap thus needs to be kept small to limit the width of the poles to obtain the desired accuracy in the machining. A small pole gap also requires smaller NI (N; number of turns and I current) for a given magnetic field ($B = NI/g$). However, the gap has to be large enough to ensure that the beam does not hit the pole pieces while traversing through the magnet. It is therefore necessary to choose the di-pole magnet gap judiciously. Finally, the magnet design needs a consideration of the fabrication aspects of the magnet since the bigger the magnet, the more difficult it is to machine it with the tight dimensional tolerances necessary to achieve the designed field uniformity.

Instead of one large single magnet, it is often advantageous to use a number of smaller sector magnets of smaller bending angles configured in the dispersion additive mode to ultimately achieve a high resolving power. For two magnets, the configurations that lead to dispersion addition and dispersion nullification (or achromatic) are shown schematically in Figure 7.4. The use of multiple di-pole magnets in dispersion addition mode offers a number of advantages. First, by making individual magnets smaller the mechanical fabrication issues become relatively less complicated. Secondly, it offers flexibility in beam

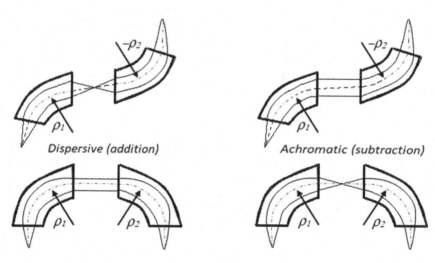

FIGURE 7.4
Dispersive and achromatic mode using two sector magnets.

Experimental Techniques

optics as it may be necessary for tuning the beam optics differently for different experiments and thirdly it helps to clean up the contaminant nuclei better by allowing slits to be put in between the di-pole magnets.

It is important to note that owing to collisions with residual gas atoms even at a pressure as low as 10^{-7} mbar, the momentum of some ions of undesired nuclei while passing through the beam line and optical elements might become equal to that of the ions of the desired isotope and therefore these ions would pass through the momentum/mass selection slit (Woll 95). Since the ions of the exotic nuclei of interest are usually lower by many orders of magnitude compared to contaminant ions, the contamination might affect subsequent measurements badly if only one di-pole magnet is used. However, if there is more than one di-pole magnet, the extent of the contamination can be reduced greatly by placing slits in between the di-pole magnets. This way the intensity of contaminants through the slits undergoes reduction in stages. Also, inside the ion source where the pressure is relatively higher, ions of a particular mass can exchange charge with other residual atoms before being accelerated at full accelerating potential at any stage along the length of the ion source. The particular mass, now having a different charge, would be accelerated with the remaining acceleration potential until the end of the ion source. Thus, a particular mass can have a distribution of energy at the exit of the ion source. Moreover, since the charge exchange can take place at any accelerating potential, the probability exists that the ratio of two masses becomes inversely proportional to their extracted energy from the ion source. In such cases, a magnet would not be able to differentiate between the masses as they would now have the same momentum. An electrostatic field, on the other hand, can differentiate between energies. So, cross contamination due to this effect can be avoided by floating two magnetic separators at different electrostatic potentials. The other option is to use an electrostatic bender and a magnetic bender in an energy-achromatic mode which can accept the energy spread of a particular mass, ensuring no loss of the radioactive nucleus of interest. Usually, after exhausting all these options to improve the resolving power, mass resolving power not exceeding a few 1000 can be achieved. To enhance the resolving power further it is necessary to reduce the emittance of the ion beam extracted from the ion source. Emittance is conserved under linear forces, and this can be achieved only by using collision or dissipative forces using a device called the Radio Frequency Quadrupole (RFQ) Cooler.

7.3.1 Radio Frequency Quadrupole (RFQ) Cooler

The cooling of an ion beam implies that the ions should occupy a small volume in phase space, which in turn means a small size beam with small transverse and longitudinal emittances. One of the most popular ways to cool the short-lived radioactive ions from an ISOL ion source is by buffer gas cooling since it is fast enough. In this cooling process, the ions lose their transverse and longitudinal energies through collisions with the buffer gas (most often helium) molecules. As a result, the ion motion is damped and in principle the ion temperature reaches, after many collisions and under thermal equilibrium, the same temperature as that of the gas, and its emittances are thus reduced. But to achieve the cooling, the ions need to be confined in all three dimensions, otherwise during the cooling process the ions would be lost by diffusion, hitting the walls of the buffer gas container. It is known that only electrostatic fields cannot create a potential minimum in three dimensions (Ernshaw's theorem). In an RFQ cooler, which is essentially a linear Paul trap (Pau 58), the three-dimensional trapping of ions is achieved by using a combination of an RF quadrupolar field which produces the transverse focusing and a static voltage that gradually decreases in the beam direction (Z) and rises at the end to create a potential minimum.

The RFQ cooler and its electrode configurations are shown schematically in Figure 7.5. The ion beam from the ion source, with energy usually in the range of 30–60 keV, is decelerated to energy below 100 eV before being injected into the RFQ cooler trap. This deceleration is necessary for the buffer gas cooling to be effective (Kim 97). However, the deceleration enhances the divergence of the beam from the ion source, typically by a factor exceeding 30 (the ratio of the square root of the initial energy, say ~60 keV, and the energy inside the trap, say ~60 eV). This needs careful design of the deceleration optics to ensure proper focussing so that most of the beam is injected into the trap through the last deceleration electrode. Once inside the trap, the transverse dimension in the trap and the transverse focusing capability should be optimally decided to accept the diverging beam and cool it with almost no loss in efficiency. The four rods in the RFQ cooler, arranged in quadrupolar geometry, are segmented in the longitudinal Z direction. Radiofrequency voltage (amplitude U_{rf}; ω typically in the range of 0.5–1.5 MHz) applied on the four rods (with a 180° phase difference between oppositely connected rods in X and Y just as in the RFQ accelerator) provides the transverse focusing. The segmented (the number of segments is typically between 15 and 25) rods insulated from each other allow, in addition to the radiofrequency voltage, application of a DC voltage that varies (decreases) from segment to segment along Z with a small gradient that helps to guide the thermalized ions towards the end of the trap. At the end of the structure a potential minimum is created by raising the potential of the last few segments of the rod and the end plate totaling about a few tens of volts. This traps the ions in the Z direction. The ions, on entering the RFQ cooler, thus lose energy due to collision in the buffer gas and are finally trapped and accumulated at the end at the position of the minimum of the trapping potential, forming a bunch of cooled ions. Finally, the voltage of the end plate is switched (as shown by dotted

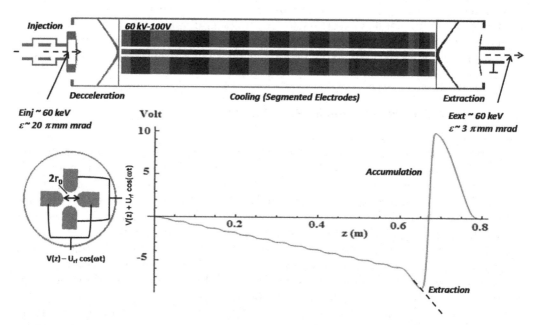

FIGURE 7.5
Schematic diagram of an RFQ cooler.

Experimental Techniques

lines in Figure 7.5) to extract the cooled ion bunch that has a typical width of a few tens of µs. A typical cooler has a total length of between 0.7 and 1 m and the typical cooling time is in the range of 1 ms to a few tens of ms. The transverse and the longitudinal emittances of the ion beam extracted from the cooler are independent of the ion source and depend upon how well the RFQ cooler parameters, such as the buffer gas pressure, RF voltage and frequency, etc., are optimized. However, the choice of injection energy into the trap (<100 eV) has a dependence on ion source emittance and has a bearing on the choice of the trap parameters including its pressure and total length. The RFQ cooler installed at ISOLDE, CERN, could achieve about tenfold decrease in emittances of a 60 keV Continuous Wave (CW) ion beam from the ion source with a trapping efficiency of about 30% (Her 01). A radial matching section (like in the RFQ Linac described in Section 6.4.1) can increase the overall capture efficiency of the input DC beam while entering the RFQ cooler (Dec 20).

The RF voltage applied to the rods produces the radial confinement, and the ion motion can be described as undergoing oscillation in a harmonic potential well called the pseudo-potential of depth V_{RF} given by:

$$V_{RF}(r) = \frac{QU_{rf}r^2}{4r_0^2}; Q = \frac{4eU_{rf}}{mr_0^2\omega^2} \tag{7.2}$$

In the above equation, Q is the dimensionless Mathieu parameter that connects the charge e and mass m of the ions with the RF voltage, the distance between surfaces of opposite rods $2r_0$, and the angular frequency ω. It can be shown that the motion remains stable for $Q < 0.91$. Such a pseudo-potential can be the minimum bounding configuration for both positive and negative ions when averaged over the complete time period of RF. This is because the quadrupolar electric field (along x or y) is weaker at the center, increasing linearly with movement away from the center. In one half of a time period, depending on the charge and polarity of the rods, ions will be pushed outward, while in the next half-cycle it will be pushed inward (by a higher electric field away from the center) towards the center. Since the electric field is increasing away from the center, the ion will feel a net push towards the center over the complete time period.

The ion will undergo a micro-motion at the frequency ω of the RF and a macro-motion of frequency ω_m, which is an oscillation in the harmonic pseudo-potential. For sinusoidal time variation of the RF voltage, ω_m can be approximated as $\omega_m = Q\omega/\sqrt{8}$ when $Q < 0.6$. Thus, the frequency of macro-motion is roughly about 1/5 of the RF frequency. For $Q < 0.6$, the micro-motion amplitude is small, and it is just a ripple over the macro-motion but for higher Q, the micro-motion amplitude becomes comparable. For optimum transverse confinement, Q should be in the range of 0.5–0.6.

The choices of buffer gas and buffer gas pressure are very important for efficient and fast cooling. The buffer gas needs to be an inert one with high ionization potential to minimize the chances of charge exchange during multiple collisions with the ions. For fast cooling of short-lived nuclei, the buffer gas molecules should be lighter than the ions and the pressure needs to be optimized. Usually, the buffer gas chosen is helium and a pressure of around 0.1 mbar results in a cooling time of around a few ms. Increasing the buffer gas pressure would result in a faster cooling rate but the pressure cannot be increased beyond a point because of voltage breakdown.

The buffer gas exerts a viscous drag to the motion of the ions thereby damping the motion. Thus, the cooling can be described by a macroscopic model based on the viscous drag experienced by the ions drifting through the gas. However, in practice microscopic calculations (Monte Carlo method simulating scattering/collision) need to be conducted for the ion cooling process in the presence of an oscillating radial and DC axial field for a proper understanding of the ion loss processes and to estimate the final ion temperatures. In such calculations, it is necessary to include the ion-atom interaction potential, for which standard datasets are available for different combinations of ions and atoms. A simulation result of a microscopic simulation for a particular ion-atom combination showing the damping of the radial motion along the length of the RFQ cooler is shown in Figure 7.6. For sufficient cooling, the drift time to traverse the RFQ cooler at a particular pressure should be more than the macro motion amplitude decay constant. At low pressure, the RF force remains effective, which forces the ion to the center. The macro motion decay time gradually decreases with an increase in pressure. After a certain pressure, collision becomes so dominant that focusing force becomes ineffective, resulting in a further increase in the velocity time constant. The decay of amplitude at different pressure and other effects of buffer gas cooling have been studied by Kellerbauer et al. (Kell 01).

The viscous drag damping assumption is valid when the ion velocity is sufficiently low, and the buffer gas atom is much lighter than the ion. For lighter ions (such as ions of ^{6}He, say) or a heavier buffer gas, say Ar instead of He, detailed simulations show that such combinations hardly produce the desired cooling. This is because of a mechanism known as RF heating. In the collision of the ion with a heavier buffer gas atom, the momentum change/change in the trajectory is drastic and would correspond to a phase jump with respect to the radio frequency field. In particular, the ion can adopt, after the collision, a trajectory that is more distant from the RFQ axis and correspond to a higher mean energy. This causes a lot of ion loss and also makes cooling inefficient. Thus, for cooling the ion should be much heavier than the buffer gas so that in each collision the ion momentum change is small.

Reduction of emittances in the RFQ cooler-buncher helps in reducing (by suitable design of the beam optics downstream of the cooler usually comprising of a di-pole magnet and a few focussing magnets) the image size to below 1 mm; often down to about 100 µm at the focal plane of the di-pole magnet. This helps in achieving a much higher resolving power of about 10,000 by using a di-pole-cooler-di-pole combination. The first di-pole with a resolving power of ~100 is used to get rid of most of the contaminants. The ion beam is then cooled in the RFQ cooler, and then the cooled ion beam is extracted and mass separated in the downstream beam line consisting of a well-designed and precisely fabricated high resolution di-pole magnet and associated quadrupole magnets and other higher-order magnets, such as sextupoles for corrections of higher-order optical aberrations.

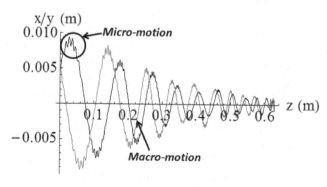

FIGURE 7.6
Damping of macro-motion for 20 eV ^{133}Cs ion with He as buffer gas.

Experimental Techniques

7.3.2 High-Resolution Separator—A Typical Example

As an example of a typical high-resolution separator, we briefly discuss the separator that has been designed and is under construction for the Advanced Rare Isotope Laboratory (ARIEL) facility at TRIUMF, Canada. The ARIEL facility will be using two primary accelerators: a 35 MeV, 100 kW, electron linear accelerator and the existing 520 MeV proton cyclotron (Bag 18) to produce neutron-rich and proton-rich radioactive nuclei. The first or the pre-separator stage (after the target station) will be used to provide the initial separation, preventing the bulk of the undesired isotopes from passing through the downstream beam line. The pre-separator consists of a 112° magnetic di-pole and a 90° electrostatic di-pole in the Nier–Johnson configuration (Sam 17). The electrostatic di-pole compensates the dispersion induced by the magnetic di-pole. The energy achromatic transport mode allows the achievement of higher resolving power even for input beams with energy spread. For focusing, electrostatic quadrupoles, rather than the magnetic quadrupoles, would be used. This would allow beam tuning independent of mass. Also, the selection in an electrostatic di-pole depends on the ion's energy, it helps to reject different mass accelerated to a different potential with charge exchange in the extraction region. Such contaminants could not be differentiated by the magnetic di-pole before the electrostatic one. The mass resolving power at the position of mass slit (width ~1.3 mm) after the magnetic di-pole is 300 while the energy resolving power at the energy slit after the electrostatic di-pole is 327 with input beam of emittance of 20π mm mrad and energy of 60 keV. Before being injected into a high-resolution separator, the beam is cooled in an RFQ cooler. The beam from the RFQ cooler has transverse emittance of 3π mm mrad (90% of the output beam) and energy width of 2 eV. A High Resolution Separator (HRS) consists of two 90° di-pole magnets (radius of curvature ~1.2 m) installed on a High Voltage platform (Mal 15). The di-pole magnets have entrance and exit angles of 27° for transverse focusing. Electrostatic quadrupoles and multipoles are introduced to provide correction up to the fifth order. The HRS magnets need to have integrated field flatness of $< 2.5 \times 10^{-5}$. For the RFQ-cooled beam, simulated resolving power is ~20,000 with a beam exit slit of dimension 120 µm. Emittance and energy width affects the achievable resolution which can be approximated as (Mal 15):

$$R_{effective} = 20000 / \left(\left(\varepsilon_x / 3.75\pi mm\ mrad \right) + \left(\Delta E / 3eV \right) \right) \tag{7.3}$$

The constraints in achieving the designed resolving power experimentally using di-poles come from various factors. Some of these are: the precision in setting the entrance and exit slit-widths, variation in peak-to-peak high voltage on the platform along with the ground variation, ripples in current power supply, flatness of pole pieces, etc. All these ultimately restrict the resolving power to a value less than the designed value and it is very difficult to attain resolving power exceeding 10,000 using sector magnets and electrostatic di-poles, etc., even after reduction of emittances in an RFQ trap. To increase the resolving power further, it is necessary to use atomic traps that are used for precision mass measurements, especially the Penning trap. This will be taken up in the section covering techniques for measurement of mass.

7.3.3 Identification of Isotopes in ISOL-Type Facilities

It is important to note that it is possible to identify a new isotope and carry out spectroscopic measurements even if the separation is not perfect. The presence of a longer-lived isobar whose decay properties are known sometimes also helps in tuning the optics of the separator. The identification of a new isotope can be carried out through its half-life (expected to be different/shorter than the other contaminants present in the focal plane), through its rather unique β-decay modes, such as β-delayed particle decay, and by studying the growth and decay of activities (β, γ, α, etc.) as the nucleus undergoes β-decay into its daughter which undergoes further decay. In fact, for super-heavy isotopes, the identification of new super-heavy isotopes/elements is carried out using a chain of α– decays in coincidence. Also, the lack of absolute purity does not affect techniques like laser spectroscopy used for measurement of nuclear moments since this technique is isotope specific. Thus, although getting an absolutely pure beam for study is always advantageous, a number of measurements (but definitely not all) can nevertheless be carried out with a limited degree of contamination and a good many properties of exotic nuclei were determined in experiments where the separation was only partial. The contamination generally results in a poorer signal-to-noise ratio and loss of precision in the measured properties, which tend to become intolerable as the nucleus becomes more and more exotic.

7.4 Separation in In-Flight Separators at Intermediate and Relativistic Energies (~50 to 1500 MeV/u)

At high energies (~50 MeV/u to a few GeV/u) the in-flight fragmentation and fission of heavy ion projectiles are the two reactions that are used for production of exotic nuclei. These reactions are discussed in detail in Chapter 2, where it is also mentioned that projectile fragment separators are used for separation of the product of interest from a very large number of other isotopes (practically all possible nuclei lighter than the projectile) that these reactions produce. In this section we discuss briefly how separation and identification of new nuclei are achieved in projectile fragment separators (PFS).

The fragments produced in projectile fragmentation/in-flight fission (after the projectile interacts with a target) fly in a narrow forward cone of a few degrees because of kinematic focusing and with a velocity close to the beam velocity. Also, as already discussed in detail in Chapter 2, the fragments have momentum spread associated with the reaction mechanism and target thickness (for in-flight fission preceded by abrasion the momentum spread of the fission fragments also contributes a little since in the rest frame of the projectile the fission fragments are produced isotropically with an average energy of ~1 MeV/u). For exotic nuclei with very low production cross-sections, it is necessary to use rather thick targets to increase the yield. This usually results in a momentum spread of the fragments typically amounting to about ±3% (energy spread double of that) or more. Also, only a good fraction of the fragments, but not all, are produced in the fully stripped ($q = Z$) ionic state. Thus, there is considerable charge state mixing from lower charge states.

The design of the ion optical system of a fragment separator should ensure: (i) efficient collection of fragments produced in the reaction and (ii) separation of the fragment of interest from other fragments including the unreacted beam/projectile. Because of very low production cross-sections of exotic nuclei, the separation has to be achieved, to the extent allowed by the measurements to be carried out in a particular experiment, without cutting down on the intensity by selecting, say, slits with optimum openings to reduce the momentum acceptance, image size, etc. The production of large numbers of fragments, most of which are produced with much larger cross-sections than the one of interest, along with rather large momentum spread and angular divergence of the fragments, makes the design of a projectile fragment separator an extremely challenging task. The presence of multiple charge states (in addition to the fully stripped one) adds further to the complications. At high energies any in-flight separator would consist of a large number of magnetic quadrupoles for focusing and a number of magnetic di-poles for separation of fragments based on their momentum ($B\rho \sim A/q.v$; B: magnetic field; ρ: radius of curvature of the di-pole; v: velocity of the ion; A: mass number; q: charge state of the fragment; $q = +Z$ for fully stripped fragments). The requirement of efficient collection of a beam with angular divergence of a few degrees demands large bore focusing magnets (quadrupoles) and also di-pole magnets with a large gap between the poles. The momentum spread of the fragments requires a large X extent of the poles of di-pole magnets (assuming beam direction to be along Z and magnetic field along Y) that makes the di-pole magnets big, bulky and costly. Further, it means the beam will occupy a large area in the dispersive plane (X–Z) inside the magnet and designing magnets with a uniform field over the entire region occupied by the beam in the X–Z plane is an extremely difficult task when the pole gap is large. The large bore of the quadrupole magnets and large pole gap of the di-pole magnets, combined with the high rigidity requirement at high energy (rigidity requirement is proportional to beam velocity and exoticity or A/Z ratio) often require the use of superconducting quadrupoles and di-poles, adding further to the cost and complications. The presence of a high radiation field in the initial section of the separator also necessitates the use of radiation-hardened materials for the construction of the magnets at the front end of a fragment separator. Thus, all the next-generation or advanced fragment separators like the BigRIPs at RIKEN (operating), ARIS at MSU for FRIB and Super-FRS at GSI for FAIR (the last two are yet to be constructed/operational) use or plan to use superconducting quadrupoles in the beam line of the separator and also a few di-poles that are superconducting. As mentioned already in Chapter 2, both the angular divergence and the momentum spread are smaller at high projectile energies. Thus, separators designed for high energies (such as the FRS/Super-FRS at GSI) allow more efficient acceptance of fragmentation/in-flight fission products with relatively smaller physical acceptances of its ion optical elements/magnets.

The A1900 fragment separator presently operating at NSCL, Michigan, USA, also uses large warm-bore superconducting quadrupoles that allow acceptance of a large fraction (up to 90%) of a large range of fragments produced in a reaction. This helps to obtain an order of magnitude higher intensity of secondary radioactive beams (Morr 03). The A1900 projectile fragment separator at NSCL is schematically shown in Figure 7.7. The separator has 24 quadrupole magnets housed in eight cryostats and four 45° di-poles. There are four

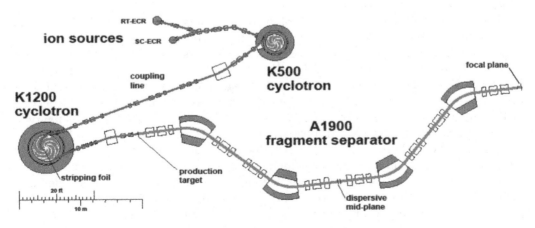

FIGURE 7.7
The A1900 fragment separator at NSCL. (Courtesy of Thomas Baumann, MSU-NSCL© all rights reserved.)

focal planes or image positions, one after each di-pole. Except for the first one which is located in the high radiation zone, the other focal planes are used for placing detectors to extract information on the beam (such as the time signal for time-of-flight measurement, position measurement, etc.) and for modification of the beam to achieve a cleaner separation (say, placement of momentum acceptance limiting slit, degrader, etc.). The final focal plane is the position where the detectors are positioned to carry out different types of measurements required for identification (such as energy loss, total energy, γ-rays from isomers, etc.) and determination of other properties of the fragments (such as excited states, β-decay half-life, β-delayed particle emissions, etc.). An analyzer beam line S800 (Baz 03), not shown in Figure 7.7, is often used after the A1900 to achieve much better separation/purification that becomes essential for the detection of very exotic fragments produced with very small cross-sections (Bau 07).

To better appreciate this discussion, it is necessary to understand the basic considerations for ion optical design of a fragment separator. It is important to note first that a single di-pole is not enough to achieve an effective separation of the fragments. This is because di-poles, being momentum analyzers, would differentiate ions of the same velocity on the basis of $A/q \sim A/Z$ of the fragments (assuming fully stripped ions). But fragmentation can produce ions of different nuclei with the same A/Z; for example, fragmentation of a ^{18}O beam at 80 MeV/u impinging on a 94 mg/cm^2 beryllium target has been found to produce ^{3}H, ^{6}He, ^{9}Li, ^{12}Be, ^{15}B, all having $A/Z = 3$ (Tho99). Also, all fragments irrespective of their A and Z are produced with a large momentum spread of the fragments. Thus, accepting all the fragments of the isotope of interest would also mean accepting some other fragments with close A/Z ratio and momentum within the same momentum window. Further, the di-pole would disperse the fragments of interest because of the momentum width around the central momentum. A second di-pole (along with quadrupoles for focusing, etc.) is therefore necessary to nullify the dispersion resulting in an achromatic transport for efficient collection of the fragment of interest. An achromatic transport essentially means that at the focal plane after the second di-pole the fragment position and angle would be independent of its momentum. Achromatic ion optics result in a smaller image size and higher resolving power independent of the initial momentum width. Thus, by using two di-poles, a momentum selection slit at the intermediate focal plane and ensuring achromatic transport some degree of separation and small image size for the fragment of interest can be ensured, but the separation would still be highly inadequate, as explained earlier, even if it is assumed that all the fragments are produced in the fully stripped ionic state ($q = Z$), which does not happen in practice.

The way to achieve a much better separation of fragments was suggested by Dufour and his co-authors in their seminal paper in 1986 (Duf 86). The authors suggested the use of a rather thick energy degrader placed at the intermediate focal plane after the momentum selection slit in between the two di-poles (Figure 7.8) which would slow down the ions differentially depending upon their A and Z. After the degrader, the momentum would not be the same for ions with the same A/Z but for a different combination of A and Z, depending on the functional dependence of energy loss on A and Z. This opens up the possibility for further separation in the second di-pole. In the energy range up to about 100 MeV/u, the momentum loss (see later for details) is such that for an aluminum degrader the ions with the same $A^{2.5}/Z^{1.5}$ would have the same momentum and would be selected by the second di-pole. A subset of ions of the same A/Z having the same $A^{2.5}/Z^{1.5}$ is thus selected after the second di-pole. This selection criterion is quite efficient for isobar separation but not so for isotonic separation. It is important to note that the idea of using a degrader to improve

Experimental Techniques

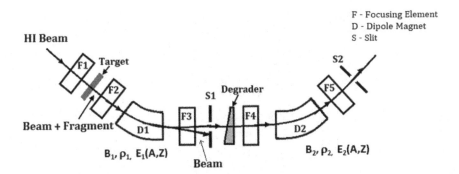

FIGURE 7.8
Schematic layout of the minimum configuration of a Projectile Fragment Separator using two di-poles and a wedge-shaped degrader placed at the intermediate focal plane between the two di-poles.

separation was an important step forward, after the initial studies at LBL using high energy ion beams from Bevalac that established the efficacy of projectile fragmentation in producing exotic nuclei and identification of new nuclei using a magnetic momentum analyzer. This new idea paved the way to use projectile fragmentation for production and identification of comparatively heavier exotic isotopes and at energies much lower than the Bevalac energies. However, use of a degrader of uniform thickness would spoil the achromatic optics because the degrader would change the dispersion. This would thus result in a much poorer resolving power. The achromatism of the optics can be preserved by using a profiled degrader of varying thickness along X (Duf 86) resulting in a smaller image size and much higher resolving power. The selection may still not be perfect and indeed in most cases it is not. The focal plane after the second di-pole in most cases still contains a good number of isotopes apart from the one of interest. The beam can be further purified by range selection which can be done by using two stacks of foils. The first set of foils with adjustable thickness can be used to stop all undesired fragments with a shorter range and the second set positioned downstream at a distance where the detectors are placed for measurements of decay radioactivity may be used to stop the fragments of interest. The remaining undesired fragments with longer range may be stopped in the beam dump located downstream at a good enough distance from the second foil so that radioactivity in the beam dump does not hamper measurements of the fragment of interest (Duf 86).

The starting point in the beam optical design is the calculation of mean energy of the fragments and the width in the energy introduced by the reaction mechanism and the finite thickness of the target. This would dictate the tuning of the first di-pole ($D1$) and slit opening at its focal plane. The projectile impinging on a target with a finite thickness can undergo fragmentation at any layer (position) of the target along the thickness of the target in the beam direction. Figure 7.9 represents the two extreme cases when the fragmentation takes place at the entry of the target and when the beam passes through all inside layers of the target without undergoing a fragmentation reaction and undergoes fragmentation just at the exit layer of the target. These two cases represent the upper (E_{2f}) and lower (E_{1f}) energy limits of the fragments of a particular isotope after the target; the energies of all other fragments of the same isotope would lie in between these two energies. The beam or the projectile with higher Z and A would lose more energy inside the target compared to the lighter fragment. Thus, the lower energy limit holds for any fragment since it is decided by the slowing down of the projectile/beam in the target.

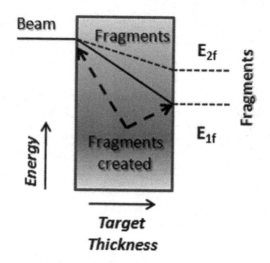

FIGURE 7.9
Energy width of the fragments due to fragmentation reaction taking place at the first (entry) layer, and the last (exit) layer of the target.

The range of fragment of given A, Z and energy E in a material can be described by following empirical relation which is applicable in the range of $Z = 4$–30 for energy ranging from 20 to 100 MeV/u:

$$\hat{R}(A, Z, E) = k \frac{A}{Z^2} E^\gamma + CA \tag{7.4}$$

where, (k, γ, C) are constants depending on the target material. For the projectile of energy E_b, the energy limits of the fragment E_f (Figure 7.9) coming out of the target of thickness (t) can be calculated to be:

$$E_{1f} = E_b\left(1 - t/R_b\right)^\gamma \ ; \ E_{2f} = E_b\left(1 - t/R_f\right)^\gamma \tag{7.5}$$

In the above equation R_b and R_f are reduced ranges ($R = R - CA$) of the beam and fragment in the target and $R_b < R_f$. Each fragment depending on its A and Z (considering fragments are fully stripped) follows a unique curve of dN/dE_f (number of fragments per unit energy interval) which is independent of the input beam energy and target thickness and is proportional to the term ($Z_f^2 / A_f - Z_b^2 / A_b$). Integrating the distribution (where dN/dE_f is the number of fragments per unit energy interval) within the energy limits would give the total number of fragments. dN/dE_f describes the distribution in momentum due to a slowing down of the fragment in the target.

However, the momentum distribution due to the reaction itself also needs to be taken into account. As discussed in Chapter 2, the projectile fragmentation inherently induces a Gaussian momentum distribution with width ($\sigma_{\parallel}$) given by Goldhaber's formula. Reaction-induced momentum distribution would not only shift the energy limits (E_{1f}, E_{2f}) of the fragments but also would add to the total width of energy distribution (σ_{tot}^2) about the mean energy of the fragments. The relations are described as follows where shift ΔE_N (from the mean energy) is due to nuclear reaction:

$$E'_{1f} = E_{1f} + \Delta E_N \, ; E'_{2f} = E_{2f} + \Delta E_N$$

$$\text{Variance}: \sigma^2_{tot} = \sigma^2_{target} + \sigma^2_{reaction} \tag{7.6}$$

The di-pole magnet after the target should be tuned for rigidity corresponding to the mean fragment energy given by ($E_f = \left(E'_{1f} + E'_{2f}\right)/2$) and should be able to accept the energy width given by the variance. The quadrupoles between the target and di-pole are usually used to transport and accept larger divergence of the produced fragments. The di-pole magnet D1 along with the slit at the focal plane should be able to accept the entire momentum width of the desired fragment. An increase in target thickness would enhance the exotic nuclei production at the cost of increased energy/momentum width which adversely affects the separation. While the di-pole is tuned for the mean energy for a particular fragment, the slit width is decided on the basis of energy width given by Equation (7.6). So, an optimum choice of target thickness is necessary to ensure maximum acceptance and transmission of the desired fragment without increasing the contamination beyond a limit set by the requirement of the particular experiment.

The next step in the design for a particular experiment is the choice of a degrader with optimum thickness to clean up the undesirable fragments with the same or a very close A/q ratio as the fragment of interest, which are transmitted through the slit at the focal plane of D1. The first di-pole would select fragments whose energy lies between $E_1(A_F, Z_F) \pm \Delta E_1(A_F, Z_F)$ or in terms of rigidity between $B_1\rho_1 \pm \Delta B_1\rho_1$. The fragments, while passing through the degrader (of thickness d) placed after the slit, would lose energy depending on their A and Z. The tuning (magnetic field) of di-pole D2 would depend on the mean energy of the fragment of interest after the degrader. After a little algebra, it can be shown that the mean rigidity of the di-pole D2 is related to the rigidity of di-pole D1 by the relation:

$$B_2\rho_2(A,Z,d) = B_1\rho_1 \left(1 - \frac{d}{k(0.1439)^{-2\gamma}} \frac{A^{2\gamma-1}}{Z^{2\gamma-2}} (B_1\rho_1)^{-2\gamma}\right)^{(1/2\gamma)} \tag{7.7}$$

for an aluminum degrader $\gamma = 1.75$. Thus, if an aluminum degrader is used the second di-pole would select ions with the same $A^{2.5}/Z^{1.5}$ ratio, while the selection by the first di-pole is governed by the ratio A/Z. It should be mentioned that the expression for the range in Equation (7.4) is not valid beyond energy of ~100 MeV/u because of relativistic effects and thus at higher energies the $A^{2.5}/Z^{1.5}$ selection criterion would not be valid. The selection criterion of the second di-pole would become velocity-dependent and would vary from more like isotonic (~$A^{2.5}/Z^{1.5}$) at lower energies up to about 100 MeV/u to isobaric just below 1 GeV/u to isotopic at around 2 GeV/u (Schm 87).

As mentioned already, the introduction of a degrader of uniform thickness would modify the dispersion, thereby destroying the achromatism of the separator which results in a poor resolving power. It can be shown that the resolving power in this case is given by:

$$\frac{A}{\Delta A} = \frac{(2\gamma - 1)}{4\gamma} \left(\frac{\Delta B_1\rho_1}{B_1\rho_1}\right)^{-1} \tag{7.8}$$

The resolving power thus depends only upon the momentum acceptance of the first di-pole. For the LISE fragment separator at GANIL, the momentum acceptance is about ±2.5% and thus for an aluminum degrader $\gamma = 1.75$, the resolving power comes out to be only about 14. However, it is possible to restore the achromatism by replacing the degrader of constant thickness (homogeneous degrader) with a wedge-shaped degrader (Figure 7.10) such that the ions lose a constant fraction of their momentum in the degrader. In the wedge-shaped degrader, the thickness $d(x)$ is varied according to energy of the fragment which has been dispersed by the first di-pole.

The rigidity of the fragment which is displaced by a distance x from the optical axis after the first di-pole magnet is given by:

$$B_1\rho_1(x) = B_1\rho_1(0)\left(1 + \frac{x}{D_1}\right) \tag{7.9}$$

D_1 being the dispersion of first di-pole magnet (Figure 7.10). The achromatism is preserved if $B_2\rho_2(x)$ after the degrader is proportional to $B_1\rho_1(x)$ for all values of x. It can be seen from Equation (7.7) that this condition can be met by requiring the product $\{d(x)(B_1\rho_1)^{-2\gamma}(x)\}$ to be independent of x. This gives the required profile of the degrader for restoring achromatism, which is given by:

$$d(x) = d(0)(1 + x/D_1)^{2\gamma} \tag{7.10}$$

$d(0)$ in the expression represents the thickness on the optical axis. It can be seen that such a profiled degrader, which is not very difficult to realize in practice, can preserve achromatism independent of A, Z, E_1, $B_1\rho_1$ and $B_2\rho_2$. It can be shown that the resolving power in this case is given by:

$$\frac{A}{\Delta A} = \frac{2\gamma - 1}{2\gamma} \frac{d(0)}{R} \frac{D_2}{X_a} \tag{7.11}$$

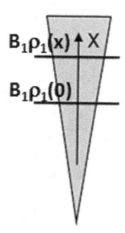

FIGURE 7.10
Wedge-shaped degrader with rigidities at the optic axis and displaced by x.

Experimental Techniques 185

In the LISE spectrometer, the wedge-shaped degrader would give mass resolving power of ~200 for a nominal beam size (X_a) of 4 mm and d (0)/R ~ 0.5. This is clearly more than an order of magnitude improvement in the resolving power compared to the homogeneous degrader case.

The use of a degrader has a few negative aspects. The most obvious one is that it results in an increase in transverse emittance of the beam, which affects the achievable resolving power. The emittances after and before the degrader roughly follow the relation: $\varepsilon = \varepsilon_0 (1 - d/R)^{-1}$, ε and ε_0 being, respectively, the emittances after and before the degrader. To keep the emittance small, a small beam spot on the target is often used. For example, GANIL uses a superconducting solenoid to bring down the beam spot on the target to about 0.2 mm on the target. A second solenoid placed after the target accepts a secondary beam of divergence ±80 mrad (Sav 96). The emittance of the secondary beam is thus about 16π mm mrad. The other negative aspect of the degrader is that it can modify the charge state of a fragment. As mentioned already, apart from the fully stripped charge state, a substantial fraction of fragments is produced with charge states of $q = Z - 1, Z - 2$, etc. The lower the projectile energy the wider the charge state distribution is. The degrader modifies the charge state further. Also, a fraction of projectile fragments undergoes projectile fragmentation in the degrader, producing lighter fragments. All these affect the level of separation achievable which is much worse than what a straightforward beam optical simulation would predict. It is possible to reduce the level of contamination to some extent by using two-stage degrader achromats. The super fragment recoil separator (FRS) at GSI is designed to have two degrader stages: the first one in the pre-separator beam line followed by the second one in the main- separator (Gei 03). The second stage can reject the cross contaminants induced due to the reaction in the wedge of the first stage. Moreover, the first achromat selects fragments at higher energy (close to 1 GeV/u) while the second achromat selects fragments at a much lower energy due to energy loss in the degrader of the first stage. The use of two degraders at two different velocities would allow two different selection criteria in the di-poles after the degraders (since the range–energy relationship depends upon the velocity) which in principle can result in a cleaner focal plane. However, this method can be gainfully employed if the projectile energy is about 1 GeV/u or higher. At lower energies in the range of 100–400 MeV/u the use of two degraders does not offer much advantage, since the nature of energy loss does not change much in this energy region. In most experiments the final focal plane comprises, in addition to the fragments of interest, a good number of fragments of unwanted isotopes which is often called a "cocktail" beam.

7.4.1 Identification of New Isotopes in the PFS Method

The identification of new nuclei is achieved by measuring the energy loss (ΔE; ΔE proportional to Z^2) and time of flight (ToF; ToF proportional to A/q) of the ions in an event-by-event mode (each event represents the measurements of ΔE, ToF, position, E, etc., of a single ion). The ToF is usually determined by recording time signals in two detectors (such as Parallel Plate Avalanche Counters, PPAC; or plastic scintillators) placed apart by good enough distance, usually a few tens of meters, in the separator beam line. The energy loss is usually measured by a series of silicon detectors (PIN diodes—intrinsic layer sandwiched between p and n type material) placed at the final focal plane. If the ToF and ΔE are plotted in a two-dimensional plot, known as a particle identification (PID) plot, fragments belonging to different nuclei can be separated and identified if the system is calibrated by identifying known nuclei. The production of fragments with multiple charge states often creates problems in the identification of new isotopes with A/q close to that of contaminant isotopes. The problem is usually severe when the new isotopes of interest are very exotic and therefore are produced with very small cross-sections. This requires a precise measurement of the position of each ion to determine its magnetic rigidity $B\rho$ (momentum) from the known magnetic fields (B_0) of the di-poles (measured using an NMR). The measurement of ToF and $B\rho$ give an accurate value of A/q, while the Z can be extracted from the measurement of ΔE and ToF. With proper calibration the PID plot (essentially Z versus A/q) leads to an unambiguous identification of new isotopes.

The technique of particle identification by measuring ΔE and ToF has been extensively used in fragment separators, such as (among others): LISE at GANIL, RIPS at RIKEN, FRS at GSI and NSCL at MSU. Continuous improvements of detectors to achieve better energy, time and position resolutions have helped in a significant way to identify and study more and more exotic fragments using this technique. Among these large numbers of experimental studies, two experiments have been arbitrarily chosen to give a glimpse of the experimental technique and what the actual PID plots look like. Figures 7.11 and 7.12 show two typical PID plots from two separate experiments. The first plot (Figure 7.11) shows data of an experiment carried out at the LISE, GANIL leading to identification, for the first time, of the doubly magic ^{48}Ni (Bl 00) and of a few other very exotic p-rich nuclei such as ^{49}Ni, ^{45}Fe and ^{42}Cr which were also observed in an earlier experiment at GSI (Blan 96). The fragmentation of a 74.5 MeV/u ^{58}Ni beam in a natural Ni target was used to produce ^{48}Ni and other p-rich exotic isotopes. The LISE3 fragment separator was used along with a Wien filter placed at the end of the LISE to purify the fragment beam further and to reduce the counting rates in the detectors. Identification of the isotope was done using a Micro Channel Plate (MCP) detector (for time signal) placed at the achromatic focal plane of the LISE spectrometer and

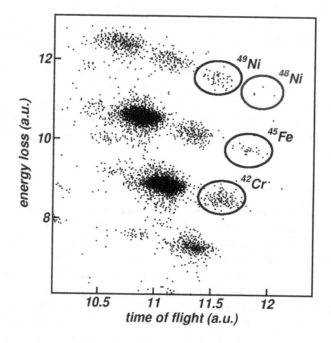

FIGURE 7.11
Two-dimensional plot based on energy loss in Si detector and ToF from MCP and Si detector at LISE3/GANIL for particle identification. (Courtesy of J. Giovinazzo, GANIL© all rights reserved.)

a silicon detector telescope (for energy loss, time and position measurement) comprising of five silicon detectors placed at the final focal plane. The first silicon ΔE detector of 300 μm thickness was used for energy loss measurement and also for the time signal for the measurement of ToF. The position information was extracted from the position-sensitive second silicon ΔE detector. The last two silicon detectors were used as veto-detectors. Overall, there was a redundancy in the measurement of both ToF (between MCP and first silicon detector and between cyclotron RF and the first silicon detector) and ΔE (in the first three silicon detectors) and an unambiguous identification of ^{48}Ni could be made.

The second particle identification plot (Figure 7.12) shows the discovery of the two most n-rich isotopes ^{40}Mg and ^{42}Al at NSCL. The experiment used the fragmentation of 141 MeV/u ^{48}Ca in a natural tungsten target and the A1900 fragment separator. The extremely n-rich nuclei are produced with very small cross-sections; in this case about 1 in 10^{15} reactions (Bau 07). Apart from very small production cross-sections, a little excitation leads to the break-up of these weakly bound nuclei by the emission of neutrons reducing their number further, often to a few per day. Unambiguous identification of these nuclei thus demands a much higher degree of purification apart from the acceptance of a large momentum spread of the fragments to increase the collection efficiency which, in turn, tends to affect the beam purity. The experiment therefore made use of the S800 spectrometer after the A1900 separator and positioned detectors for energy loss, timing, etc., measurements at the end of the S800 beam line.

The event-by-event mode of determination of a number of ΔE signals (in seven silicon PIN diodes positioned at the final focal plane after S800), two ToF measurements over flight lengths of 46 and 21 m (using time "start" signals from plastic scintillators; one located at

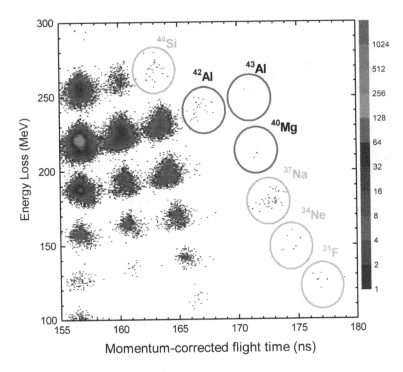

FIGURE 7.12
Particle identification plot leading to the discovery of ^{40}Mg and ^{42}Al. (Courtesy of Thomas Baumann, MSU-NSCL© all rights reserved.)

the final A1900 focal plane and the other at the object point of S800 and the stop signal from one of the seven silicon PIN diodes), and the position signals from the PPACs placed at the dispersive focal plane of S800 (for determination of momentum of the fragments) allowed an unambiguous identification of all nuclei including ^{40}Mg and ^{42}Al, using initial calibration of the entire system with the primary beam and the location of gaps in the two-dimensional particle identification spectrum corresponding to known unbound nuclei.

The separator-spectrometer mode, as accomplished at NSCL by putting together the fragment separator A1900 and the spectrometer S800, is an integral feature of the BigRIPS separator at RIKEN, which was designed to be a two-stage separator (Ku 03, Ohn 08, Ku 12). In fact, the ARIS fragment separators at FRIB and the Super-FRS at FAIR, GSI, which are being built at present are also designed to be multi-stage separators (refer for example to Ku 16), which allows a lot of flexibility in choosing different optical modes, which is required to optimize conditions for different experiments.

The BigRIPS is presently the most versatile fragment separator; it has been operating since the first quarter of 2007 and has been able to produce a large volume of front-line research outputs in RIB physics (including the discovery of a large number of new n-rich isotopes) taking advantage of the availability of presently the world's highest intensity heavy ions from ^{48}Ca to ^{238}U at energies up to 400 MeV/u, delivered by the chain of cyclotrons at the RIKEN RI Beam Factory (RIBF). The BigRIPS is designed for very high angular (80 mrad horizontal and 100 mrad vertical) and momentum (6%) acceptances and can collect at around 50% of the fission fragments produced by in-flight fission of a 345 MeV/u ^{238}U beam. The BigRIPS (Figure 7.13) uses 14 large-aperture superconducting quadrupole

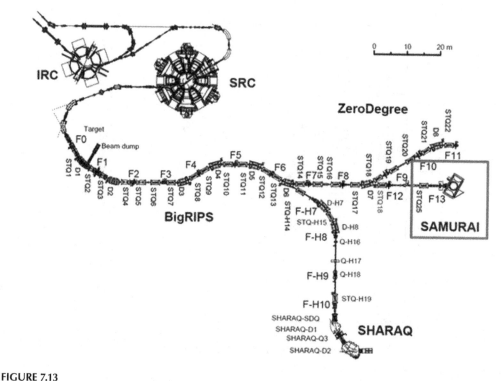

FIGURE 7.13
The schematic layout of BigRIPS at RIKEN along with the zero-degree spectrometer, SHARAQ and the SAMURAI. (Courtesy of Osamu Kamigaito, RIKEN© all rights reserved.)

triplets (STQ), six room temperature di-poles (D), and contains seven focal planes (F). The first stage is used to produce and separate fragments using an achromatic degrader, etc., while the second stage is used as an analyzer with high momentum resolution to analyze and identify the fragments. The momentum resolution is high enough to allow identification of fragments with different charge states without measuring their total kinetic energies. The energy loss measurements are carried out using silicon detectors placed at the achromatic focus F7. The $B\rho$ measurements are carried out using position and angle information at two position-sensitive PPACs placed at the F3 achromatic focus and another two similar PPACs at the F5 dispersive focus. The ToF is measured by extracting time information from two plastic scintillators, one placed at F3 and the other at F7. Isomeric tagging from the delayed γ-rays emitted by some fragments offers a cross-check for the identification of isotopes based on a PID plot. The BigRIPS could identify exotic nuclei with production rates as low as one per day (Ohn 08).

It should be mentioned that a fragment separator not only allows separation and identification of new isotopes, it also allows, through correct tagging of the fragments, nuclear reaction studies with secondary exotic beams, accurate determination of mass of the exotic fragments by injecting the energetic fragment beam into storage rings and various kinds of decay studies with stopped beams. For these studies, the entire fragment separator acts as a tagged (identified) source of secondary exotic nuclei with given energy, energy width and angular spread.

7.5 Measurement of Mass

Mass measurement has a long history starting from Aston in 1920 and these measurements have played a key role in our understanding of atomic nuclei. The initial development of nuclear physics (e.g., near constancy of binding energy per nucleon, shell effects, etc.) mainly required determination of masses of stable nuclei and with only moderate precision of ~10^{-5}. However, with the advent of the field of exotic nuclei, it was realized that mass formulae were not good enough to predict the masses of these nuclei and masses of exotic nuclei needed to be experimentally determined to fine-tune the theory for reliable prediction of masses of these nuclei. However, there are three challenges to overcome. First, the exotic nuclei are short-lived and are difficult to produce in enough numbers, requiring high sensitivity of the techniques to be employed. Secondly, the level of precision required to address the physics of exotic nuclei is a few orders of magnitude higher. The level of precision required depends upon the physics issue being addressed and these are listed in Table 7.1. The principal physics issues in exotic nuclei include reliable prediction of the limits of stability, especially the neutron drip line for heavier ($A > 80$) isotopes (these nuclei would be difficult to produce and hence measure experimentally in near future), understanding of the structure of exotic nuclei such as the evolution of shell structure, chalking out the probable nucleo-synthesis paths in explosive stellar environments, and the testing of fundamental symmetries such as the CVC hypothesis. The accuracy needed for these measurements as listed in Table 7.1 are in the range of 10^{-7} to 10^{-9}. Along with mass determination of exotic nuclei, determination of masses of stable nuclei and also of particles (say proton) and anti-particles (say anti-proton) remain very important for a wide variety of fundamental studies such as the double β-decay, tests of CPT invariance, metrology, etc. The degree of precision necessary for these studies is very high (~10^{-9}–10^{-12}).

TABLE 7.1
Precision Required in Mass Measurements for Different Studies

Physics Problems	Accuracy ($\delta m/m$)
Shells, sub-shells, pairing	10^{-5} to 10^{-6}
Nuclear deformation, halos	
r process, rp process, separation of isomers, nuclear models	10^{-7} to 10^{-8}
Weak interaction, CVC hypothesis, CKM unitarity	$<10^{-8}$
Double β decay, fundamental constants, QED, atomic binding energies	10^{-9} to 10^{-11}
Metrology, fundamental constants, CPT	$\leq 10^{-11}$

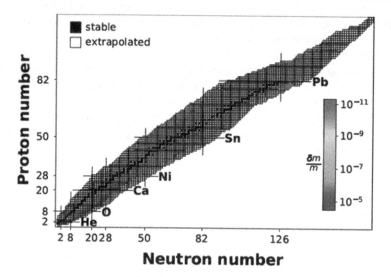

FIGURE 7.14
Nuclei for which mass has been measured using different techniques are shown in the N–Z chart along with the relative mass uncertainties. (Courtesy of Klaus Blaum, GSI/FAIR© all rights reserved.)

At present, the masses of about 3500 nuclides are known with various degrees of precision. These are shown in the N–Z chart (Figure 7.14). The measurements were carried out using different techniques. Some of these experimental methods will be discussed in sections 7.5 and 7.6. It is hoped that the next generation of RIB facilities providing more intense beams would allow measurement of masses of another 1000 exotic nuclei using these techniques. Also, continuous improvements in the techniques would result in measurements with improved precisions.

7.5.1 Indirect Methods for Mass Measurement of Exotic Nuclei

Indirect methods involve determining masses through nuclear decay, such as β or α decay, and through nuclear reactions. These are called indirect methods since these only measure mass differences. In direct methods, strictly speaking, masses are also not determined directly and so the distinction between indirect and direct methods is somewhat arbitrary. In the direct method, masses are only determined with respect to the mass of a reference nucleus whose mass is assumed to be precisely known. The reference mass is also measured in the same experiment and it is used for accurate calibration of quantities, often the magnetic field, leading to a precise determination of the mass of the nucleus of

Experimental Techniques

interest. It should be mentioned that calibration using an accurately known reference mass allows the magnetic field to be determined with much higher precision than is possible by any direct measurement technique, say, using an NMR probe.

7.5.1.1 Q_β and Q_α Measurements

Measurement of Q_β through β end-point energy determination has been used extensively for mass determination of exotic nuclei. Q_β measurements are always done by stopping the ion beam of interest in a foil after it is separated from other reaction products (contaminants) produced in the nuclear reaction. The measurement can be carried out either in an ISOL-based or fragmentation-based facility. In this technique the β particles are usually measured from the decay of the parent in coincidence with γ transitions in the daughter from the levels fed by the β decay (as schematically shown in Figure 7.15a). Addition of β end-point energy (extracted from the analysis of the experimental β spectrum) with the energies of γ transitions in coincidence (leading to the ground state of the daughter) gives the Q_β value. This technique is usually known as β-delayed γ spectroscopy. The mass of the parent nucleus is then obtained by adding the mass of the daughter, which is not determined in the experiment, to the experimentally determined Q_β. This constitutes a drawback since for exotic nuclei the mass of the daughter (which is itself an exotic isotope) may not be precisely known. There are other difficulties too. Some exotic nuclei have the strongest β feeding to the ground state of the daughter. In such cases the presence of other contaminant β-unstable isotopes/isobars/isotones would spoil the β spectrum and would lead to erroneous β– end-point energy determination. An accurate determination of Q_β is thus not possible for such nuclei unless a contaminant-free separation is achieved, which is difficult in practice. Also, even for favorable cases where there is a strong β feeding to a low-lying state in the daughter that de-excites by emission of one or more γ-rays to the ground state of the daughter, the complete decay scheme should be known, otherwise determination of Q_β could be erroneous. For example, there could be β feeding to higher-lying excited states connected by γ cascades which would distort the low energy part of

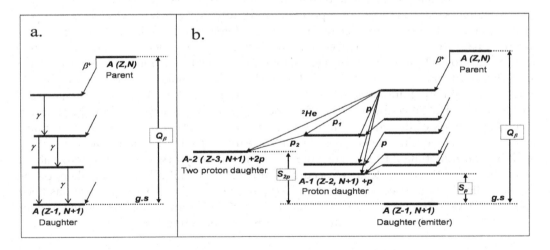

a. Beta-delayed gamma-ray spectroscopy   b. Beta-delayed proton and two-proton emission

FIGURE 7.15
Schematic representation of β-delayed γ and proton decay.

the β spectrum, introducing error to the end-point determination. Another complexity arises because of the need to determine the β detector (often plastic scintillators) response function accurately, which must be convoluted to determine the end-point energy of the continuous β spectrum. This also affects the accuracy of measurement. As such, in favorable cases the precision achievable in Q_β measurement is mainly limited by the energy resolution of the β detector (plastic scintillator) and the uncertainties in its energy calibration and is in the range of 50–100 keV (~10^{-6}).

The β decay of exotic nuclei close to drip lines, because of their high Q_β values often populates excited states in the daughter that are particle unbound. β decay to particle unbound states leads to β-delayed p, $2p$, α, etc., emissions for decay of p-rich nuclei (schematically shown in Figure 7.15b) and β-delayed n, $2n$, $3n$ emissions for decay of n-rich nuclei. Apart from mass measurement, the β-delayed γ and particle decay give us a wealth of information about excited states in the daughter including the particle-unbound states at higher excitations. It also allows accurate determination of β-decay half-life using pulsed beam, where the activity is produced during beam "on" and γ-rays/protons/neutrons, etc. resulting from the de-excitation of excited states in the daughter nucleus fed by the β-decay, are measured as a function of time during the beam "off" period. Study of β-decay of exotic nuclei is thus an invaluable tool for studying the properties of exotic nuclei. However, because of the complexities mentioned above and with the availability of other more accurate and precise techniques, Q_β measurements are no longer the principal source of information for nuclear masses of exotic nuclei.

Near proton drip line α decay modes are the dominant ones for heavier nuclei ($Z > 72$) and several chains have been identified and linked with known mass. Determination of Q_α in heavier neutron-deficient/p-rich nuclei on or beyond the proton drip line results in quite precise determination of mass since in this case there is no sharing of energy with neutrinos and the α energies can be accurately measured using Si detectors (refer for example to Dav 01). In many cases there are sequential α decays preceded by proton decay (Coulomb delayed or β-delayed). The proton decay leads to a daughter nucleus which undergoes α decay and the subsequent daughters also undergo α decays. Ultimately the sequence of α decays leads to a daughter close to the stability line whose mass is precisely known. From the measured kinetic energies of α particles and protons the masses of a large group of nuclei connected through the decay chain can be determined precisely. One such example (Dav 08) is the Coulomb-delayed proton decay of highly p-rich ^{167}Ir (half-life ~30 ms) which leads to ^{166}Os that decays by α emissions and subsequent daughters also decay by α emission until ^{150}Er (^{166}Os $\rightarrow$ ^{162}W $\rightarrow$ ^{158}Hf $\rightarrow$ ^{154}Yb $\rightarrow$ ^{150}Er) is reached. ^{150}E, being a particle bound nucleus (still quite exotic since ^{162}Er is the most p-rich stable isotope), undergoes β^+/EC decay to ^{150}Ho. Since the mass of ^{150}Ho and Q_β value of the decay of ^{150}Er are precisely known, the mass of the entire chain of isotopes can be determined. This is a standard technique employed in super-heavy element research for identification of a new isotope and for determination of its mass through a chain of α decays. Accurate knowledge about any member of the chain leads to accurate information about all other members of the decay chain through precisely determined α particle energies. This technique has been used extensively in the super-heavy element research program at GSI and Dubna (see for example Hof 01 and Oga 01), and in other laboratories including RIKEN.

7.5.1.2 Missing Mass Method

As mentioned already, nuclear reactions also provide ways of determining masses indirectly. The two techniques that are used most often are the "missing mass method" and

Experimental Techniques

the "invariant mass method" (refer for example to Pen 01). In the missing mass method (Figure 7.16) a suitable transfer reaction of type $A(a, b)B$ is chosen, where A represents the target, a represents the projectile, b represents the ejectile and B represents the unknown exotic nucleus whose mass is to be determined. Thus, the exit channel has two products, an ejectile whose mass is known and an exotic nucleus whose mass is to be determined. The mass of the unknown exotic nucleus (B), which could be particle-bound or unbound, is determined by determining the Q value of the reaction and the momentum of its complementary partner b (the ejectile) using momentum and energy conservation. The necessary information about B is thus obtained without measuring its final states, thereby avoiding the uncertainties and complications associated with such a measurement. A typical example would help to explain the method more clearly. To determine the mass excess of weakly bound ^{15}B, the multi-nucleon transfer reaction: ^{14}C (a) + ^{13}C (A) = ^{12}N (b) + ^{15}B (B) + Q, was used (Kal 00). The ejectile ^{12}N is particle-bound and has no excited states. This is an important criterion for choosing a particular reaction for the study. Thus, measurement of the energy/momentum spectrum of ^{12}N reflects the level scheme (ground and excited states) of the residual nucleus ^{15}B as shown in Figure 7.16. The background or the continuum in the energy spectrum resulted from in-flight decay of excited ^{13}N* and ^{14}N* via ^{13}N* → ^{12}N + n and ^{14}N* → ^{12}N + n + n. The separation, identification, measurement of angle, measurement of energy of the ejectile (^{12}N) and the determination of the reaction Q value are done using on-line recoil/reaction product analyzers/magnetic spectrometers in combination with detectors measuring energy loss, energy, position, etc., placed at the focal plane. Since all the three masses barring the one to be determined are known precisely, the measurements yield an accurate mass excess.

The accuracy in mass determination should be better than a few tens of keV since a precision of this level is needed to decide if the nucleus is particle-bound or unbound. This requires the target to be thin to allow good energy resolution since the ejectile is emitted with low energy. Use of a thin target in turn puts a higher demand on the projectile's intensity for adequate statistics. This becomes a limiting factor especially when unstable radioactive projectiles are used in a RIB facility. The projectile energy of 24 MeV/u used in the experiment was rather high for multi-nucleon transfer reactions

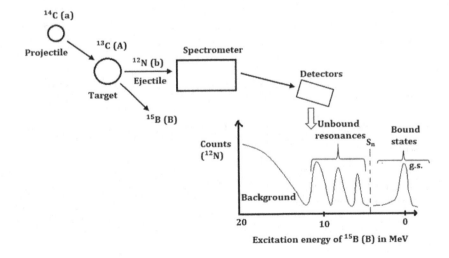

FIGURE 7.16
Schematic representation of the missing mass technique for a typical case.

but was necessary because of the rather high negative Q value of the reaction chosen. It should be noted that the use of relatively high energy for the projectile leads to a momentum transfer mismatch for reactions with low angular momentum transfer, resulting in low cross-section for the transfer reactions. Usually, the incident energy range for the application of the missing method technique lies typically between 10 and 30 MeV/u and, if possible, a reaction that needs the projectile's incident energy of around 10 MeV/u is chosen. However, this is not possible always since a reaction where the complementary partner (b) is particle-stable and has no excited states (or excited states at high excitations) is necessary. For determining masses of light n-rich bound and unbound isotopes like $^{4-6}$H, ^{9,10}He, ^{10,11}Li, ^{13}Be, $^{13-16}$B, the complementary partners chosen are ^{8}B, ^{9}C, ^{12,13}N, ^{13,14}O, ^{17}Ne and ^{20}Mg. Although the technique allows determination of masses and study of a number of light n-rich drip line and beyond drip line nuclei, its scope is rather limited because of the lack of availability of suitable complementary partner nuclei across the nuclear chart. Also, this technique is not directly applicable to PFS-type RIB facilities where the fragment (projectile) energies are too high. To use the technique in PFS-type facilities it is necessary to slow down the fragments to about 20 MeV/u for proper momentum matching (Burg 14).

7.5.1.3 Invariant Mass Spectroscopy

The other method called the "invariant mass method" (Figure 7.17) is often employed for drip line nuclei, especially for light n-rich nuclei on or beyond the drip line (for a recent review refer to Nak 17). The first experiment utilizing this technique helped to determine the mass excess of ^{9}He (Seth 87).

In this technique, the nucleus of interest (B), which may be particle-stable or unbound (beyond the drip line) is studied directly. In the reaction, usually a projectile fragmentation, the nucleus of interest is produced in an excited unbound state above the break-up threshold of a particular exit channel ($B \rightarrow C + x$) consisting of a projectile like residue (C) and one or more particles (x), most often neutrons in the case of exotic n-rich nuclei for which the last or the last two neutrons are only loosely bound. The production of the particular unbound state would thus lead to decay of the state into a recoiling nucleus (the projectile-like fragment) and the light particles. The invariant mass of the particular unbound state of the parent nuclide undergoing decay is constructed by measuring the

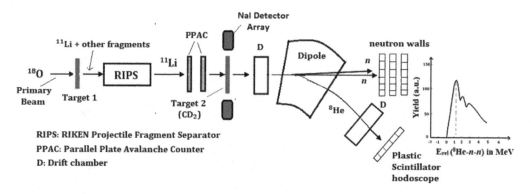

FIGURE 7.17
Schematic representation of the set up used for Invariant Mass spectroscopy of unbound ^{10}He using RIPS at RIKEN.

kinematics of the recoiling decay product and decay particle(s). The invariant mass M^* of an unbound state is given by (in the unit velocity of light c = 1)

$$M^* = \sqrt{\left(\sum E_i\right)^2 - \left|\sum p_i\right|^2} \; ; (E_i, p_i) \text{ is the four momentum of particle } i. \quad (7.12)$$

In practice, in an experiment a quantity called relative energy E_{rel} is derived, which is the energy above the break-up threshold of $B \rightarrow C + x$. E_{rel} is invariant under a Lorentz transformation, and is given by the difference between the invariant mass of the unbound state and sum of the rest masses of exit channel particles (C and x):

$$E_{rel} = M^* - \sum \text{masses of decay products (projectile-like fragment and neutrons)} \quad (7.13)$$

E_{rel} represents, in the center of mass of the decay system, the total kinetic energy, that is, the decay energy. The excitation energy of the unbound state relative to the bound ground state of nucleus B is given by (assuming the unbound state in the parent decays into a bound excited state of the recoiling nucleus that decays by γ emission with energy E_γ; E_γ = 0 if the unbound state in the parent nucleus decays into the ground state of the recoiling nucleus):

$$E_x = E_{rel} + S + E_r \; ; S \text{ being the neutron separation energy } (S_n, S_{2n}, etc.) \quad (7.14)$$

For example, if the nucleus B is ^{16}C produced in an excited unbound state above the break-up threshold (^{15}C + n), the exit channel decay products will be ^{15}C (C) and a neutron (x). The excitation energy of the unbound state in ^{16}C in this case is equal to $E_{rel} + S_n$, if the decay leads to the ground state of ^{15}C; S_n being the separation energy of a neutron from ^{16}C. But if ^{15}C has a bound excited state and the decay leads to that state then it is necessary to add the energy difference to calculate the excited energy of the unbound state in ^{16}C. Since the bound excited state in ^{15}C would decay by emitting a γ-ray, the γ needs to be detected and its energy measured and added to it to get the excited energy of the decaying state in ^{16}C. However, if the nucleus B is unbound in its ground state (like ^{9,10}He) and the exit channel heavy fragment (say ^{8}He) has no excited states, only E_{rel} is of interest.

The invariant mass method is independent of the incident beam energy and is ideally suited at high and intermediate energies using-PFS type facilities. An appropriate secondary beam is produced using a PF reaction at high energy and the separated/tagged secondary beam (the projectile fragment of interest), acting as the projectile produces the unbound excited state of the nucleus of interest through nuclear reaction in a secondary target. A great advantage is that a thick target can be used (approximately a few hundred mg/cm² or even thicker depending upon the secondary beam intensity) to compensate for the low intensity of secondary RIBs. Also, since the beam velocity is high, the exit channel products are produced in a narrow forward cone allowing high geometrical efficiency for their detection. However, in this method all the particles in the exit channel including γ-rays from the de-excitation of the recoiling nucleus, if any, need to be measured. This requires large detector arrays with high angular resolutions for detection of neutrons as well as for γ-rays (usually an array of NaI detectors is placed surrounding the target to detect γ-rays from the decay of the excited projectile fragment (C)). A typical experimental set-up for invariant mass spectroscopy studies at RIKEN using RIPS is shown in Figure 7.17. One of the first experiments that was carried out at RIKEN using a similar set-up was the invariant mass spectroscopy of unbound ^{10}He (Kor 94). ^{10}He is the lightest double magic nucleus but it is unbound. In the experiment ^{10}He was produced in the fragmentation (one proton removal channel) of β-unstable n-rich drip line isotope ^{11}Li using a thick CD_2 target. ^{11}Li was produced by projectile fragmentation of 100 MeV/u ^{18}O beam using a ^{9}Be target. The RIKEN projectile fragment separator, RIPS, was used to separate the fragmentation products and to select and tag the ^{11}Li (through energy loss and time of flight). The tagging is important since the beam obtained after separation/selection using RIPS did not contain only ^{11}Li but also admixtures of other beams. The fragmentation of ^{18}O resulted in ^{11}Li beam intensity of about 2×10^4 /s at an energy of ~61 MeV/u at the focal plane of RIPS. The ^{11}Li beam was then allowed to impinge on a CD_2 target (the secondary target) of thickness 390 mg/cm^2 to produce ^{10}He. The decay products of ^{10}He (^{8}He and two neutrons) are then detected in coincidence by using a scintillation counter hodoscope for detection of ^{8}He and neutron walls consisting of arrays of plastic scintillators for detecting the neutrons. A large di-pole magnet with large pole gap ~30 cm, placed after the secondary target, was used for efficient acceptance of the decay products, providing the necessary resolution and for steering away the charged ^{8}He nucleus from the neutrons. The neutrons fly in a narrow forward cone, being unaffected by the magnet, and hit the neutron walls. Two drift chambers, one before the magnet and the other after the magnet, allow determination of the trajectory of ^{8}He. The energy of the relative motion E_{rel} was constructed from the measured energies and angles of ^{8}He and two neutrons in coincidence. A strong peak was obtained at 1.2 MeV with an error of about 300 keV (Figure 7.17) indicating a resonant state in ^{10}He at 1.2 MeV above the break-up threshold, ^{8}He + n + n.

7.5.2 Direct Methods of Mass Measurement of Exotic Nuclei

The separation of masses in di-pole magnet-based mass separators or spectrometers allows direct determination of mass in addition to spectroscopic measurements on exotic isotopes, leading to determination of decay half-lives, excited states, decay modes and Q_β values, etc. Direct measurement of mass is always carried out with respect to known masses. The accuracy of mass measurement therefore depends on, in addition to the precision of the di-pole magnetic field and the voltages applied in the ion source extractor, electrostatic di-poles, etc., the accuracy with which the reference mass is known. At ISOLDE, the masses of a number of n-deficient and n-rich isotopes of alkaline earth elements Rb and Cs have been determined by the mass spectrometric technique (Eph 79). The technique comprises of stopping of the 60 keV mass separated ions in a tantalum tube, and then re-ionizing and extracting the ions with a potential of about 9 kV and focusing them using electrostatic and magnetic sector magnets. The mass of the exotic nucleus of interest is determined by making its ions and the ions of reference mass follow the same trajectory through intermediate and final slits by changing the extraction voltage and the voltage of the electrostatic sector magnet but keeping the magnetic field of the di-pole magnet constant. In such a case, the ratio of masses would be equal to the ratio of voltages: $m_1/m_2 = V_2/V_1 = U_2/U_1$, where V and U represent the extractor and the electrostatic di-pole voltage respectively. The voltages need to be stable for precise measurements and it was possible to determine the masses of nuclei around $A \sim 100$ with about 100 keV accuracy (accuracy of 1 in 10^6).

Direct mass measurements have also been carried out using a cyclotron which is an excellent mass separator/spectrometer, as already mentioned in Chapter 2. Using the cyclotrons CSS1 and CSS2 at GANIL, the masses of many neutron-deficient exotic nuclei in the range of A = 64–80 have been determined. The CSS1 cyclotron was used to accelerate the projectile beam to a few MeV/u that produces neutron-deficient exotic nuclei through compound nuclear fusion reactions in a target. The evaporation products of selected A/q were injected into the CSS2 cyclotron and were accelerated almost up to the extraction radius where the ions were intercepted by a silicon $\Delta E - E$ detector telescope placed radially inside the cyclotron for measurement of energy loss, the remaining energy and for extracting the time signal of the ion's arrival. The recording of ToF (measured with respect to cyclotron RF), ΔE and E was carried out in an event-by-event mode. The long flight length (~1 km) inside the cyclotron allowed a long ToF of approximately several µs resulting in an excellent time resolution ~3×10^{-5} (Gome 06). The masses are then determined using the simple relation: $\delta(m/q)/(m/q) = \delta t/t = \delta\varphi/\varphi$. The ions of different m/q would have different arrival time (δt), which corresponds to a phase difference $\delta\varphi$. A number of reference ions, all of the same A/q, were used for the mass measurements. The method allowed determination of masses of exotic species within about 200 keV. The CSS2 cyclotron is coupled in frequency with the CSS1 cyclotron and this makes the change of frequency during the experiment difficult. This is avoided by using the more recent CIME cyclotron, which can accept ions directly from the SPIRAL ion source and allows acceleration of a much wider range of masses. In this technique the m/q of the ion of interest is determined with respect to the reference ion by the difference of the frequencies used for acceleration: $\delta(m/q)/(m/q) = \gamma^2 \delta f/f$, where γ is the Lorentz factor. The precision in mass measurement would be of the same order (1 in 10^6) as in the ToF technique.

Direct measurements of projectile fragments produced in the energy range of 50–100 MeV/u have been carried out at GANIL using an α-shaped spectrometer that selects the high energy fragments and transports the fragments to a high-resolution spectrometer, SPEG. The mass is deduced using the relation: $B\rho = \gamma m_0 v/q$. Identification of fragments has been done using ΔE, E and ToF. The ToF is measured by measuring the time interval between two detectors placed 82 m apart, one near the target and the other at the final focal plane of the SPEG spectrometer. The typical flight time is ~1 µs that has allowed a resolution of ~2.10^{-4} in ToF measurement. The magnetic rigidity $B\rho$ has been derived from two position measurements, one at the dispersive focal plane of the analyzer magnet and the other at the final focal plane of the SPEG. An accuracy of 10^{-4} could be achieved in the rigidity (momentum) measurement. The mass calibration has been obtained from masses of known nuclei that are measured simultaneously. The measurements have resulted in a mass accuracy in the range of $2-4 \times 10^{-4}$. The technique is highly sensitive and allows a quick mapping of mass surface away from the β-stability. For example, using ^{48}Ca projectile beam of 60 MeV/u and a Ta target, the masses of more than 40 very exotic n-rich nuclei from ^{23}N to ^{47}Cl have been determined using the SPEG spectrometer at GANIL (Jur 07).

Direct measurements using magnetic spectrometers including a cyclotron can only reach a precision of ~10^{-6}, which is useful for the fine-tuning of mass formulae. However, as mentioned earlier, a precision of ~1 keV (better than 10^{-8}) is needed for exotic nuclei in order to decide on the limits of stability and on the actual nucleo-synthesis paths under explosive scenarios. This is achieved using Penning traps in ISOL-type facilities and storage rings in PFS-type facilities. A comparatively recent technique known as Multiple Reflection Time of Flight (MR-ToF) also allows high-precision mass measurement in ISOL-type facilities. These techniques are briefly discussed in the next section.

7.5.3 Mass Separation and Measurement in Paul and Penning Traps

"A single trapped particle floating forever at rest in free space would be the ideal object for precision measurements" is an idea introduced by Hans Georg Dehmelt, the inventor

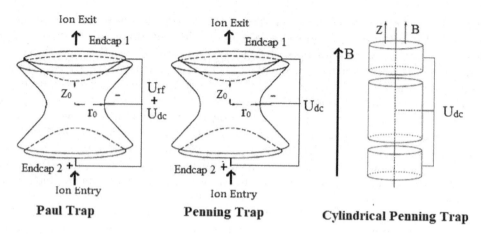

FIGURE 7.18
Schematic of Paul trap (RF field) and Penning trap (uniform magnetic field).

of the Penning trap who shared one half of the 1989 Noble Prize with Wolfgang Paul, the inventor of the other trap known as the Paul Trap. The trapping of ions confined in a small space in three dimensions achieved using a combination of DC and RF fields (as in the Paul trap) or a combination of DC field and magnetic field (as in the Penning trap) can create conditions very close to those described by Dehmelt. As described by John Pierce (Pie 49), the trapping of ions needs a force proportional to the distance of the ion from the center of the trap. This would thus require a harmonic potential which has been realized by using a ring electrode and two end caps, all with hyperbolic shapes. The application of a DC voltage between the ring and end cap electrodes with superimposed RF confines the ions in three dimensions in a Paul trap. In a Penning trap, the confinement is attained by a combination of a strong and homogeneous magnetic field and a relatively weak quadrupolar electric field. The two traps are shown schematically in Figure 7.18. There is another variant of electrode geometry for the Penning trap consisting of a stack of cylindrical electrodes, which gives rise to a quadrupolar potential in the lowest order with higher-order terms almost negligible if the ions occupy a small space around the center of the trap. This configuration allows easy injection and extraction of ions and is easier to fabricate and align (Gab 89).

7.5.3.1 Paul Trap

A Paul trap (Pau 90) is a three-dimensional analog of a radio frequency quadrupole mass filter (QMF). A QMF provides confinement only in the vertical and lateral directions whereas the Paul trap also confines the ions in the direction of ion beam injection and extraction. In a QMF two-dimensional trapping is achieved by using four rods arranged in quadrupolar symmetry and excited by DC and RF voltages. The potential in QMF is given by:

$$\varphi(x,y) = \frac{(U_{dc} + V_{rf} \cos(\omega_{rf} t))(x^2 - y^2)}{2r_0^2} \quad (7.15)$$

Depending on the DC and amplitude of RF voltage, the transverse motion of ions remains stable for a particular mass-to-charge ratio or for a range of mass-to-charge ratios. The ions for which the transverse motion is stable are not lost and drift along the axis of the QMF

Experimental Techniques

to reach the detector. For a QMF, the stability plots are expressed in two-dimensional diagrams with axes representing the following parameters:

$$a = \frac{4zU_{dc}}{mr_0^2\omega_{rf}^2}, \quad q = \frac{2zV_{rf}}{mr_0^2\omega_{rf}^2} \tag{7.16}$$

On the q axis (with DC voltage zero), the QMF acts as a high pass filter allowing all mass greater than a minimum mass (corresponding to $q_{max} = 0.91$). With increasing DC voltage, the q range for stable motion shrinks (a triangular stability region with its base on the q-axis) and the QMF acts as bandpass filter with gradually decreasing mass range Δm. So, keeping frequency constant U_{dc} and V_{rf} can be scanned along a fixed slope ("mass scan line") allowing transmission of the masses within the stability region (Figure 7.19).

The three-dimensional trapping or confinement of ions in a Paul trap is realized in a configuration consisting of a ring electrode with two end caps shaped as hyperbolic electrodes (shown in Figure 7.18). The ideal potential form in the case of a Paul trap can be expressed as:

$$\varphi(r,z) = \frac{\left(U_{dc} + V_{rf}Cos(\omega_{rf}t)\right)\left(r^2 - 2z^2\right)}{d_0^2} \tag{7.17}$$

$$d_0^2 = \sqrt{0.5r_0^2 + z_0^2}$$

In the above equation, U_{dc}, V_{rf}, ω_{rf} are applied DC voltage, amplitude and frequency of RF voltage respectively. The stable motion in r or z in this case can also be expressed in terms of dimensionless parameters a_z and q_z, which are related to the applied potentials, frequency and most importantly the charge-to-mass ratio of the ions. The mass range of the stored ion can thus be selected on the basis of the ratio between the DC voltage and amplitude of the RF voltage. Usually, in order to have favorable initial conditions, the ions

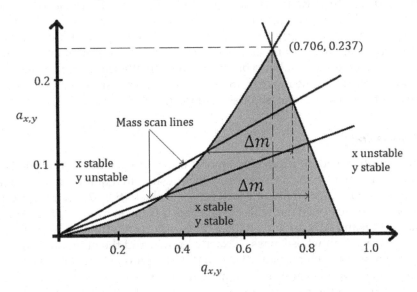

FIGURE 7.19
Operating stability region in QMF represented in an (a–q) plot.

are first centrally trapped within a small region through collisional cooling with a buffer gas (helium at ~10^{-3} torr), which reduces the ion's average kinetic energy by ~ 0.1 eV. The typical time for an ion cloud to get centrally trapped is in the range of 1–30 ms. Mass spectra are usually recorded by operating the trap in mass selective axial instability mode. In this mode the amplitude of RF voltage is gradually ramped causing the instability of different ion species (as q_z exceeds 0.91), which finally gets ejected from the trap through the exit hole in the end cap, in order of the increasing mass-to-charge ratio (Mar 97), to reach the detector. In this mode no DC voltages are applied on the end cap electrodes and trap is thus operated along the q axis. The ions ejected are then detected by placing an electron multiplier outside the trap. The ToF is recorded to identify the mass of the ions. It should be mentioned that the voltage ramp rate should be slow enough to properly resolve the masses. The other technique of ejecting the ions in the Z direction out of the trap is that of resonant excitation. The trapped ions have uncoupled motion in the radial and axial directions oscillating with frequencies dependent on the mass-to-charge ratio of the ions. Depending on the ratio of RF and DC voltage the ions are first brought to the mass selective stability line. The end-cap electrodes are then excited by a small RF voltage (di-polar mode) with frequency in resonance with the axial frequency of the ion. The ions will pick up energy and if the resonant signal is strong enough the amplitude of oscillating ions will increase sufficiently for the ions to leave the trap through the hole. The frequency can be swept gradually, ejecting ions of increasing mass-to-charge ratio. A mass resolving power close to 10^6 is achievable in a Paul trap.

Paul traps as mass filters have been extensively used for studies in rest gas analysis, analytical chemistry, biochemistry, environmental sciences and also for studies leading to the development of precise atomic clocks and basic studies in quantum mechanics, such as the observation of entanglement of atomic states. However, precision mass measurement of exotic nuclei using Paul traps is limited in scope because of the lack of stability of the oscillating electric field. Paul traps were, however, used in high precision mass measurement as a preparation trap before a Penning trap. The cooling of ions in a small volume makes the ion transport to the next trap very efficient. However, because of ion loss at the entry and exit due to high RF fields the combined injection cooling–ejection efficiency in a Paul trap is very low and is typically ~10^{-4}. The low efficiency acts as a hindrance in the measurement of masses of exotic nuclei because of a very low rate of production for these nuclei. Thus, Paul traps are now replaced by the RFQ cooler, described in the previous section, which is also a kind of Paul trap but has an overall efficiency that is higher by orders of magnitude, typically in the range of 15–30%. This is a very important gain as far as the measurement of masses of exotic isotopes is concerned.

7.5.3.2 Penning Trap

A Penning trap (the readers may refer, among others, to the articles by Bla 06 and Bla 10) measures mass in a frequency domain, which results in a quantum jump in accuracy and precision in mass measurement breaking a deadlock (Bol 87). In a magnetic field B, the cyclotron frequency of an ion with charge state q is given by $\omega_c = qB/m$. Simultaneous determination of B and ω_c is difficult, but the ratio of cyclotron frequencies of two different ions (one, a reference ion whose mass is exactly known and the other, the ion of an exotic isotope whose mass is to be measured) is equal to the ratio of their masses, provided the charge state of the two ions and the magnetic fields are the same. The mass can thus be measured accurately if the reference mass is known to a high degree of accuracy and the magnetic field is calibrated using the reference ion. Precise determination of the mass of

Experimental Techniques

the reference ion is thus a pre-requisite. The development of a carbon cluster ion source at CERN for the ISOLDE (Bla 02) and subsequent availability of ions of carbon clusters throughout the mass range of particle stable nuclei in steps of 12 mass units has provided a unique solution to this problem. As carbon is the reference atom for the atomic masses, there is no question of uncertainty in the reference mass. Also, availability of carbon cluster ions in steps of 12 mass units provides the choice of a reference ion that is within six mass units of the ion whose mass is to be measured. The precision in mass achieved in Penning traps has now reached a level better than 10^{-11} for stable nuclei which can be confined and cooled for a long time. The precision achieved for short-lived exotic nuclei is however much less, typically in the range of 10^{-7} to 10^{-9}. But precision of this level is quite often sufficient for nuclear structure physics and for determining the routes of different nucleo-synthesis processes in stars. This is the reason that in all the ISOL-type RIB facilities Penning traps are used for carrying out the most precise mass measurements of exotic nuclei.

As mentioned already, in a Penning trap the ions are confined in three dimensions by the combination of a strong magnetic field and a rather weak static electric quadrupole field. The magnetic field provides the radial and the static quadrupole field provides the axial confinement. Since static fields are applied, there is no micro-motion and consequently no resultant heating of the ions in a Penning trap unlike in a Paul trap. The ion motion in a Penning trap (Bro 82, 86, Klu 13) is composed of three independent oscillations, one along the direction of the applied magnetic field called the axial oscillation and the other being the modified cyclotron and magnetron oscillation along the radial directions, as shown in Figure 7.20. The axial and radial components are completely decoupled, giving rise to two independent equations of motion. The axial motion is a harmonic oscillation about the trap center with a frequency dependent on the ion mass and potential difference (U) applied between the ring electrode and the end caps.

The characteristic axial motion frequency is given by the following relation:

$$\omega_z = \sqrt{qU/md^2}\,;\, d^2 = \frac{1}{2}\left(z_0^2 + \frac{r_0^2}{2}\right) \tag{7.18}$$

The ion would move around in a circular orbit about the uniform magnetic field with a cyclotron frequency given by: $\omega_c = qB/m$. The electric field in the axial direction is towards the center of the trap (creating axial oscillations), which would result in a field away from

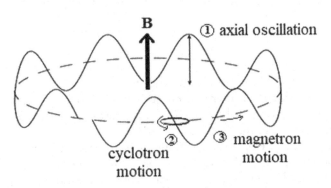

FIGURE 7.20
Three independent oscillatory motions of ions in a Penning trap.

the center in the radial plane (x–y) obeying Gauss's law. The ion motion would thus be subjected to crossed E and B fields. As the ion is dragged outwards from the trap due to the electric field, it accelerates increasing the cyclotron orbit. The magnetic field would try to curve the ion towards the center; the decelerating motion (since the electric field is directed outwards) would induce reduction of the cyclotron orbit. Thus, the center of the cyclotron motion (guiding center) would experience a drag due to $E \times B$ drift. Consequently, there would be a slow magnetron motion around the center of the trap. The drift of guiding center would also slightly reduce the cyclotron motion. The characteristic frequencies of modified cyclotron motion (ω_+) and slow magnetron motion (ω_-), both of which are perpendicular to the axial motion, are given by:

$$\omega_\pm = \frac{1}{2}\left(\omega_c \pm \sqrt{\omega_c^2 - 2\omega_z^2}\right); \omega_c = \omega_+ + \omega_- \tag{7.19}$$

Stability condition requires the frequency to be real, which would give a relation between the applied magnetic field and potential difference. If the potential difference between the electrodes creates a larger drift, the motion becomes unstable and cannot be contained radially by the magnetic field. The cyclotron frequency is larger than the axial motion and the frequencies are in the order: $\omega_+ > \omega_z > \omega_-$. In the first-order approximation, it can be shown that magnetron motion is independent of the mass of the trapped particle and is given by:

$$\omega_- = \frac{U}{2d^2 B} \tag{7.20}$$

Typical frequencies for an ion of $A/q \sim 85$ in a trap with $z_0 \sim 5.5$ mm, $r_0 \sim 6.38$ mm, $U \sim 10$ V, and $B \sim 7$ T can be calculated to be 1.260 MHz (ω_+; modified cyclotron), 106.6 kHz (ω_z; axial) and 4.509 kHz (ω_-; modified magnetron) (Bla 06).

Real trap electrodes deviate from ideal shapes. This is because openings need to be kept for injection and extraction of the ions and also the hyperbolic electrodes are truncated, not extending to infinity. Deviations from an ideal quadrupolar potential trap lead to frequency shifts or broadening of resonances that affect the resolving power and hence the precision of mass measurement. Field inhomogeneity and misalignment of the trap axis with respect to the magnetic field also induce inaccuracy in mass measurements. Also, ion motion would be affected by the Coulomb field if there is more than one ion in the trap, affecting the resolving power. Brown and Gabrielse (Bro 82) have shown that ω_c can be expressed as:

$$\omega_c^2 = \omega_z^2 + \omega_+^2 + \omega_-^2 \tag{7.21}$$

The relation (7.21) is referred to as the invariance theorem and it makes ω_c independent of trap misalignments since field imperfections get canceled to first order. The same relation also allows the determination of the mass-to-charge ratio of the stored ions by measuring the cyclotron frequency ω_c of the trap.

Apart from minimizing trap imperfections, it is important to restrict the oscillation amplitudes of the ions to the minimum possible so that the ions are trapped in a small volume and experience less of the field imperfections; for example, magnetic field homogeneity of $\sim 10^{-8}$ is achievable only over a small region of 1 cm^3. The trapping within a small region is achieved by ion cooling. The reduced emittances of cooled ions also help in the

Experimental Techniques

efficient transport of ions to the next Penning trap; the second Penning trap is often used to achieve higher precision in mass measurement.

The cooling techniques are varied, such as buffer gas cooling, resistive cooling, laser cooling, evaporative cooling, sympathetic cooling, etc. Of these, the mass selective buffer gas (usually helium at a pressure of 10^{-4} mbar) cooling is widely used for mass measurements of exotic nuclei. Buffer gas cooling can remove unwanted species such as isobaric contaminants from the stored ion cloud (Sav 91). Buffer gas cooling, as mentioned already, can be understood in terms of a viscous drag. In a Paul trap such as the RFQ cooler, buffer gas cooling is possible owing to the three-dimensional harmonic well potential created by the electrodes. So, collision with the buffer gas extracts energy from the ion, pushing it towards the center of the trap. In the case of a Penning trap, however, the same electrical quadrupole potential that provides the axial confinement also generates a radial repulsive field pushing the ion away from the center. The outward drift actually creates the slow magnetron motion which can be confined due to the presence of the magnetic field turning the ion back towards the center, in case no buffer gas is present. The magnetron motion has a small kinetic energy part and a large electric potential energy with the total energy being maximum at the center of the trap. Thus, removing energy from the ion due to collision with buffer gas would push the ion gradually away from the center and ultimately to the walls. Thus, to achieve buffer gas cooling in the trap, the radial blow-up of the stored ions needs to be constrained. This is done by excitation of the ions using an external RF quadrupolar field with the frequency of $\omega_c = \omega_+ + \omega_-$. The RF field of frequency ω_c is applied azimuthally on the ring electrode that is segmented into four sections as shown in Figure 7.21. The RF field couples the modified cyclotron motion with the magnetron motion and the magnetron motion is converted into the cyclotron motion. The time required for conversion is inversely proportional to the applied RF voltage.

Due to collision with the buffer gas molecules, the energy of the cyclotron motion would continuously decrease and because of coupling this would reduce the magnetron radius and the ions would be pushed towards the center of the trap. The time taken for conversion of the motion should be less than the time taken for the blow-up of the magnetron radius due to collision. The reduction of energy would continue until the ions are in thermal equilibrium with the temperature of the buffer gas. Since the frequency of the applied quadrupole excitation is ω_c, the cooling is mass selective (trapping and cooling only the desired mass). Thus, the buffer gas cooling technique in a Penning trap can get

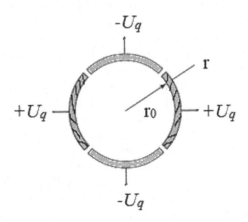

FIGURE 7.21
Quadrupolar excitation with RF applied to a four segmented ring electrode.

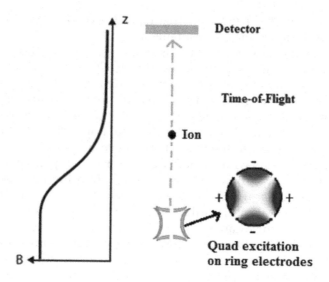

FIGURE 7.22
Time-of-Flight Ion Cyclotron Resonance technique represented schematically and showing the position of the trap and the detector, and the magnetic field variation along the released ion path.

rid of contamination from less exotic and much more abundantly produced isobars and the resolving power can reach as high as 10^6. It should be mentioned that the application of a di-pole RF field between two opposite segments of the ring electrode tuned to one of the eigen frequencies (ω_+ or ω_-) may be used to resonantly excite the particular eigen motion whose amplitude would therefore increase. This can be used to radially eject the unwanted ions and to attain purification.

For measurement of masses of exotic nuclei using Penning traps, two techniques are used. These are: time of flight ion cyclotron resonance (ToF-ICR) and Fourier transform ion cyclotron resonance (FT-ICR). In the ToF-ICR technique, schematically shown in Figure 7.22, the ions stored in the trap are excited by an external RF quadrupole field applied in the four-fold segmented ring electrode (as described while discussing buffer gas cooling) with a frequency equal to the cyclotron frequency for a time period T_{RF}. The RF excitation converts the magnetron to cyclotron motion accompanied by an increase in the radial energy of the ions. The ions are then released from the trap by lowering the trapping potential of the exit electrode. The ions then pass through the magnetic field gradient to reach the detector (micro-channel plate) where the magnetic field is very low. While traversing from the high magnetic field region inside the trap (typically 3–9 T) to low magnetic field at the position of the detector (~0.0001 T), the radial energy is converted into longitudinal energy because of the coupling of the magnetic moment associated with the radial motion with the magnetic field and the ions get accelerated along Z in the falling magnetic field ($F = -\mu_r \,\text{grad}\, B$). The ions are detected in the micro-channel plates and the ToF is recorded. The minimum ToF from the trap to the detector would correspond to the maximum radial energy (which gets converted into longitudinal energy) of the ion which happens when $\omega_{rf} = \omega_c$, that is the quadrupolar RF field is in resonance with the cyclotron frequency of the stored ions. The resonance spectrum, characterized by a typical dip in the ToF, is generated by varying the excitation RF frequency over a few tens of Hertz around the expected cyclotron frequency and repeating the process of extraction of ions from the trap and ToF measurement. Usually about 200 ions are needed to observe

Experimental Techniques

the resonance (Bla 13). A typical resonance spectrum for ^{82}Zn, an exotic n-rich nucleus is shown in Figure 7.23 (courtesy of Klaus Blaum).

The resonance line width is inversely proportional to the time period of RF excitation (Fourier limited by the T_{RF}). Thus, a higher T_{RF} is preferable to attain a narrower line width (or higher resolving power). It should be mentioned that for exotic nuclei T_{RF} is limited by the half-life of the isotope to be studied. Even for long-lived or stable nuclei T_{RF} cannot be arbitrarily long since it is difficult to keep a magnetic field stable over a long period of time (achievable relative field instability ~10^{-8} per hour). The resolving power of a Penning trap is given by: $R = m/\Delta m = \omega_c /\Delta \omega_c \sim f_c T_{RF}$. Thus, higher resolving power can be attained by increasing the cyclotron frequency, that is, by increasing the magnetic field or the charge state or both. Thus, accuracy of mass measurement is given by:

$$\delta m \propto \frac{1}{Bq\sqrt{N}T_{RF}}, \tag{7.22}$$

where N is the number of ions measured in a single resonance. For an excitation time (T_{RF}) of 1 s and cyclotron frequency of 1 MHz, a resolving power $R \sim 10^6$ is obtained. This is sufficient for isobaric separation and sometimes isomeric separation, depending upon the energy difference between the ground and the isomeric states. If the statistics are good enough for the data points around the resonance peak, the resonance peak can be fitted well, leading to a determination of the centroid of the peak (f_c which is directly related to the mass to be determined) with an accuracy better than about 10% of the FWHM, depending upon how well the resonance line shape is known. In the case of determining the sum frequency through quadrupole excitation mode, the centroid can be determined with an accuracy of 1% of FWHM since the line shape is well known. In this case an accuracy that

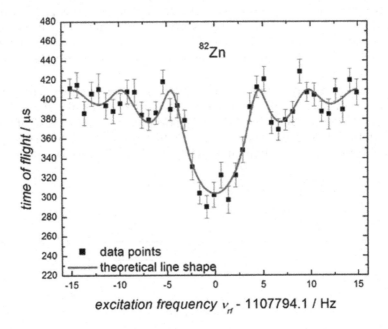

FIGURE 7.23
A typical resonance spectrum using the ToF-ICR technique. (Courtesy of Klaus Blaum, GSI/FAIR© all rights reserved.)

is two orders of magnitude higher than the resolving power is thus obtained, that is 1 in 10^8 or at the level of 10^{-8}, that is to a precision of about 1 keV for a nucleus of $A \sim 100$.

To extract the atomic mass of the measured ion, it is necessary to calibrate the magnetic field by using a reference ion and account for the electron mass. The cyclotron frequency of the reference ion is measured twice: once before the determination of the cyclotron frequency of the ion of interest and once after. This is done to eliminate drifts in the magnetic field. A complete measurement thus involves determination of three cyclotron frequencies, one for the ion of interest and the other two for the reference ion. Any uncertainty in the mass of the reference ion adds to the uncertainty in the determination of mass. The use of ions of carbon clusters: C_n^{+1}, with n up to 20 to calibrate the magnetic field has helped in removing the uncertainties related to reference mass.

The other technique of mass measurement is the FT-ICR, shown schematically in Figure 7.24. The ToF-ICR is a destructive technique where the ions are lost from the trap. But the FT-ICR technique is a non-destructive one that relies on observation of image charges induced on the segmented electrodes by the stored circulating ion. Time-dependent image currents will be produced if a tank circuit tuned at the same frequency as that of the oscillating ion is connected across the segments of the ring electrode. The current signal is amplified using a low noise amplifier and the mass spectrum is generated by fast Fourier transform (FFT) which would show a peak at the oscillation frequency of the ion. In the FT-ICR technique there is no need to re-load the trap, which makes mass determination much quicker than the ToF-ICR technique and also the frequency range can be fully scanned. In principle, mass determination is possible with a single ion and thus this technique is ideally suited for detection of isotopes of super-heavy elements (SHE) which have an extremely low production rate and at the same time have half-lives in the range of a few hundred ms and higher. However, precision mass measurement is carried out with only a few ions in the trap to avoid frequency shift owing to the space charge. So, the induced current is very small (a few fA) and this constitutes the major challenge in the FT-ICR technique due to the presence of electronic noises. The signal-to-noise ratio can be improved by using a high-quality factor (Q) tank circuit for a stronger signal, cooling the tank circuit to liquid helium temperature for the reduction of noise, and increasing the charge state and accumulation time of the ion to increase the resolving power.

A good number of Penning traps are now operating at RIB facilities across the globe and many more are in the planning stage. The traps that are operational include: the ISOLTRAP at ISOLDE, CERN (Bol 96); the SHIPTRAP at GSI (Marx 01), the JYFLTRAP at the University of Jyvaskyla (Jok 06); the LEBIT at NSCL, MSU (Schu 07), the Canadian Penning Trap (CPT) at Argonne (Cla 03) and the TITAN at TRIUMF (Dil 06). The masses of several hundred

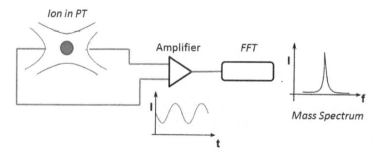

FIGURE 7.24
Schematic representation of the FT-ICR technique for mass measurement.

exotic nuclei have been determined using these Penning trap facilities. Among these, the ISOLTRAP and TITAN trap facilities receive beams from thick target-type ISOL facilities, whereas for the other four installed in in-flight facilities, slowing down of the ion beam is needed. This is achieved by using the IGISOL (Ays 01) technique where low energy ions are stopped in low pressure (~200 mbar) helium gas and are then accelerated to about 30 keV for mass separation in a di-pole magnet. After separation, the ion beam is fed first into an RFQ cooler and then into ion traps, as in the conventional Penning traps facilities at thick target ISOL facilities. For high energy projectile fragments, the fragments need to be slowed down first in thick adjustable degrader foils before entering the gas stopper cell to bring the ions to rest in the gas (Sav 03). It should be mentioned that the IGISOL technique was used to measure precisely the masses of fission fragments $^{95-100}$Sr, $^{95-101}$Y, $^{98-105}$Zr, $^{101-107}$Nb and $^{102-110}$Mo (Hag 06), all of which, except strontium, are very difficult to produce with appreciable intensity in thick target ISOL facilities.

The shortest half-life isotope, for which the mass is measured using a Penning trap, is so far ^{11}Li, which has a half-life of ~8.8 ms. The mass of ^{11}Li was determined using the TITAN at ISAC, TRIUMF to an accuracy of 690 eV (Smi 08). To increase precision, charge breeders are added in the SMILETRAP (Ber 02), in the TITAN at TRIUMF for radioactive species (Ett 11 Sim12) and also in the HITRAP facility at GSI (Her 10). Similarly, gain in precision has been achieved in the NSCL Penning trap by using a high magnetic field of 9.4 T. It was possible, using the trap, to determine the mass of the short-lived radionuclide ^{38}Ca^{2+} ($T_{1/2}$ ~ 440 ms) with an accuracy of 280 eV, that is, $\delta m/m \sim 8 \times 10^{-9}$ (Bol 06). Also, octupolar excitation, where the ring electrode is segmented into eight sections radially, has been proven to yield an order of magnitude gain in resolving power compared to the conventional excitation and was used to measure the mass difference of ^{164}Er (candidate for neutrino-less double electron capture) and ^{164}Dy with a resolving power of 2×10^7, an order of magnitude gain over quadrupolar excitation for the same time period of observation (Eli 11).

The ISOLTRAP at ISOLDE in CERN was the first Penning trap to be used for mass measurement of exotic nuclei and masses of more than 200 exotic isotopes have been determined using the facility. The ISOLTRAP facility has been updated a number of times since its installation in 1987, mainly to increase the capture efficiency in the trap of a mass-separated continuous beam from the thick target ion source ISOL, which forms the front-end of the mass measurement set-up. The increase in efficiency has allowed determination of masses of more exotic and shorter-lived nuclei. The developments leading to the up-gradation of the ISOLTRAP through the years have been nicely reflected in the paper by D. Lunney (Lun 17). The ISOLTRAP in its present form is schematically shown in Figure 7.25. The reaction products are ionized in the ion source, extracted at 30–50 keV from the ion source, separated in the ISOLDE high resolution separator and are then slowed down before entering the RFQ cooler, which as explained already, is a linear Paul trap with high capture efficiency. The RFQ cools, accumulates and bunches the continuous beam from the ISOLDE facility. The ion bunches are then decelerated to about 100 eV and fed into a preparation Penning trap for removal of isobaric contaminations (shown by the dotted line in Figure 7.25). The preparation trap is a cylindrical trap where a mass selective buffer gas (helium at ~10^{-4} mbar) cooling technique is employed using a quadrupolar RF excitation. The typical magnetic field is about 5 T. The cooling helps to localize the ions of interest within a small region around the center. A mass resolving power exceeding 10^5 and suppression of isobaric contamination exceeding 10^4 can be obtained in the trap. After sufficient ions are trapped, the ions are extracted by switching the trap potential and the ions enter the hyperbolic-shaped high precision Penning trap for the mass measurement. The precision trap operates under high vacuum. In the precision trap, ions are once again

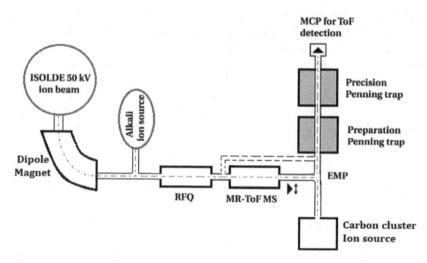

FIGURE 7.25
The schematic of the ISOLTRAP at ISOLDE.

manipulated for a finite time (T_{RF}) using quadrupolar RF excitation and then are extracted from the trap and the mass measurement is carried out using the ToF ion cyclotron (ToF-ICR) technique. As the reference ion, a carbon cluster ion source is often used as shown in Figure 7.25. Depending upon the RF excitation time (T_{RF}), mass measurement precision of ~10^{-8} has been obtained for short-lived exotic nuclei.

7.5.3.3 MR-ToF and Measurement of Mass

The Multiple Reflection Time of Flight (MR-ToF) is a new device in the ISOLTRAP in between the RFQ cooler and the preparation Penning trap introduced almost a decade ago. Thus, instead of directly injecting the ions from the RFQ cooler into the preparation Penning trap, the ions are now first fed into the MR-ToF and then into the preparation Penning trap. As a device, the MR-ToF can be used both for suppression of isobaric contaminations and for precise determination of mass. So, it is a mass separator cum spectrometer.

As the name implies, the MR-ToF (Woll 90) uses ToF for mass separation as well as for mass measurement. If all ions travel the same distance, the resolving power $R = m/\Delta m = t/2\Delta t$. Thus, any ion detector placed after the MR-ToF would be able to discriminate between the masses, if the difference in ToF is greater than the time-width of the individual mass (Δt). To achieve high resolving power the ToF needs to be made as large as possible and Δt as small as possible. In practice, the resolving power is limited since time width cannot be reduced beyond a limit and flight time also cannot be made very long because it needs a very long travel distance in high vacuum. The MR-ToF provides the solution by increasing the path length and hence the time by multiple reflections in between ion-optical mirrors, as shown in Figure 7.26. The ion travels back and forth a few thousand times in between the electrostatic ion-optical mirrors, increasing the path length and the ToF to a few tens of ms. For a time width of a few tens of ns, a resolving power exceeding 10^5 can therefore be achieved. About 50% of the ions are lost during a typical storage time of a few tens of ms and the loss is mainly due to the collision of ions with the neutral gas atoms inside the device.

Experimental Techniques

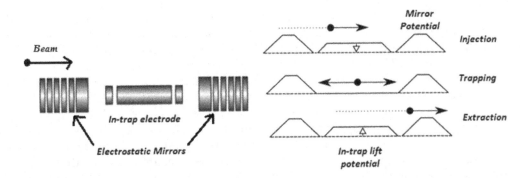

FIGURE 7.26
Schematic layout of an MR-ToF with in-trap lift potential configuration during trapping and injection/extraction.

The mechanism of storing ions in the MR-ToF trap is explained in Figure 7.26. The trap has two mirrors (comprising of a number of electrodes) kept at a suitable potential, one at the ion entrance and the other at the ion exit. In between, there is a drift tube whose voltage can be quickly switched. Ion bunches from the ion source, or better, from the RFQ cooler, with kinetic energy more than the potential of the entrance mirror electrode enters the trap and then enters the central drift tube which is kept at a certain potential and is decelerated. During the passage of the ions through the drift tube its potential is suddenly (in a time of approximately a few tens of ns that is much faster than the ion passage time of approximately a few μs) switched to zero. This will make the total energy of the ion lower than the potential of the mirror electrodes and thus ions will be trapped and undergo multiple reflections. After sufficient storage time the ions are ejected from the trap by lifting the potential of the central drift-tube.

The switching of the in-trap potential of the drift-tube for trapping offers a number of advantages over the usual method of lowering the potentials of the mirror electrodes during injection and extraction. This requires the switching of potentials of several mirror electrodes at the entry and the exit, which is operationally more complex. Also, switching affects the stability of the potentials on which the resolving power critically depends. The switching of a single potential thus minimizes this problem. Also, the kinetic energy of the ions inside the trap in this case becomes independent of the kinetic energies of the beam upstream in the beam line before the trap and downstream after the trap. It decouples the two, and as long as the input energy is sufficient to surpass the barrier height of the mirror, any kinetic energy for the input beam can be used. The other important advantage is that it allows for optimization of the resolving power by changing the in-trap voltage (Wie 15).

The time width in the MR-ToF has contributions from the time width of the input beam and also due to fluctuations in turnaround time at the electrostatic mirrors. Increasing the number of passes helps in reducing the effect of fluctuations in the turnaround time. The electric potential on mirror electrodes and the drift length in between needs to be designed properly to achieve an isochronous condition as well as to reduce aberrations (Horn 18). The time and energy width of the input beam can be reduced by using an RFQ cooler before the MR-ToF, as in the ISOLTRAP. In the ISOLTRAP, the MR-ToF is used in combination with a Bradbury–Nielsen gate (Bru 12) which acts as a fast electrical shutter allowing selection of the ion of interest. With a separation time (ToF) of a few tens of ms, it has been possible to achieve a mass resolving power exceeding 10^5, sufficient for separating isobars.

Instead of using a Bradbury–Nielsen gate to separate the nuclide of interest at the end of the MR-ToF, if all the ions are allowed to exit the trap and be detected in an ion detector, usually a MCP, placed at a distance recording the ToF, the mass of the nuclide can be determined. The MR-ToF used for mass measurement has specific advantages over the Penning Trap Mass Spectrometry for more exotic nuclei that are produced with lesser intensity and are shorter-lived. The MR-ToF does not require any cooling and can utilize about 50% of the ions injected into it. The observation time required is of the order of few tens of milliseconds giving access to shorter half-life nuclides. Moreover, it is broadband and non-scanning in nature. It does not a require pre-separator to remove the contaminants. The isobaric contaminants can simultaneously be measured along with the nuclei of interest as long as the total number of ions does not exceed a limit beyond which space charge starts to play a dominant role. On the contrary, the isobaric contaminants can provide calibration for mass measurement of unknown nuclides. The mass resolving power in a Penning trap increases linearly with the time of observation, so it can reach an extremely high value for stable nuclides but for exotic nuclei the time constraint limits the resolving power, as already mentioned. Experimental study (Wolf 13a) at the ISOLTRAP has shown that the mass resolving power (for $A/q \sim 90$) in a Penning trap increases linearly from 10^3 to 10^6 when the observation time is increased from 1 to around 800 ms. While for the MR-ToF resolving power increases linearly from 10^4 to 10^5 for observation times of 1 to 20 ms after which it saturates. This shows that for shorter-lived nuclei ($t_{1/2} < 100$ ms), the MR-ToF can give competitive resolving power. At TITAN, TRIUMF, the masses of neutron-rich nuclei $^{51-55}$Ti (lowest half-life is about 560 ms for ^{55}Ti) have been measured independently using the MR-ToF and the Penning trap. The measured masses are found to be in good agreement with each other (Lie 18). While the Penning trap leads to more precise measurements, the MR-ToF measurements are also precise enough to make it competitive for precision mass determination for nuclei that are even shorter-lived and more exotic (produced with lesser intensity). In fact, the MR-ToF has already been operating or under installation in most of the RIB facilities in the world. Lastly, compared to a Penning trap which requires cryogenic infrastructure, the MR-ToF is more compact and mobile and can be used in different environments. A mobile MR-ToF measuring 1.2 m in height and 0.8 m in width and with a resolving power of ~100,000, has already been used for analytical mass spectrometry and can also be used for varieties of other in-situ applications (Dick13).

7.6 Mass Measurements in Storage Ring

At high ion energies, storage rings offer excellent opportunities for precision mass determination. As in a Penning trap, the mass measurements in storage rings also utilize the relation between the frequency of revolution of the stored ions and the mass-to-charge (m/q) ratio, that is, masses are determined from measurement of revolution frequencies. For determination of masses of exotic nuclei produced in projectile fragmentation or in-flight fission reactions storage rings are added at the end of projectile fragment separators. To store charged ions for a considerable time (for many revolution cycles) the vacuum inside the storage ring should be $\sim 10^{-11}$ mbar. Most of the mass measurements in storage rings have been carried out so far using the Experimental Storage Ring (ESR) at GSI that

Experimental Techniques 211

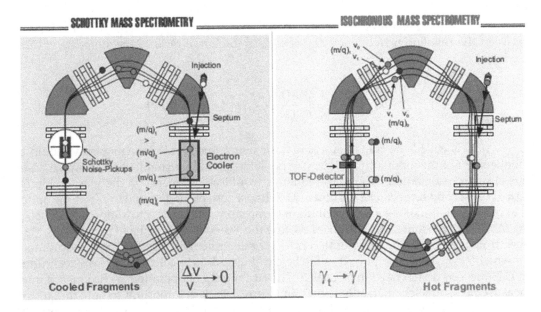

FIGURE 7.27
Schematic of SMS and IMS. (GSI/FAIR© all rights reserved; source https://www.gsi.de/en/work/research/appamml/atomic_physics/experimental_facilities/esr.htm.)

accepts high energy (~500 MeV/u) and highly charged fragments from the FRS fragment separator. For many years, the ESR (Fra 87) has been the only storage ring used for mass measurements of exotic nuclei (Rad 97). At present there are two storage ring facilities; the second one being the experimental cooler-storage ring (CSRe) at the Institute of Modern Physics, Lanzhou, China (Yan 16), where the beam from the fragment separator RIBLL2 is injected into the CSRe for mass measurement.

A storage ring basically consists of a number of di-pole and quadrupole magnets, forming the so-called lattice. For any experiment the lattice is tuned to accept a certain magnetic rigidity window around a central rigidity. The ESR is shown schematically in Figure 7.27. It contains six 60° di-pole magnets and a number of quadrupole triplets and doublets to keep the ions in their trajectory. The maximum rigidity of the ESR is 10 Tm. The revolution frequency (f) of an ion with velocity ($v = \beta c$) around the storage ring with an orbit circumference R is given by $f = \beta c / 2\pi R$. Deviation in revolution time would thus depend on a fractional change in orbit circumference and velocity. For ions varying in momentum (dP/P), change in velocity ($d\beta/\beta$) is given by: $(d\beta/\beta) = (1/\gamma^2)(dP/P)$, where γ is the Lorentz factor. The radial deviation from the designed orbit due to a change in momentum would depend on the dispersion function of the magnetic arrangements. The relative orbit difference (dR/R) for a relative momentum change (dP/P) can be expressed as: $dR/R = \alpha_p(dP/P)$, where α_p is called the momentum compaction factor, which is constant for a given ion-optical setting. Relative change in revolution frequency can thus be expressed as (Hol 14):

$$(df/f) = (d\beta/\beta) - (dR/R) = \left(\frac{1}{\gamma^2} - \alpha_p\right)(dP/P) = \eta(dP/P) \qquad (7.23)$$

Since dP/P can be expressed as: $(dP/P) = (\Delta(m/q)/(m/q)) + \gamma^2(\Delta v/v)$, the relation between the revolution frequency and the mass-to-charge ratio of the stored ions takes the form:

$$(df/f) = -\left(\frac{1}{\gamma_t^2}\right)\left(\frac{\Delta(m/q)}{(m/q)}\right) + \left(1 - \frac{\gamma^2}{\gamma_t^2}\right)(\Delta v/v) \quad (7.24)$$

$\gamma_t = 1/\sqrt{\alpha_p}$ is called the transition energy, for which the revolution frequency would be independent of the momentum and γ_t can be determined by using ions of known masses. It is clear from Equation (7.24) that frequency would only depend on the mass-to-charge ratio if the second term can be minimized to almost zero or made equal to zero. One option is to reduce the initial velocity distribution of the stored ion beams to almost zero by cooling ($\Delta v/v \to 0$) the ions. This option (of cooling the stored ions) is utilized in Schottky Mass Spectrometry (SMS). The other option is to operate/tune the ring at the ion optical mode for which $\gamma = \gamma_t$. This is employed in the Isochronous Mass Spectrometry (IMS) technique. In IMS the ions do not need to be cooled. The two modes of operation/techniques of mass measurements are schematically shown in Figure 7.26. It is important to note that a storage ring can accept m/q width of about ±1% which means it can simultaneously store and circulate ions with a range of m/q values within the acceptance limit.

7.6.1 Schottky Mass Spectrometry (SMS)

As mentioned, SMS requires beam cooling. Ions circulating in a ring can be cooled in various ways: using laser excitations, using electronic feedback (stochastic cooling) and using Coulomb interactions with an electron beam with almost the same velocity as the ion beam. "Beam cooling" basically means reduction of inherent emittance through interactions with non-Louivillian devices using non-conservative forces. Cooling should also ensure no loss in the number of particles, thus compressing the same number of particles within a smaller phase space volume enhancing the density of particles. Laser cooling is based on resonant absorption of photons by bound electrons followed by isotropic re-emission of the photons with a net momentum along the direction of the laser. Another counter-propagating laser corrected for the Doppler effect can reduce the velocity fluctuation and cool the beam longitudinally. Low-lying excitation states of electrons are required for the resonant absorption of lasers, which exist mostly for moderately charged ions. For highly charged ions, laser cooling is possible only for those having H-like or Li-like electron configuration due to the presence of hyperfine splitting of the ground state (Kla 94). Laser cooling, although very effective at relativistic energies (Win 15), is not universal and often the other two cooling techniques are employed.

Stochastic cooling was invented by Van Der Meer at CERN in the 1970s to reduce the energy spread and divergence of colliding antiprotons and protons creating W and Z bosons, for which he was awarded Noble Prize in 1984 along with Carlo Rubbia (https://www.nobelprize.org/uploads/2018/06/meer-lecture.pdf). It works on the principle of self-correction of the trajectory around the ring. The ion stored in the ring would undergo betatron oscillations (in presence of focusing fields) about the ideal orbit due to deviation of the position or momentum while injecting into the ring. Cooling is required to damp this oscillation. The deviation of an ion path from the ideal orbit is sensed with pick-up electrodes of the passing ion and the correction is applied through a kicker providing a deflecting force proportional to the error in position at pick-up (Figure 7.28). The difference

Experimental Techniques

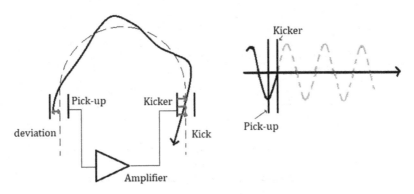

FIGURE 7.28
The pick-up and kicker arrangement in stochastic cooling.

in signal on the pick-up electrodes placed on either side of the trajectory registers the deviation from the central orbit. With the distance between the kicker and pick-up plates being an odd number of betatron wavelengths, the kicker corrects the angle depending on the position at the pick-up position. If the particle is at the crest of oscillation while passing through the pick-up, it will pass through zero position in the kicker but at an angle (Figure 7.28). The kicker will deflect and correct this angle and the particle will traverse the ring along the central orbit. For particles crossing the pick-up at positions other than the crest, correction will be partial, but after a few revolutions the oscillation will be gradually damped (Moh 83).

Stochastic cooling is more suitable for cooling beams with a larger momentum spread ($\Delta p/p \sim 10^{-3} - 10^{-2}$). This acts as a pre-cooling mechanism before going for electron beam cooling which is not suited for a large momentum spread.

Electron cooling was first proposed by G. I. Budker in 1965 for the cooling of hadron beams. It is the most popular cooling mechanism installed (Bud 75) at most of the storage ring facilities. In this, hot ions exchange momentum with collinear cold electrons, moving with almost the same velocity, by Coulomb collisions. In order to have the electron moving with the same average velocity as that of an ion, the kinetic energy of the accelerated electron would be m_e/M_i time of the ion (m_e and M_i are the mass of the electron and ion respectively). So, a much lower energy electron accelerator is required. After allowing the electron and ion to move for a few meters, electrons are extracted at the end of the cooling section (Figure 7.29) and ions are returned to interact with a fresh batch of cold electrons. Repeating this procedure would ensure that the ions reach thermal equilibrium with the temperature of the cold electrons.

Electron cooling can produce beams of very small momentum spread ($\Delta p/p \sim 10^{-6}$) and high phase density. The momentum spread depends upon the stored ion beam intensity. At high beam intensities, the cooling is offset by heating effects such as intra beam Coulomb scattering, which increases with the number of particles. For example, a fully stripped ionized uranium beam of intensity 10^7/s can be cooled to have a momentum spread of 10^{-5}, which can be further reduced to 10^{-6} if the intensity is about 10^3 (Lit 15). The cooling time required is generally a few seconds but is more if the initial velocity spread of ions is more. For example, for initial velocity spread of $\sim 10^{-2}$, the cooling time required is in the range of 10–20 seconds. The cooling time restricts the use of this technique to the measurement of masses of nuclei with half-lives of the same order. Stochastic cooling (reducing momentum width to 10^{-4}) followed by electron cooling can reduce the net cooling time

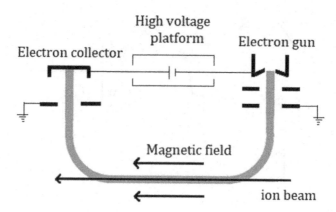

FIGURE 7.29
Schematic layout of electron cooler facility.

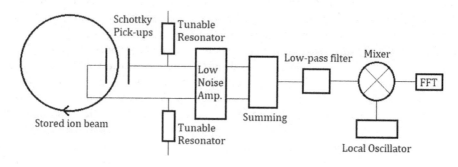

FIGURE 7.30
Schottky diagnostics data acquisition system.

to a few seconds, thus allowing measurement of masses of exotic nuclei with half-lives of approximately a few seconds. The combined cooling technique is often employed in the ESR at GSI (Ste09).

Schottky mass spectrometry is a non-destructive technique based on the Schottky noise induced by the ions revolving around the storage ring on two pick-up electrodes placed inside the beam tube (Figures 7.27 and 7.30). In every revolution, the ions induce mirror charges on the pick-up electrodes. Once the beam is cooled, the ions have almost the same velocity and their revolution frequencies will depend only upon their m/q values. The bending radius of ions with different m/q values (within the m/q acceptance width of the ring) will be different depending on the momentum compaction factor of the designed storage ring and thus the revolution frequencies will be different. The Fourier Transform (FT) of Schottky noise will show up as highly resolved peaks in the frequency spectrum, corresponding to different m/q values of the ions that are simultaneously stored in the ring (Figure 7.31). The masses of unknown fragment ions can be determined by calibrating the momentum compaction factor $\alpha_p \, (= 1/\gamma_t^2)$ using the Schottky peak positions of the known masses.

The Schottky frequency spectrum consists of lines around every harmonic of the revolution frequency of the stored particles. The width of the peaks depends upon the longitudinal momentum spread of ion species which can be of the order of 10^{-6} at low beam

Experimental Techniques

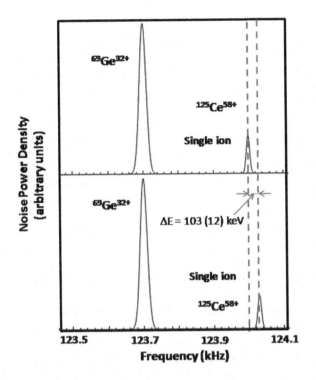

FIGURE 7.31
Ground and isomeric state of ^{125}Ce resolved in the SMS technique.

intensities. For a small number of ions intra-beam scattering becomes negligible, and in such cases the width of the Schottky line depends on the stability of magnet power supplies (Schl-Th 97). The area in each Schottky peak is directly proportional to the number of ions, the square of charge and the square of revolution frequency. Thus, for higher charge states and at relativistic energies (higher revolution frequency) Schottky diagnosis is highly effective.

The revolution frequency of the ions is in the range of a few MHz. For example, the ESR has a circumference of 108 m and a 400 MeV/u beam has a revolution frequency of 2 MHz. The Schottky noise induced by a revolving ion is recorded in two capacitive plates in the time domain. To observe the frequency peak several thousand revolutions are recorded over a time span as the induced signals are very small. A higher harmonic of the fundamental revolution frequency is recorded, which would yield the required resolution over a shorter time. This is especially required to resolve isobars. Usually, the 34th harmonic of the revolution frequency has been used for mass spectroscopy at the ESR, GSI (Gei 01). Tunable resonators are connected to the plates in order to increase the signal intensity corresponding to a particular harmonic. Signals are amplified in a low (thermal) noise amplifier and are summed. The high frequency (MHz) is then reduced to the frequency range of kHz by mixing the signal with a local oscillator frequency (Figure 7.30). This is done for two reasons. Firstly, noise from other harmonics can be removed and data (above a frequency) corresponding to only one harmonic selected. Secondly, the signal needs to be digitized using a data acquisition system before doing FFT. According to the Nyquist theorem, the sampling rate should be twice that of the signal frequency. So, reduction of the frequency helps in matching with the working range of the FFT system.

The signals are random in nature and power is not constant in time. In order to achieve precise measurement, averaging of the signal in the time domain is required for which a low pass filter is used before mixing (Figure 7.30). This would improve the signal-to-noise ratio of the spectrum. Finally, the FFT of the processed signal is analyzed to identify the unknown mass based on the calibrated reference mass. A typical bandwidth of the Schottky spectra consisting of frequency peaks corresponding to different fragments is in the range of 100–300 kHz (sometimes more up to ~900 kHz), which is enough to record data for a different mass-to-charge ratio that can be simultaneously stored in the ring. The achieved mass resolving power was 700,000 (Att 02). Relative mass accuracy of 10^{-7} has been achieved at the ESR.

The ESR acceptance is $\Delta(m/q)/(m/q) = \pm 1.2\%$. This allows simultaneous mass measurements of several nuclides. Masses of 15 nuclides were obtained simultaneously using SMS from fragmentation of a ^{152}Sm projectile. The SMS technique allows single ion sensitivity, which makes possible the determination of low-lying isomeric states. For example, the long-lived isomeric state and ground state of ^{125}Ce ($t_{1/2}$ (g. s) = 9 s) differing in energy by 103 keV could be resolved in the frequency spectrum as shown schematically in Figure 7.31 (Sun 07). Further, since the peak area (of a Schottky peak) is proportional to the number of nuclei, mass resolved half-life measurement can be conducted and the reaction cross-section of a particular reaction using Schottky spectroscopy can also be determined (Irn 95, Bos 96). For example, the half-life measurement of Cosmo-chronometer ^{187}Re ($t_{1/2}$ ~ 5 × 10^{10} yr for the atom) in the ESR using this technique revealed dramatically that the fully charge-stripped ^{187}Re^{+75} has a half-life of only about 33 yr (Bos 96). Generally speaking, the high vacuum level in the storage ring (better than 10^{-10} mbar) helps to observe the high charge state heavy ions for a longer duration of time and allow decay studies of nearly stable highly charged nuclei (Lit 11), and experiments of astrophysical interest (Lit 13) using an internal windowless gas target. It should be mentioned that while for mass determination cocktail beams are a desirable feature allowing in-situ calibration, the half-life determination and reaction studies using secondary beams using SMS require purer beams.

High resolving power, the ability to determine a large number of masses simultaneously, the ease of calibration using the same frequency spectrum (calibration does not need a separate run) and single ion sensitivity have made the SMS an extremely versatile technique. Masses of nearly 300 new isotopes/isomers have been measured using the SMS technique at the ESR, GSI (Yan 16 and references therein). At the CSRe, separate Schottky noise was also clearly observed for ^{12}C^{6+} and ^{16}O^{8+} ions, which demonstrated the potential of carrying out mass measurements using SMS at the CSRe (Wan 18).

7.6.2 Isochronous Mass Spectrometry (IMS)

In IMS, the ring is operated at transition energy and in this ion-optical mode a higher velocity ion of a given m/q would traverse a longer orbit compared to a lower velocity ion of the same m/q, thus having the same revolution frequency. The transition energy for the ESR is $\gamma_t = 1.41$ while that for the CSRe is 1.39 (Lit 10). IMS would not require beam cooling, allowing mass measurement of short-lived nuclides with half-lives below 1 ms. The revolution time for each ion for each turn is recorded by a ToF detector time-stamped with the ion passing through it. The ToF detection system comprises a thin carbon foil (thickness ~20–30 µg/cm^2) and two MCPs placed on two sides of the foil. MCPs record the secondary electrons (delta electrons) emitted by the passage of an ion through the carbon foil (in coincidence to reduce the background). The resolution of mass measurement is

Experimental Techniques

TABLE 7.2
Comparison of Four Precision Mass Measurement Techniques for Exotic Nuclei

Technique	Beam Energy	$T_{measurement}$	$m/\Delta m$	$\delta m/m$	Mode
Penning Trap	1–10 keV	1 sec	10^6–10^7	$\leq 10^{-7}$	Narrowband
Schottky MS	~500 MeV/u	10 sec	10^6	10^{-7}	Broadband
Isochronous MS	~500 MeV/u	100 μsec	10^5	10^{-6}	Broadband
MR-ToF-MS	110 keV	10 msec	$\geq 10^5$	$<10^{-6}$	Broadband

determined by uncertainty in detector timing since it is usually more than the time spread of the ions in isochronous mode. The energy loss of ions passing through the thin foil is less, allowing multiple turns (typically a few hundred revolutions each taking about 500 ns in the ESR) to be completed and simultaneously recorded before the ion is lost. The masses of unknown nuclei are determined from the ToF spectrum by comparing the ToF with that of known masses stored simultaneously. Although the MCP signal is above the noise level for a single turn, for precise measurement of revolution a frequency of about 100 turns is usually recorded. The revolution time is recorded with a high sampling rate digital oscilloscope. Averaging time stamps over multiple turns allows the achievement of higher precision in ToF and a mass resolving power of ~10^5 and a relative mass accuracy of ~10^{-6} can be obtained. Since the hot beam has a larger $\Delta(m/q)/(m/q)$, the IMS spectrum has a bandwidth that is about six times broader than that of the cooled SMS spectrum. Using the IMS technique at the ESR, a 17 μs isomer in ^{133}Sb had been identified (Sun 10). It is possible to combine the IMS with SMS by placing a pill box cavity-type resonator in air (outside the vacuum tube of the ring), coupled to the vacuum tube electromagnetically via a ceramic gap, designed to pick up the Schottky noise induced by the ion. This allows for precise measurement of revolution frequency in a short time, enhancing precision in mass measurement using IMS (Nol 11). The combined isochronous–Schottky mass spectrometry based on a resonant Schottky cavity has recently been applied at the CSRe to determine the half-lives of fully ionized ^{49}Cr and ^{53}Fe (Tu18).

About 1000 nuclides have been measured in the FRS-ESR facility out of which masses of about 350 exotic nuclei were determined for the first time. About 300 of these new masses were determined using SMS or derived by Q values or α decay and masses of the remaining 50 were determined using the IMS technique (Bos 13). Unlike Penning trap mass spectrometry, storage ring mass spectrometry is broadband and allows simultaneous measurements of a number of masses. Penning trap spectroscopy, however, can determine masses more precisely, roughly by an order of magnitude. Table 7.2 summarizes the important parameters of the four most precise mass measurement techniques employed for determining masses of exotic nuclei.

7.7 Measurement of Ground State Properties of Nuclei Using Laser Spectroscopic Techniques

Apart from the mass, the other important ground state properties of nuclei are their spin, magnetic moment, electric quadrupole moment and the charge and the matter radii. On-line atomic spectroscopy techniques using lasers have been used extensively since the 1980s for the study of these ground state properties (except the matter radii which need to be studied using nuclear reactions). It has already been mentioned in Chapter 1 that it was the measurements of spin, magnetic moment (Arn 87) and quadrupole moment (Arn 92) of ^{11}Li using laser spectroscopic techniques that ruled out the deformation of ^{11}Li. This provided strong support in favor of the halo structure of ^{11}Li, which was most surprisingly found to have an unusually large matter radius (Ta 85). A major experimental program to study the ground state properties of exotic nuclei using on-line laser spectroscopy was initiated at ISOLDE in the 1980s and continues (with many improvements/modifications/diversifications of the basic technique) to be a major program at ISOLDE. Subsequently, all the major laboratories across the world pursuing RIB science such as IGISOL, JYFL; ISAC, TRIUMF; NSCL, MSU; RI Beam Factory, RIKEN; etc., have taken up similar programs. These efforts have together resulted in a large number of publications reporting ground state properties of about 1900 exotic nuclei (Cam 16). A number of excellent review articles have also been published from time to time, covering these techniques and the findings. Interested readers may consult, among others, the comparatively recent review articles by Blaum et al. (Bla 13), Campbell et al. (Cam 16) and Neugart et al. (Neu 17) and the references therein for an overview of laser spectroscopic techniques, present status and future outlook.

The atomic spectral lines carry in them information about the charge radii, ground state spins, magnetic and quadrupole moments of atomic nuclei. Although the electrons are far away from the nucleus, the size and shape of the nuclear charge distribution, nuclear mass and nuclear moments perturb the electronic (atomic) energy levels. The transition energy or the resonance frequency (ν) of an electronic transition changes from one isotope (A) to another (A'). This is called the isotopic shift. The isotopic shift has two components. The first component is called the mass shift which arises due to a change in the nuclear mass (due to a change in the neutron number between the two isotopes) which changes the center of mass motion of the nucleus and hence the kinetic energy of the electron. The second component is called the field shift or volume shift which arises from the finite size of the nucleus. The electrons, especially the s electrons have finite probability of being inside the nucleus and would not see the nucleus as a point-like charge which results in a small decrease in the binding energy. The field shift is proportional to the change of the mean square nuclear charge radius, usually denoted by $\delta <r_c^2>$. The net isotopic frequency shift is given by:

$$\delta\nu_{IS}^{AA'} = \nu^{A'} - \nu^A = K(M_{A'} - M_A)/M_A M_{A'} + C_F \, \delta<r_c^2> \qquad (7.25)$$

The constant K in the mass shift term consists of two parts: a normal mass shift K_N that depends on the change in the reduced mass and which is usually the dominant part, and a specific mass shift K_s arises due to correlations between the electrons. K_N is always positive for a heavier isotope but K_s can be positive or negative. The net mass shift for a heavier isotope (in an isotopic chain) is generally positive (the transition frequency increases) for s to p transitions. C_F in the second term is called the field shift constant which is proportional to the change of probability density of the electron inside the nucleus when transition takes place from an initial state to the final state, say for example from the s to p state. In an isotopic chain where the charge radius increases for a heavier isotope, the field shift for the s to p transition would be negative (transition frequency would decrease) and if the increase in the charge radius is large enough, the mass and the field shifts might almost cancel each other, resulting in a very small isotopic shift. The near cancellation cannot happen for very light nuclei where mass shift is a few orders of magnitude higher than the field shift. The mass shift is large for very light

Experimental Techniques

nuclei since the addition of a single neutron brings a large relative change which is not true once there are many neutrons, as in the case of heavy nuclei. The mass shift for very light isotopes (say Li) is about a few tens of GHz, while the field shift is very small; for Li it is about 10,000 times less. As A increases, the mass shift decreases but the field shift increases since it has a $Z^2/A^{1/3}$ dependence. The magnitude of the two shifts becomes nearly equal around $Z \sim 25$.

The calculation of a normal mass shift is straightforward but that is not true for the specific mass shift. The specific mass shift can be calculated with high spectroscopic accuracy only for systems up to three electrons. For systems with more electrons the calculations are not so accurate. The contributions of mass and the field shift can also be separated out by following the King plot (King 84) procedure which requires the accurate values of charge radii of at least three stable isotopes using some other technique. Once the mass shift is separated out $\delta <r_c^2>$ can be extracted from the isotopic shift measurement. To extract the value of nuclear charge radius ($R_{A'}$) from $\delta <r_c^2>$, a reference isotope (usually a stable isotope) is used whose charge radius (R_A) is exactly known or determined from an electron scattering experiment to obtain the charge radius using the relation: $R_{A'} = (R_A^2 + \delta <r_c^2>)^{1/2}$.

The nuclear spin ($I > 0$) of the ground state of a nucleus gives rise to a magnetic di-pole moment that interacts with the magnetic field produced by electrons at the center of the nucleus. This interaction couples the nuclear spin (I) and the total angular momentum of the shell electrons (J) to the total atomic angular momentum $F = I + J$. The allowed eigen values of F lie between $|I - J|$ and $I + J$. This interaction leads to hyperfine splitting of atomic spectral lines with energy shift proportional to the magnetic moment μ of the nucleus and the magnetic field produced by the electrons $B_e(0)$ at the center of the nucleus:

$$\Delta E_m = A/2 \times C \tag{7.26}$$

where $A = \mu B_e(0)/IJ$ is the hyperfine coupling constant and $C = [F(F + 1) - I(I + 1) - J(J + 1)]$. Thus, the energy difference between two hyperfine levels of the same multiplet is given by: $\Delta E_m(F + 1) - \Delta E_m(F) = A(F + 1)$, which for $\mu > 0$ ($\mu < 0$) increases (decreases) with F. The magnetic interaction is strongest for s electrons that have the largest spin density inside the nucleus.

The magnetic hyperfine levels get shifted in case the nucleus is deformed (the shift is different for prolate and oblate). The presence of deformation results in an electric quadrupole moment and the electric field gradient induced by the electrons interacts with the quadrupole moment, resulting in a shift of magnetic hyperfine levels. The shift is given by:

$$\Delta E_e = B\left[3C(C+1) - 4I(I+1)J(J+1)\right] / \left[8I(2I-1)J(2J-1)\right] \tag{7.27}$$

where $B = e Q_s \partial^2 V_e / \partial z^2$.

In Equation 7.27, B is the electric quadrupole interaction constant, $\partial^2 V_e/\partial z^2$ is the electric field gradient (V_{zz}) at the nucleus and Q_s is the spectroscopic quadrupole moment, which because of selection rules is zero for $I = \frac{1}{2}$. Q_s is the quantum mechanical measured quantity in the laboratory frame which is related to but not the same as the intrinsic quadrupole moment Q_0 given by $Q_0 = \int d^3r\, \rho(r)(3z^2 - r^2)$, defined with respect to body-fixed axes. For axially symmetric nuclei Q_0 can be deduced from Q_s using the relation $Q_0 = Q_s(I+1)(2I+3)/I(2I-1)$ and from Q_0 the deformation parameter can be deduced. In addition to $I > 1/2$, the angular momentum of the electronic states J has to be greater than $1/2$ (Equation 7.27) for the electric hyperfine structure to be observed.

The magnetic moment and quadrupole moment of an unknown isotope are determined by comparison with a reference isotope whose moments are known, by using the relations:

$$\mu = \mu_{ref}\left(A/A_{ref}\right)\left(I/I_{ref}\right); \text{ and } Q_s = Q_{ref}\left(B/B_{ref}\right) \quad (7.28)$$

The spin of a nucleus can be determined from the number of hyperfine components in the atomic spectrum (for $I < J$) or can be extracted directly from relative energy differences of hyperfine levels and relative intensities of transitions from these levels. It should be noted that the magnitude of hyperfine splitting is of the order of 10–1000 MHz, which is about eight orders of magnitude smaller than the atomic/ionic transitions, which are in the visible range and about ~10^{14-15} Hz. Very high spectroscopic resolution, better than 10^{-7} to 10^{-8}, is thus required for studying the hyperfine splitting.

A number of laser spectroscopy techniques have been employed in the studies of exotic nuclei. These techniques are of varied nature and often experiment-specific. In the following sections only three basic techniques are discussed. For other techniques and studies carried out using those techniques, readers may consult the references cited at the beginning (Bla 13, Cam 16 and Neu 17) and the references therein.

7.7.1 The Collinear Laser Spectroscopy (CLS) Technique

The collinear laser spectroscopy (CLS) technique was first used at the TRIGA reactor, Mainz (Sch 78) and within a few years a CLS beam line was built at ISOLDE in the 1980s, which has since then been used extensively for studies on exotic nuclei. The technique is based on the detection of photons resulting from fluorescence decay of excited atoms or ions resonantly excited by a narrow bandwidth laser. The components of a typical CLS set-up are schematically shown in Figure 7.32. The exotic nuclei produced in a reaction are

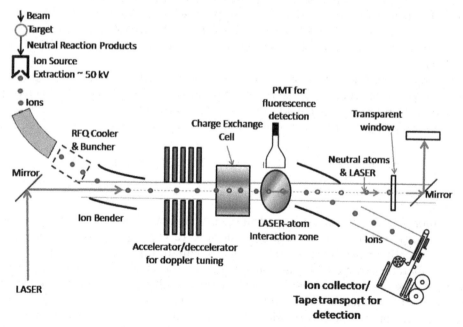

FIGURE 7.32
Schematic diagram showing components of a typical CLS set-up.

first ionized in an ion source and then mass separated. The mass separated ion beam of a few tens of keV (typically ~50 keV) is then allowed to interact with a narrow bandwidth (~1 MHz) CW laser beam propagating along or opposite to the ion direction. The laser resonantly excites the ions and the photons from the fluorescence decay are detected using a photomultiplier tube, positioned suitably with respect to an ellipsoidal reflector used to enhance the light collection efficiency. The key point in this technique is the acceleration of ion beams to a few tens of keV that provides both the spectral resolution necessary for studying the hyperfine structure and the sensitivity necessary for studying exotic nuclei that can only be produced in small numbers (low intensities).

The resonant frequency of an electronic transition would be Doppler shifted in the laboratory frame. In the ion source where the temperature is typically 2000 K or more, the velocity distribution of ions would be governed by Maxwell–Boltzmann distribution.

Laser spectroscopy study on such an ensemble of ions would not reveal the hyperfine structure because the Doppler width in this case is typically in the range of a few GHz, which is much higher than the frequency differences between the hyperfine multiplets that are in the range of ten to a few hundred MHz. Acceleration of ions to about 50 keV (by using the extraction potential of 50 kV) reduces the Doppler broadening drastically to below 10 MHz, which is of the same order as the natural width of the spectral lines (natural lifetime of the states are typically approximately a few nanoseconds). This reduction happens because the energy spread ($\delta E = mv\delta v$) of ions does not change under electrostatic acceleration but the velocity increases by a factor of 100 or more compressing δv by the same factor. It should be mentioned that apart from Doppler broadening, which is the main concern in CLS, laser power should be chosen judiciously to keep the broadening to the minimum possible since increasing the laser power means an increase in the rate of stimulated emission that effectively shortens the lifetime of the upper state, thereby increasing the line width. The reduction of Doppler width down to natural width not only allows the necessary spectral resolution but also allows all the ions to be resonantly excited simultaneously by the laser beam, making the technique highly sensitive.

The collinear geometry (superimposition of laser and the ion beam) also allows sufficient interaction time. It should be mentioned that the ion optics and relative alignment of the laser and the ion beams should ensure that the laser beam overlaps efficiently with the ion beam to minimize any loss of sensitivity on account of a less-than-perfect overlap of the two.

The hyperfine states can be resonantly excited by the laser, one by one, by varying the laser frequency over a few GHz and detecting the fluorescence decays from the excited levels by the PMT. The recordings of fluorescence photons give rise to a spectrum as a function of laser frequency. A typical hyperfine splitting of the electronic energy levels $S_{1/2}$ and $P_{3/2}$ for a nuclear spin of 5/2 and the fluorescence spectrum as a function of relative frequency is shown schematically in Figure 7.33. The figure is not to scale since it shows both the hyperfine splitting of electronic levels and the electronic transitions that differ by about seven orders of magnitude in frequency/energy.

In actual practice, the Doppler tuning for resonant excitation to different states is often not done by scanning the laser frequency since it gives rise to systematic errors due to non-linearity in the laser scan and problems associated with the instability of the laser frequency. Instead, the laser frequency is kept fixed (locked to a fixed frequency) and the ion velocity is varied which is equivalent to changing laser frequency in the ion's rest frame. Thus, by changing the ion velocity slightly the rest frame laser frequencies corresponding to excitations to different hyperfine multiplets can be tuned. This Doppler tuning is done by changing the velocity of the ions using accelerating/retarding potential before the

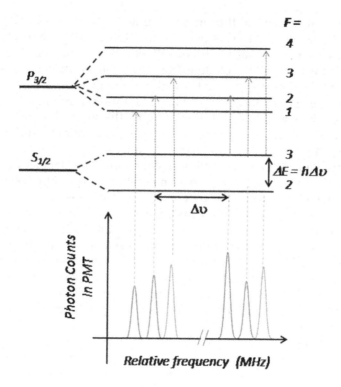

FIGURE 7.33
Hyperfine splitting of electronic states $S_{1/2}$ and $P_{3/2}$ for nuclear spin $I = 5/2$. The frequency spectrum resulting from fluorescent decay of resonantly excited hyperfine levels is also shown.

fluorescence detection zone, as indicated in Figure 7.32. In Figure 7.32 a charge exchange cell (CEC) is shown that is used to convert the singly charged ions into neutral atoms. The charge exchange is done in collision with alkali vapor atoms (usually Na or K) at low pressure ~10^{-6} mbar. The process is quite efficient (efficiency ~70%) and the charge exchange does not lead to any appreciable degradation in the beam quality (size, divergence, etc.). The conversion of ions into atoms is often done since transitions in the ionic system have higher excitation energies compared to atomic systems; requiring CW lasers in the UV range (say up to 200 nm) that involves more cost and effort. However, it is also quite common to carry out laser spectroscopy on singly charged ions. Ti:sapphire and dye lasers with frequency doubling and quadrupling arrangements are most often used to carry out spectroscopy on ions. If a CEC is used, photon detection should be conducted, as shown in Figure 7.32, immediately after the CEC for efficient collection of fluorescence photons. The ions (or remaining ions after neutralization if a CEC is used) can be deflected using a bender and collected on a tape transport system for decay spectroscopic studies.

Despite all its advantages, the sensitivity in CLS is limited by the background photons produced by the scattering of laser photons from apertures and other components of the beam line (also due to dark counts in PMT that can be reduced to about 150 count/s by cooling the PMT). Also, the solid angle for photon detection remains rather poor. The scatter background is often a few thousand counts per second, adversely affecting the signal-to-noise ratio. To get enough fluorescent photon counts above the background, a beam intensity exceeding ~10^{4-5} ions/s is usually needed. This often does not allow measurements to be carried out on very exotic species that are produced with lower intensities.

Experimental Techniques 223

The background could be reduced by about three to four orders of magnitude by using a bunched beam from an RFQ cooler buncher, described earlier, positioned upstream of the detection region in a CLS beam line as shown in Figure 7.32 (in dotted line). The addition of the cooler buncher can make it possible to study exotic species produced even with intensities as low as 100 ions/s (Nie 02). As mentioned earlier, an RFQ cooler requires a minimum of a few ms for cooling. The RFQ cooling accumulation time (t_{ac}) used for most experiments using CLS is typically around 100 ms (but for shorter-lived nuclei, a cooling time $t_{ac} \sim t_{1/2}$, which could be as small as a few ms is used; in such cases cooling might not always be optimum resulting in comparatively higher beam emittances). The width (t_w) of the cooled ion bunches extracted from the cooler is typically around 5 to 20 μs. The resonance fluorescence signal is detected only during the time window of the ion bunch and this leads to the reduction of background by the ratio t_{ac}/t_w, which is 10,000, assuming an accumulation time of ~100 ms and bunch width of 10 μs. The RFQ cooler also reduces, both the longitudinal and transverse beam emittances as mentioned earlier. These respectively allow reduction of Doppler broadening to lower than the natural width for most ionic and atomic transitions and better overlap of the laser and the ion/atomic beam. An RFQ ion cooler was first used in CLS at the IGISOL facility at Jyvaskyla; nowadays it is an integral part of a CLS beam line in all the laboratories. CLS combined with an RFQ cooler has been extensively used for studies of ground state and isomeric states of a large number of exotic isotopes. For example, at Jyvaskyla, the technique was used to study ground and isomeric states of a large number of yttrium isotopes in the range of $^{86-102}$Y. Magnetic di-pole and electric quadrupole moments, as well as the differences of charge radii, were deduced. These helped in establishing a shape transition from oblate to prolate shape at around $N = 60$ (Che 07). At ISOLDE, a change in the shell structure was revealed between $N = 40$ and 50 through studies on n-rich Cu and Ga nuclei (Fla 09, Che 10).

7.7.2 The Collinear Resonant Ionization Spectroscopy (CRIS) Technique

In this technique the sensitivity is greatly increased by combining the highly efficient resonant collinear laser excitation with highly sensitive ion detection. In this technique a second (or sometimes third) laser is used to excite the atom stepwise to the continuum to ionize it, as discussed in Chapter 2 in the laser ion source section. The resonant optical excitation is a highly selective process, and the detection of optical resonances is done by counting ions, which is almost 100% efficient and almost background-free. The technique uses multiple pulsed lasers as in the resonant ionization and requires a variety of pulsed and CW lasers with frequency doubling, etc. (Gro 17), to cater to various ionization schemes. The resonant ionization spectroscopy in ion source, is, however, limited in resolution by Doppler broadening of about 4 GHz and is used for producing high quality ion beams of isobaric purity. But the CRIS technique is not limited by Doppler broadening.

Resonant excitation of a continuous beam from an ISOL facility using pulsed lasers means inefficient use of a beam only for the duration of the pulse. An RFQ cooler buncher is therefore used, which along with its other advantages, allows matching of its accumulation and extraction time with the duty cycle of the high-power laser. All ions can thus be resonantly excited with almost 100 % efficiency. A typical CRIS set-up is schematically shown in Figure 7.34. The ion bunch is neutralized in a charge exchange cell and then the neutral atoms enter a differential pumping region where ions that are not neutralized (usually about 30% since a CEC can neutralize typically about 70% of the ions) are removed by using electrostatic deflection. The neutral atoms then enter the interaction zone where it is resonantly ionized by two or three collinear laser beams. The vacuum in the interaction

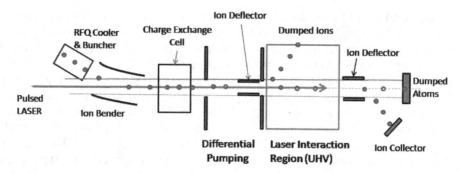

FIGURE 7.34
Schematic diagram showing a typical CRIS set-up.

region and in the rest of the beam line downstream is maintained at a level better than 10^{-8} mbar to minimize non-resonant re-ionization of the atoms in collision with residual particles in the vacuum. The length of the interaction region is kept in order to accommodate an entire ion bunch which can be ionized by the overlapping laser pulses before entering the detection region. Typically, an interaction zone length of 1–1.5 m is needed, depending on the time width of the ion pulse extracted from the RFQ cooler (typically ~5 μs) and the beam energy (typically ~30 keV) after the RFQ cooler. For the high-resolution hyperfine spectrum, the laser used for excitation to the first excited level should be of narrow width; typically, about 10–20 MHz or less (the second step leading to ionization may be a non-resonant step). However, in many cases measurements, including those of isotopic shifts, might not require very high resolution. In those cases, it is enough to use a laser with a width of ~1 GHz for the first step excitation. The larger bandwidth also helps in locating the resonances quickly.

The resonantly ionized ions are then deflected into the ion detection region and detected by an MCP ion detector. Alternately, the ions can be transported to a decay spectroscopy station (not shown in Figure 7.34). The CRIS technique is highly sensitive and allows measurements on exotic nuclei produced with yields of fewer than 100 ions/s. At ISOLDE, the CRIS technique has been used to determine the isotopic shifts, magnetic and quadrupole moments of a long chain of Fr isotopes (Gro 15, Fla 13). These studies have produced a wealth of information about Fr isotopes. For example, through electric quadrupole moment measurement it has been established that the ground state of ^{219}Fr (half-life ~21 ms) is deformed, contrary to the expectation of a spherical shape based on the shell model. In the chain of Fr isotopes studied, ^{214}Fr is the isotope with the shortest half-life (half-life ~5 ms) for which the isotope shift has been measured using CRIS (Smi 16). At the end of the CRIS beam line, a decay spectroscopy (laser-assisted decay spectroscopy) measurement station allows study of the decay of the isomers and the ground state of isotopes selected by CRIS. The set-up at ISOLDE, for example, has been used to measure α decay from ground and isomeric states of a number of Fr isotopes (Lyn 14, Lyn 16, Smi 16).

Resonant ionization spectroscopy (RIS) requiring low resolution can be carried out at the ion source itself. It is usually done in cases of very exotic species where the production rate is very low, about 1 ion/s. In measurements requiring high resolution, RIS is also used in the ion source to get from the ion source isobarically pure ion (the so-called laser ion source mode) beam comprising of a chain of isotopes of a particular element (such as Ca, for example) which are also produced in the same nuclear reaction. High-resolution laser spectroscopy using the CLS technique can then be carried out on a bunched (using an RFQ cooler) ion beam to measure the isotopic shift of the chain of isotopes that gives information about the evolution of nuclear structure (nuclear charge radius) as a function of the neutron number. A rather recent experiment on Ca isotopes ($^{40-52}$Ca) carried out at ISOLDE using RIS followed by CLS revealed a large increase in the nuclear charge radii for Ca isotopes (especially for ^{52}Ca) beyond ^{48}Ca, that put into question the earlier suggested magicity at $N = 32$ (Rui 16). This experiment is an example where resonant ionization, bunching and CLS are all used to make the measurements possible.

7.7.3 Optical Pumping Using Collinear Laser and β–NMR

A typical set-up for the combined technique of optical pumping using collinear laser and β-detected NMR is schematically shown in Figure 7.35. A good degree of nuclear polarization can be obtained by optical pumping of atoms/ions to one of the hyperfine structure levels using circularly polarized light, since due to hyperfine interaction the polarization is shared by the coupled atomic system of electronic and nuclear spin. A weak longitudinal magnetic field provides the quantization axis along which both the electronic and nuclear moments are aligned. After the optical pumping zone, the spins are rotated adiabatically in the direction of the magnetic field of the NMR magnet. The atoms/ions are then implanted in a suitable host crystal placed in the strong magnetic field of an NMR magnet. During implantation in the crystal, the spin of the valance electron and the nuclear spin get de-coupled in the strong magnetic field of the NMR magnet. The nuclear polarization is then detected through β decay asymmetry as a function of laser frequency (actually Doppler tuning voltage). This is a much more sensitive technique than fluorescent photon detection and historically this technique was used first at ISOLDE for measurement of nuclear spin, magnetic di-pole moment and electric quadrupole moment of ^{11}Li, ruling out any large deformation of ^{11}Li (Arn 87, Arn 92). As shown in Figure 7.35, a charge exchange cell is often used to neutralize the ions into atoms and the atoms are polarized by optical pumping using circularly polarized laser. The β decay asymmetry is measured as the difference in the number of β particles emitted parallel and anti-parallel to the magnetic field divided by the total number of β particles. The β particles are counted using two plastic scintillators placed 180° apart. Thus, the asymmetry parameter (a) is given by: $a = (N_0 - N_{180})/(N_0 + N_{180})$. RF coils are placed around the host crystal that can be put on for NMR spectroscopy.

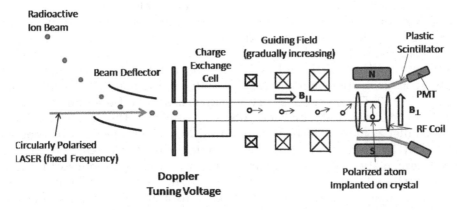

FIGURE 7.35
Schematic diagram of a β–NMR assisted collinear laser spectroscopy set-up.

The laser resonance corresponds to maximum polarization and therefore maximum asymmetry in β counting between the two β-detectors. The spin can be determined/inferred from the hyperfine spectrum (β asymmetry as a function of Doppler tuning voltage).

For the measurement of magnetic moment or the g factor, the crystal used for implantation should be one with a cubic lattice that has no electric field gradient at its lattice site. This ensures that the measurement is not affected by the existence of quadrupole moment, if any. The RF field is then applied and at resonance the RF frequency (Larmor frequency ω_L) matches with the Zeeman splitting. The RF will induce transition between adjacent magnetic sublevels, which destroys the polarization produced by the laser beam by changing the magnetic sub-level population. Thus, a change in β decay asymmetry at resonance can be observed. The g factor can be calculated from the relation: $\omega_L = g \mu_N I B / \hbar$, for which the magnetic field needs to be accurately known. The magnetic field is usually calibrated by performing an auxiliary experiment for another isotope for which the nuclear magnetic moment is precisely known. In the case of ^{11}Li, the magnetic field was calibrated by making measurements on ^{8}Li.

For the measurement of quadrupole moment, the atoms of the isotope are implanted in crystal with non-cubic lattice. The interaction of the electric field gradient with the quadrupole moment would lead to shifting of the magnetic sublevels and an NMR spectrum would show transitions at equidistant RF frequencies (leading to loss of β asymmetry). Since the frequency difference is proportional to the quadrupole moment the quadrupole moment Q_s can be deduced. For the case of ^{11}Li, by measuring the frequency difference for both ^{9}Li and ^{11}Li, the ratio Q_s (^{11}Li)/Q_s (^{9}Li) was found to be 1.14, which excludes any large deformation and suggests a spherical structure. At ISOLDE, the spin and magnetic moment of ^{31}Mg (half-life ~250 ms), an isotope in the island of inversion, was determined using this technique. Contrary to expectations based on shell model calculations, its ground state spin was found to be 1/2 with positive parity. The measured magnetic moment and spin show that 2p–2h intruder configuration dominates the ground state wave function (Ney 05). One of the shortcomings of this technique is that it does not allow very accurate determination of isotopic shifts because the process of optical pumping in a varying magnetic field (Figure 7.35) complicates the line shape.

It should be mentioned that apart from lasers projectile fragmentation reactions can also produce a sizable (although smaller in magnitude than the laser polarization) nuclear spin polarization, which is ~10% for a few nucleon removal fragmentation reactions (Asa 90). The β–NMR technique is thus employed for measurement of magnetic and quadrupole moments of exotic isotopes produced in projectile fragmentation. For example, static moments were measured for a chain of Al nuclei ($^{30-33}$Al) using LISE at GANIL and RIPS at RIKEN (Borr 02, Uen 05, Shi 12). The observation of a large electric quadrupole moment for ^{33}Al (half-life ~42 ms) signified the erosion of shell gap at $N = 20$. This method is very sensitive, has no half-life limitation and measurement can be carried out with beam intensity ~1–10 pps. The degree of polarization becomes smaller as more nucleons are removed from the projectile. The problem can be avoided by two-step production; for example, instead of producing ^{33}Al directly through fragmentation of ^{48}Ca (primary beam), ^{33}Al can be produced through: ^{48}Ca (primary beam) → Target → ^{34}Al → Target → ^{33}Al. In the two-step process, since ^{33}Al is produced ultimately through a one-neutron removal reaction, a sizable polarization can be obtained.

7.8 Matter Radii of Drip Line Isotopes through Measurements of Interaction Cross-Sections

It was through measurements of the interaction cross-section of light exotic nuclei at Bevalac, LBL that the halo structure in [11]Li was first conjectured (Ta 85). The measurement of the interaction cross-section at 790 MeV/u allowed extraction of the root mean square matter radius of [11]Li. The interaction cross-section σ_I is defined as the total cross-section for the process of nucleon removal. The total reaction cross-section is obtained by adding it to the cross-section for inelastic excitations to bound states. At high energies exceeding about 200 MeV/u, the interaction cross-section is practically equal to the reaction cross-section since the contribution from inelastic scattering is negligible. In the experiment (Ta 85) interaction cross-sections of different nuclei including [11]Li were measured in a transmission-type experiment. In a transmission-type experiment, the interaction cross-section is basically determined by counting the fraction of the beam particles (say [11]Li) that were incident on the target survived after the target.

In the [11]Li experiment, [11]Li along with other fragmentation products were produced in the reaction of a high energy [20]Ne beam with a Be production target (for the production of other Li and Be isotopes studied in the experiment [11]B projectile was used instead of [20]Ne). The fragments were then allowed to interact with a secondary target, called the reaction target, placed downstream. Different reaction targets (Be, C and Al) were used in separate beam runs. The reaction target produced further fragmentation of the secondary beam consisting of [11]Li and other fragments, but a large fraction of the beam would pass through the target without undergoing reaction. Identification of fragments before and after the reaction target was done by measuring ToF, magnetic rigidity and energy loss. For the momentum analysis and efficient collection of fragments after the reaction target a very large acceptance di-pole magnet placed downstream of the reaction target was used. For the identification of fragments before and after the reaction target, two sets of plastic scintillators (for time signal and energy loss signal) and multi-wire proportional counters (MWPC; for observing ion track) were used (one set placed upstream and the other placed downstream of the reaction target). The experimental arrangement used is schematically shown in Figure 7.36. Any event where the same nuclide was observed before and after

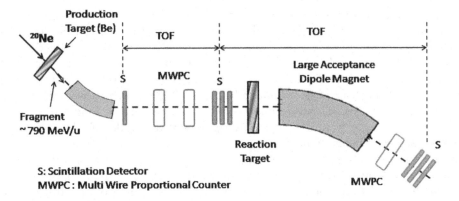

FIGURE 7.36
Schematic experimental set-up for interaction cross-section measurement using transmission method (Ta 85).

the reaction target was counted as a transmission event. The interaction cross-section was determined using the relation:

$$\sigma_I = (1/N_t) \ln(\eta_0/\eta) \tag{7.29}$$

where N_t is the number of target nuclei/cm², η is the ratio of the number of ions of a particular nuclide detected in the detector after the target to that of the number of ions of the same nuclide before the target with the target in position, and η_0 is the same ratio for an empty target run. The deviation of η_0 from the expected value of unity was mainly due to nuclear reactions in the detectors, which also act as targets. In the LBL experiment η_0 was close to 0.95. The value of η would depend upon the target thickness; the larger the thickness, the smaller η will be. It is important to note that the transmission method is very efficient for determining an interaction cross-section at high energies because of kinematic focusing and because high energies allow the use of a thick enough target ~1 g/cm².

It is known that an interaction cross-section essentially saturates for beam energy exceeding 200 MeV/u. So, σ_I is expected to represent a well-defined nuclear size. The interaction nuclear radii were defined (Ta 85) by the relation:

$$\sigma_I = \pi \left[R(p) + R(t) \right]^2 \tag{7.30}$$

where $R(p)$ and $R(t)$ are the interaction radii of the projectile and the target respectively. The separability of the two radii assumed in this expression was verified experimentally by determining σ_I of a given projectile for three different targets. This was done for a number of projectiles and it was established that the projectile's interaction radius is independent of the target chosen. Thus, for the same target, if the interaction cross-section for an exotic nucleus comes out to be much larger than the neighboring isotopes, it would indicate a large matter radius for the exotic isotope. This was found to be the case for ¹¹Li. A rough estimate of the matter radius using the above definition (Equation 7.30) of interaction radii can be determined to see, for example, how that compares with the radius expected on the basis of the $r = r_0 A^{1/3}$ law. For example, in the ¹¹Li case, the interaction cross-section was determined to be 1040 ± 60 mb compared to 796 ± 6 mb for ⁹Li and the approximate interaction radius R (¹¹Li) was deduced to be 3.14 ± 0.16 fm, which is much larger than the matter radius that would have been expected on the basis of $r = r_0 A^{1/3}$.

For deduction of rms radius of matter distribution from σ_I, a relatively simple Glauber-type calculation was used since at high energies the nuclear reaction is dominated by nucleon–nucleon collisions. It was observed however (Ta 85) that an effective value of about 80% of the free nucleon–nucleon interaction cross-section gave the best fit to the experimental values for the light nuclei (isotopes of He, Li and Be) studied in the experiment. The rms radius of ¹¹Li came out to be 3.27 ± 0.24 fm, which is much larger than expected from $r = r_0 A^{1/3}$. This was interpreted as due to either a long tail in the matter distribution originating from the weak binding of neutrons (halo) or a large deformation, which was later established to be a halo.

The transmission technique of measuring radii through measurement of the interaction/reaction cross-section is very convenient as well as efficient (at high energies exceeding about 200 MeV/u) and has been applied in the study of a number of drip line or near drip line light n-rich nuclei. In particular, it has led to the revelation of thick neutron skin in ⁸He (Ta 92) and in neutron-rich Na isotopes (Suz 95) and to the discovery of a two-neutron halo in ²²C (Tana 10, Tog 16). The halo in ²²C was discovered using the RIPS fragment

Experimental Techniques

separator at RIKEN and it was unique in the sense that the measurements were done at quite a low energy of 40 MeV/u. In the experiment, reaction cross-sections were measured for ^{19,20,22}C. The measured cross-sections for ^{19,20}C were found to be around 800 mb. But a large enhancement was observed for the cross-section of ^{22}C, which was found to be 1338 ± 274 mb. The rms matter radius of ^{22}C extracted using a Glauber calculation was 5.4 ± 0.9 fm, which means a halo much larger than ^{11}Li. The validity of the Glauber calculation is questionable at this low energy and the cross-section data also had statistical uncertainty of about ± 20%. The measurement was subsequently repeated at a higher energy of 235 MeV/u at the Riken RI Beam factory with much higher statistics that resulted in a much more accurate determination of the cross-section (1280 ± 23 mb) and the rms radius resulting from a Glauber calculation yielded a radius of 3.44 ± 0.08 fm. The result still confirms the existence of a two-neutron halo in ^{22}C as inferred from the earlier experiment but yielded much more precise data. The transmission experiments also revealed enhancement of cross-sections in very n-rich Ne and Mg isotopes that are attributed to quadrupolar deformation (Nak 17 and references therein). Availability of higher beam intensities at energies ~200 MeV/u or higher (as expected from future RIB facilities and already available at the RI Beam Factory at RIKEN) would allow pushing these studies to heavier and more exotic nuclei.

7.9 Measurement of Half-Life of Exotic Nuclei

The measurement of β-decay half-lives is important for a number of reasons. First of all, they carry the signature of the gross features of nuclear structure and experimental determination of the β-decay half-lives of exotic nuclei allows refinements of nuclear theories, such as the gross theory of β-decay. Also, as discussed in Chapter 1, the β-decay half-lives of n-rich nuclei and p-rich nuclei play a crucial role in deciding the stellar nucleo-synthesis paths in the hot r and rp processes and hence in the resulting abundance pattern. Also, the experimental determination of ft values for a number of $0^+ \to 0^+$ transitions would enable the testing of the CVC through constancy of the corrected ft, that is, the Ft values. For this, it is necessary to have very precise determination of half-lives of certain exotic nuclei, as discussed in Chapter 1.

The mass difference between the parent and the daughter or the Q_β value is the most important parameter that determines the β-decay half-lives and thus exotic nuclei with large Q_β values have short half-lives, usually in the range from a few ms to a few 100 ms; sometimes even shorter than 1 ms for nuclei on the drip lines. Besides Q_β, β decay transition probabilities (β strength functions) to states in the daughter also play an important role in determining the half-lives. Short half-lives of exotic nuclei require on-line techniques to experimentally determine their half-lives. Also, the separation of the nuclide of interest from other reaction products is essential since the nuclear reactions which are mostly used for the production of such exotic nuclei, such as the spallation, projectile fragmentation and fission also produce a large number of other nuclei. Any meaningful measurement of half-life using decay spectroscopy thus requires a clean or at least a good degree of separation. In all cases, the beam of exotic nuclei after separation must be stopped, usually in a catcher foil/aluminized mylar tape for carrying out the decay measurements or in active detectors, for example a Double-Sided Silicon Strip Detector (DSSD), which are used at much higher ion energies in a PFS facility.

The large energy window available for β-decay of exotic nuclei results in β-decay populating particle-unbound states in the daughter and thus in addition to β-delayed, γ emission, β-delayed particle emissions are also observed. The half-life may be measured either by detecting the β particles from the decay or by detecting any of these radiations: gamma, proton, alpha, neutron, etc., as a function time.

In the ISOL method, a much better degree of separation can be achieved (as discussed already) compared to the PFS method. Historically, therefore, the study of decay spectroscopy, including half-life determination, was mainly carried out in ISOL-based facilities. The basic experimental arrangement in the ISOL method comprises of production, ionization, separation and deposition of separated exotic nuclei on a foil or aluminized Mylar tape positioned at the detection station where the spectroscopic measurements are carried out. In heavy ion recoil separators, the reaction products after separation having energies of a few MeV/u need to be slowed down by using a foil degrader. For measurement of a half-life the beam needs to be pulsed on and off. The basic set-up used for half-life measurement is shown schematically in Figure 7.37. The activity is implanted on an aluminized Mylar tape during the beam-on time (collection time of the activity), which usually does not exceed about five times the expected half-life of the nuclide of interest (five half-lives are enough to saturate the activity). The beam is then pulsed off and the tape is moved quickly to a detection station for β counting in a multi-scaling mode (recording the ADC channel and time for each β counting event). The counting time is usually kept about ten times the expected half-life, to have enough data points for accurate fitting of the half-life curve. For low statistics experiments, which are rather common for exotic nuclei, the method of maximum likelihood is used for fitting the half-life curve. The sequence of activity collection, tape movement and detection are repeated several hundred times or more as required for sufficient statistics.

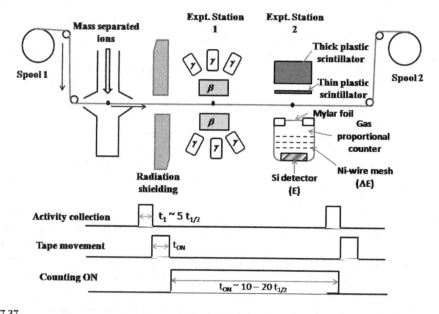

FIGURE 7.37
Schematic set-up showing the detectors for β counting, the tape transport system and the detector arrangements for β-delayed γ and proton, and $Q_β$ measurements.

Experimental Techniques 231

The detector used for β counting is either a gas proportional counter or a plastic scintillator. Such detectors can provide almost 4π solid angle coverage and 100% efficiency for β detection and are therefore suitable for half-life studies of exotic nuclei. However, since the β energy spectrum is a continuous one, it has no isotope selectivity. Thus, any undetected impurity nuclide would also contribute to the β counting introducing error to the half-life measurement. However, if the β-delayed γ-rays are also detected in coincidence with β detection, the selectivity is gained if the γ-ray energies are known. Conversely, γ detection can be used to qualify the level of impurity in the implanted beam. The detector used for γ measurement is a large volume Compton suppressed HPGe clover detector. Since γ detection efficiency is rather poor a large number of detectors/γ detector array is used for adequate solid angle coverage. Similarly, for delayed neutron measurement, a large number of neutron detectors is used (not shown in Figure 7.37) providing enough solid angle coverage. The neutron detectors used are ^{3}He and BF_3 proportional counters embedded in a paraffin/polyethylene moderator. The moderator thermalizes the neutrons increasing their capture cross-sections many-fold, and a total neutron detection efficiency of about 40% can be achieved.

For delayed proton/α measurement gas proportional counters as ΔE detector and a silicon detector as E detector is used, as shown in Figure 7.37. The detector is placed at Station 2, physically at a distance from Station 1 to facilitate the positioning of detectors of different types close to the collection spot. In this case, the tape, after collection of activities, is moved to Station 2. A single station for all types of measurements can also be used by replacing one detector set-up with another required for the particular experiment. The β particles would lose much less energy in the gas compared to the protons/alphas. The same is also true for the silicon detector whose thickness is chosen to fully stop only the protons/alphas. Thus, in a two-dimensional ΔE–E plot, the β particles would all crowd at low energies and protons/alphas would be well separated. Removal of β events through a software cut would result in a clean proton/ α spectrum and the half-life can be determined by plotting proton/ α peak areas as a function of time. Figure 7.37 also shows the combination of a thin and a thick plastic scintillator. Such a detector telescope is used for Q_β measurement. The coincidence between ΔE and E is required for γ rejection (since γ-rays hardly lose any energy in the ΔE detector) that allows recording of a clean β spectrum from which the end point energy can be extracted.

The half-lives of a large number of p-rich and n-rich exotic isotopes have been determined using the above techniques in ISOL-type facilities such as ISOLDE, IGISOL, ISAC and elsewhere. These include measurements on super-allowed $Z = N$ p-rich positron emitters (Bal 01, Kan 05, Gri 08, Fin 11, Dun 16a), such as ^{74}Rb ($t_{1/2} \sim 65$ ms) and half-lives of a large number of n-rich nuclei along or near the r process nucleo-synthesis path, such as ^{99}Kr, ^{130}Cd, ^{150}Cs (Berg 03, Lic 17, Dun 16b), to name just a few.

In many cases the decay measurement is carried out after a Penning trap that allows measurements free from isobaric contamination. Recently, the MR-ToF in combination with a cooler buncher has been proved to be an exciting alternative for isobaric separation and the measurement of half-lives (Wolf 16). In a proof of principle experiment, the half-lives of a number of short-lived nuclei: ^{27}Na ($t_{1/2} \sim 440$ ms), ^{97}Rb ($t_{1/2} \sim 244$ ms), ^{99}Rb ($t_{1/2} \sim 78$ ms), ^{100}Rb ($t_{1/2} \sim 76$ ms), ^{100}Sr ($t_{1/2} \sim 311$ ms), ^{103}Sr ($t_{1/2} \sim 103$ ms), ^{129}Cd ($t_{1/2} \sim 306$ ms) were determined using an RFQ cooler buncher and MR-ToF combination. The ions were stored for different accumulation times in the RFQ cooler during which they decayed to different extents according to their half-lives. After a certain accumulation time the ions were injected into the MR-ToF, where the remaining ion species (which did not decay in the RFQ cooler) were mass separated and counted. The number of ions of a given nuclide counted as a function of accumulation time in an RFQ cooler gives the half-life. The half-lives determined (as mentioned in the parentheses) following this technique agree well with the half-lives of the same isotopes measured using other techniques, except for ^{129}Cd where there is some significant deviation. The MR-ToF offers a very simple and efficient way of measuring the half-lives of exotic species.

The PFS method allows the measurement of the half-lives of exotic isotopes of a wide range of elements because it is independent of the chemical nature of the isotope. However, the purity of the beam is a concern in PFS facilities, especially for decay spectroscopic measurements. As has already been mentioned in Section 7.4, the ion beam, even after separation in a fragment separator, contains a good number of fragments of different isotopes (cocktail beam). This shortcoming has largely been overcome by ensuring selectivity in the β-detection method by implanting the fragments in an active detector and detecting the β particles resulting from the decay of fragments in correlation with the identified fragments. The active detector used is a DSSD that detects the fragments as well as the β particles resulting from the decay of the fragments. The active detector usually comprises multi-layer strip detectors (DSSD) and is placed in the β-detection station situated downstream of the PFS separator. Prior to implantation in the DSSD, the fragments are identified on the basis of ΔE, ToF and $B\rho$ measurements with the ΔE detector placed in the immediate upstream of the DSSD. Each DSSD is divided into a number of horizontal strips on the front side and vertical strips on the back, each with a width, usually of about 1 to 3 mm. Its position and timing can be deduced from the signal generated by the implanted isotope. As the isotope undergoes β-decay it generates a signal at the same position (pixel) in the same DSSD layer. Such correlated events are considered to be true events and the decay curve can be generated by detecting the β-particles as a function of time after the implantation of ions. By placing a γ array around the DSSD, the β-delayed γ-rays can be measured. Similarly, the β-delayed neutrons can be measured by using a neutron array around the DSSD. Thus, both the ISOL and the PFS techniques can be used to determine the short β-decay half-lives of exotic nuclei and each can be used to cross-check the half-lives determined using the other technique.

In the last decade or so, half-lives of a large number of n-rich nuclides (exceeding 150), of interest to the r process nucleo-synthesis have been determined in PFS facilities (Per 09, Nis 11, Lor 15, Wu 17). A majority of these were first half-life measurements for very exotic n-rich nuclei, e.g., ^{139}Sn ($t_{1/2} \sim$ 130 ms); ^{132}Ag ($t_{1/2} \sim$ 28 ms); ^{115}Nb ($t_{1/2} \sim$ 23 ms); ^{103}Rb ($t_{1/2} \sim$ 23); ^{100}Kr ($t_{1/2} \sim$ 7 ms).

7.10 Coulomb Excitation and Study of Exotic Nuclei

Coulomb excitation has been extensively used to probe the shell structure of exotic nuclei. The technique has been especially used to study the nuclei in and around the island of inversion and those in and around the conventional shell closures ($N =$ 20, 28, 50, 82). As discussed in Chapter 1, Section 1.1.3, Coulomb excitation probes the shell structure through the determination of the excitation energy of the first $J = 2^+$ excited state, E (2^+), and the transition probability of 0^+ to 2^+ excitation, B ($E2$), for even–even nuclei. To reiterate it should be mentioned again that E (2^+) and B ($E2$) give a measure of deformation of a nucleus. For deformed nuclei, the first excited state would be much closer to the ground

Experimental Techniques

state and the B ($E2$) value would be much more enhanced compared to a magic nucleus in which the 2^+ state is at much higher excitation and hence is difficult to excite.

Experimentally the excitation energy of the 2^+ state and the cross-section of excitation can be determined by exciting the nucleus using inelastic scattering or nuclear reaction. If the excitation is done electromagnetically such as in a Coulomb excitation experiment, B ($E2$) can be deduced from the measured cross-section, since the cross-section is approximately proportional to B ($E2$). The advantage of the Coulomb excitation over the nuclear excitation is that the theory of electromagnetic interaction is well developed and B ($E2$) can be determined in a model-independent way.

In a Coulomb excitation (COULomb EXcitation or COULEX) experiment involving a radioactive nuclide, a target (often high Z) is bombarded with a beam of the radioactive projectile. The virtual photons created by the target-projectile interaction excite the projectile (through the projectile absorbing a virtual photon). The target nuclei would also get excited by absorbing photons. If the excited level in the projectile is bound, it decays by emitting γ-rays. The maximum excitation energy of the projectile in a COULEX experiment increases with beam velocity and decreases with the distance of the closest approach (Bert 07). At sub-barrier projectile energies $v/c < 0.1$, the maximum excitation energy is limited to a few MeV, while at much higher energy exceeding 100 MeV/u, excited states exceeding 10 MeV can be excited in principle. Thus, in a COULEX experiment done at sub-barrier energies (often called a "safe" Coulomb excitation experiment) to determine E (2^+) and B ($E2$), there will be no contribution from nuclear excitation and no corrections would be required for possible feedings from the higher-lying excited levels, which might also be excited at higher projectile energies.

Pioneering studies on exotic nuclei using a COULEX experiment at sub-barrier energies have been carried out at HRIBF (Rad 02, Rod 05) and at ISOLDE (Nie 05, Hur 07, Wal 07, Ced 07). These studies have been aimed at studying the shell structure of exotic nuclei at and around the shell closures (N = 20, 50, 82). The experimental set-up used in all these experiments, basically comprised of two parts: first, the production, ionization, separation and acceleration of the exotic nuclide of interest, and second the Coulomb excitation of the nuclide of interest (the radioactive projectile) in interaction with a secondary target and measurement of emitted γ-rays resulting from the decay of the excited projectile in coincidence with the projectile. The basic components of the set-up used at ISOLDE for the determination of E (2^+) and B ($E2$) of p-rich ^{110}Sn are shown schematically in Figure 7.38.

The radioactive ^{110}Sn was produced in the reaction of 1.4 GeV proton beam with a 2.7 g/cm^2 LaC$_x$ target. The reaction products were transferred to the ion source by the process of diffusion and effusion as discussed earlier in Section 2.7. In the ion source Sn atoms were ionized to 1^+ charge state using a laser ionization technique. After ionization the Sn ions were extracted, and mass separated in a separator di-pole magnet. The ions were then accumulated, cooled and bunched in a gas-filled Penning trap. From the Penning trap the ions were fed into an EBIS charge breeder, where it was charge-bred to a 27^+ charge state, for which a confinement time of 98 ms was used. It should be noted that for shorter-lived nuclei with half-lives of approximately tens of ms, a lower confinement time needs to be used which would result in a lower charge state after charge breeding. The high charge state ions from EBIS were extracted in about 100 μs long pulses and were fed into the REX-ISOLDE linac system for post-acceleration. A pure ^{58}Ni target of 2 mg/cm^2 was used for the Coulomb excitation of the 2.82 MeV/u ^{110}Sn beam. For detection of γ-rays from the de-excitation of Coulomb-excited states in the projectile as well as the target, a set of 24 HPGe detectors with 144 segments (each with six segments), called the MINIBALL array,

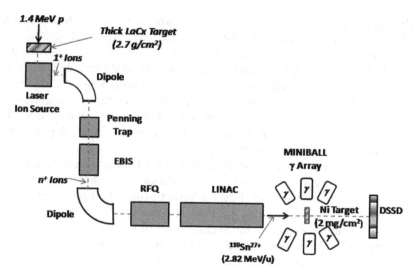

FIGURE 7.38
Schematic set-up for a safe COULEX experiment at ISOLDE.

was used surrounding the Ni target in close geometry. The total photo peak efficiency obtained with the array was around 10% for 1.3 MeV γ-ray. For detection of the scattered ^{110}Sn beam as well as the target nucleus ^{58}Ni a disc-shaped DSSD was used 30.6 mm downstream of the target. The detector (DSSD) had 16 annular strips on the front and 24 radial strips on the back. The detector was further sub-divided into four separate quadrants. The DSSD was used for recording the energy and the angle of the scattered particle. At any given angle the scattered beam (^{110}Sn) and the target nuclei (^{58}Ni) could be well separated on the basis of their energy. The energies and angles of the γ-rays detected in coincidence with those of the scattered charged particles allowed the reconstruction of the kinematics of each event, which in turn allowed for Doppler shift correction and good resolution for γ detection. It should be noted that at 2.82 MeV/u ($v/c \sim 0.08$), the Doppler shift uncorrected γ spectrum would contain only very broad peaks. The analysis of the experiment data resulted in an energy of 1211.9 keV for the first 2$^+$ excited state (almost constant across the tin isotopes as expected from the seniority scheme) and a $B(E2)$ value of $0.220 \pm 0.022\ e^2b^2$, which is large enough to conclude that ^{110}Sn is a deformed nucleus and that the trend in the $B(E2)$ values in even Sn nuclei is not symmetric with respect to the mid-shell number $A = 116$ (in between the two doubly magic nuclei ^{100}Sn$_{50}$ and ^{132}Sn$_{82}$).

However, in PFS facilities, where incident energies are high the COULEX experiments are carried out at much higher projectile energies. The projectile is Coulomb-excited using high Z targets like Au, Pb, etc. At these energies, the projectile would also be excited because of nuclear excitation. The contribution from nuclear excitation is minimized by making measurements at large impact parameters, that is, at small scattering angles. Further, the contribution of nuclear excitation can be extrapolated by conducting additional measurements using a low Z target such as carbon, for which Coulomb excitation probability is much smaller. Often, the $B(E2)$ value of a stable isotope, which is well known, is also determined in the same experiment and the $B(E2)$ value of the exotic nuclide is determined relative to that of the stable isotope. High projectile energy allows the use of much thicker secondary targets with target thickness of approximately a few hundred mg/cm^2, which is a few hundred times thicker than the typical target thickness of a few mg/cm^2 used

Experimental Techniques

in sub-barrier energy experiments. The increase in the yield by more than two orders of magnitude allows experiments to be carried out with low intensity radioactive projectiles, making possible COULEX experiments on more exotic nuclides that are produced with low intensities. Because of this important advantage, the COULEX experiments for exotic nuclei are increasingly being done nowadays at PFS facilities (refer, among others, to Sch 96, Iwa 01, Iwa 05, Sor 02, Gua 13, Doo 14).

As an example of a typical experimental set-up used for a COULEX experiment at PFS facilities, the one used at GANIL (Sor 02) for the study of ^{66,68}Ni is shown schematically in Figure 7.39. The experiment used a ^{70}Zn primary beam at an energy of 65.9 MeV/u for the production of ^{66,68}Ni nuclides. The primary target used for the production was also a Ni target. The fragments of interest (^{66,68}Ni) were separated using the LISE separator. The Coulomb excitation of the selected projectiles was induced by a 220 mg/cm^2 thick lead target placed at the final focal plane of LISE. The γ-rays were measured by an ensemble of four segmented clover detectors placed around the secondary lead target in close geometry resulting in a total photo peak efficiency of about 4% (arrays comprising of NaI (Tl) detectors are also used for γ-ray measurements because of their higher intrinsic efficiency). An x–y set of two PPACs placed before the secondary lead target followed by two drift chambers after the target were used for particle tracking. The particle tracking ensured that the beam (^{70}Zn) and the fragments of interest ^{66,68}Ni had almost the same beam profile, ensuring in turn that the measurements for all these nuclides were carried out under the same geometrical conditions. The projectiles/fragments were identified using a ΔE-E Silicon telescope that comprised of two annular silicon detectors, the first one ΔE followed by the second one E. The particle detectors cover an angle from 1.5 to 6° in the reference frame of the projectile. Nuclides emerging at smaller angles were detected in a plastic scintillator placed downstream of the annular ΔE-E silicon telescope. The scintillator gave the total number of fragments (say of ^{68}Ni) impinged on the secondary target (since a very small fraction undergoes reaction) while the number scattered in the silicon detector gives the number of fragments that underwent Coulomb excitation, and the ratio was used for calculation of 0$^+$ to 2$^+$ excitation cross-section. The measurements were carried out with ^{70}Zn beam at reduced intensities and for ^{66,68}Ni. The B ($E2$) of ^{70}Zn was used as a reference to determine the B ($E2$) values for ^{66,68}Ni. Thus, the B ($E2$) for ^{66}Ni and ^{68}Ni were determined by normalization to the known B ($E2$) of ^{70}Zn and also by determining the cross-sections for 0$^+$ to 2$^+$ excitation. The emission angles of the γ-rays detected in the germanium detectors

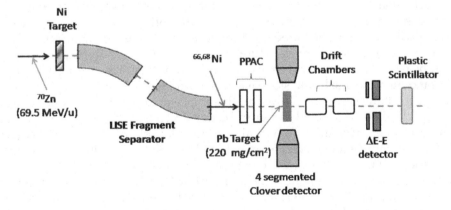

FIGURE 7.39
Typical set-up for a COULEX experiment at PFS facilities.

were used to correct for the Doppler shift, and to obtain suitable energy resolution. The comparison of the Doppler corrected and uncorrected spectra allowed the disentanglement of γ-rays originating from the target and the projectile. It was found that the E (2+) for ^{68}Ni was quite high (=2033 keV) and the B ($E2$) value was small and only 255 $e^2 fm^4$. The corresponding values of E (2+) and B ($E2$) for ^{66}Ni are about 1425 keV and 600 $e^2 fm^4$ respectively. Thus, there is a clear signature of a shell gap at $N = 40$. However, the $2n$ separation energy and other evidence such as low excitation energies of E (2+) in ^{70}Zn (+2p; 885 keV) and ^{66}Fe (−2p; 573 keV) suggest that the shell gap in ^{68}Ni gets eroded very fast.

7.10.1 Coulomb Break-Up

As mentioned in Chapter 1 (Section 1.1.2), large Coulomb break-up (or Coulomb dissociation) cross-section connected to soft $E1$ excitation resulting from the low binding energy of the halo neutron (s) is a signature of halo nuclei. Coulomb break-up is just a Coulomb excitation in which the excitation populates continuum states in the halo nuclide and the excited nucleus decays through particle emission along with the core (^{19}C* → ^{18}C + n; ^{11}Li → ^{9}Li + n + n, etc.). At the high beam energies typically used in PFS-type facilities and for high Z targets, it is established that $E1$ transitions mainly contribute to Coulomb excitation (and hence in the subsequent break-up) and especially for n-rich halo nuclei with an extended tail, the $E2$ contribution to the Coulomb excitation is negligibly small and can be ignored. Thus, the Coulomb break-up is almost entirely due to $E1$ excitation. Coulomb break-up studies have been extensively carried out on light n-rich nuclei at all the major PFS facilities to identify the existence of the halo and to study its structure. At typical projectile energies (a few tens to several 100 MeV/u) used in PFS separators, the calculated number of virtual $E1$ photons, created due to the rapidly changing electromagnetic field as the projectile approaches a high atomic number Pb target, shows an energy dependence that decreases exponentially with the energy of the photons (Bert 09). The number of $E1$ photons is therefore pretty large at lower photon energies in the range of 0–6 MeV. This makes the Coulomb break-up a sensitive tool for probing the soft $E1$ excitation in halo nuclei that is expected to occur at an excitation energy of about 1 MeV. As the projectile energy increases, say from 50 to 500 MeV/u, the virtual photon energy spectrum extends to higher energies reaching the GDR region. To study PDR ($E^* \sim 10$ MeV) and GDR ($E^* \sim$ 15–20 MeV) in neutron-skin nuclei comparatively higher projectile energies are therefore preferable.

The information to be extracted from a typical break-up experiment consist of the Coulomb dissociation cross-section, the nature of the soft di-pole excitation, the neutron separation energy and the shell model orbital occupied by the halo neutron(s). The other important property of n-rich nuclei studied by Coulomb break-up experiments is the pygmy di-pole resonances in n-rich nuclei with neutron skins. The pygmy resonance and its importance in nuclear astrophysics have been discussed in Chapter 1.

The experimental technique followed in these measurements is most often the invariant mass spectroscopy discussed earlier in Section 7.5.1.3 and therefore the experimental set-up used is identical/similar. A typical experimental set-up used at RIKEN (Nak 94, Nak 99) for Coulomb dissociation experiments is shown in Figure 7.40. The example of the break-up reaction ^{19}C* → ^{18}C + n (carried out at ^{19}C beam energy of 67 MeV/u) is taken to explain the experimental technique, which is the same for all other break-up studies at PFS facilities. The exotic nuclide of interest (^{19}C), after production and separation in a Projectile Fragment Separator (with its energy, position and angle determined event by event), is

Experimental Techniques

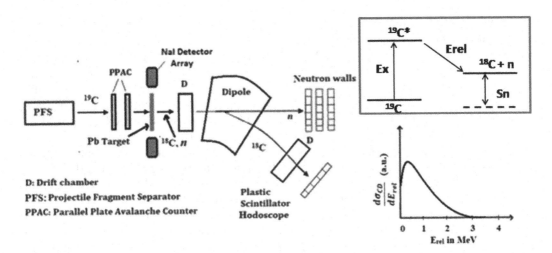

FIGURE 7.40
Schematic set-up for a Coulomb break-up experiment.

excited using a secondary target of high Z (^{208}Pb). The excited nuclide would decay into a heavy fragment (^{18}C) and light particles (a neutron). The large acceptance di-pole magnet bends the charge particles (^{18}C) and the identification of ^{18}C is done by ΔE, ToF and $B\rho$. The plastic scintillator hodoscope placed downstream of the large acceptance di-pole magnet is used for measurements of ToF between the target and the hodoscope as well as the energy loss ΔE. The particle tracking information is derived from the drift chamber placed downstream just after the magnet. The momentum vector of the heavy fragment ^{18}C is determined from the ToF and the particle trajectory in the drift chamber. The neutrons move undeflected through the magnet and are detected in the plastic scintillators forming the neutron walls. The neutron momentum vector is determined by the ToF and the position of its detection in the scintillation counters. The invariant mass, actually the relative or the decay energy E_{rel}, is obtained from the four momenta of the heavy fragment (^{18}C) and the light particle (n). The excitation energy of the projectile (^{19}C) is given by: $E_x = E_{rel} + S_n + E_\gamma$. A NaI γ-detector array placed around the target is used to measure γ-rays from the excited bound state of the heavy fragment. In many cases, for exotic nuclei close to the drip line, there are no bound excited states in the fragment (as also for ^{18}C). In such cases no γ-rays would be detected in the γ-detector in coincidence with the neutron and the excitation energy is just $E_x = E_{rel} + S_n$.

The break-up would have contributions from nuclear as well as Coulomb interactions. To obtain the Coulomb break-up cross-section, the nuclear contribution to the cross-section needs to be subtracted from the experimentally determined cross-section. This is done by performing two separate experiments, one using a light target like carbon (C) and the other using a heavy target (Pb, Au, etc.). The assumption is made that for the carbon target, the entire break-up cross-section is from nuclear interaction. The nuclear contribution for the break-up reaction in the case of the Pb target can then be well approximated by scaling the cross-section for carbon appropriately for the nuclear size effect (Nak 09). The cross-section due to pure Coulomb interaction is deduced by deducting the nuclear contribution of the cross-section from the experimental cross-section. At high energies the Coulomb part usually dominates, and the Coulomb cross-section can thus be deduced quite reliably by employing this method.

The energy spectrum for extracted pure energy differential Coulomb dissociation cross-section ($d\sigma_{CD}/dE_{rel}$ vs E_{rel}) in the case of $^{19}C^* \rightarrow {}^{18}C + n$ (Nak 99) is shown in Figure 7.40, which shows a peak at about $E_{rel} \sim 300$ keV. The nature of the energy spectrum is typical and is more or less the same for other halo nuclei (a peak or strong enhancement in soft E1 excitation just above the neutron separation energy, which corresponds to excitation energy $E_x \sim 1$ MeV, or even less, depending upon S_n or S_{2n}). The total Coulomb break-up cross-section can be obtained by integration (area under the curve in the figure), which is expected to be large. In the case of ^{19}C, it was found to be ~1 b.

In the equivalent photon model (Ber 88), the experimentally determined energy differential Coulomb break-up cross-section $d\sigma_{CD}/dE_x$ or $d\sigma_{CD}/dE_{rel}$ is proportional to the number of virtual E1 photons at the excitation energy E_x and to the reduced transition probability $dB(E1)/dE_x$. So, for halo nuclei for which S_n and ground state wave-function are known, the value of $dB(E1)/dE_x$ can be deduced, and hence energy integrated total $B(E1)$ from the experimental spectrum. The $B(E1)$ values usually come out to be much higher than that the strongest E1 transitions between the ground state and a bound excited state. The deduced $dB(E1)/dE_x$ spectrum can also be compared with the theoretical predictions based on direct break-up models. For most halo nuclei, the agreement between the theoretical prediction and the experiment has been found to be satisfactory enough to suggest that the soft di-pole resonance in neutron halo nuclei is not arising predominantly out of a slow collective oscillation of the halo neutron against the core but is due to a direct break-up process.

Conversely, for the cases (for example ^{19}C) where S_n and ground state wave function are not known from any other experiment, the energy differential cross-section spectrum can theoretically be deduced by calculating $dB(E1)/dE_x$. The calculation in this case is done by assuming different possible ground state configurations and S_n values and the theoretical spectrum is compared with the observed cross-section spectrum in each case. Such an analysis carried out for ^{19}C indicated that the halo in ^{19}C (Nak 99) is a predominantly s wave halo. The best fit required an S_n value of about 530 keV. In the same experiment S_n was also extracted from the angular distribution data of $^{18}C + n$ (in the c.m. system) and the value of $S_n = 530 \pm 130$ keV was consistent with the value deduced from the relative energy spectrum.

The pygmy resonance in a number of n-rich nuclei, such as $^{130,132}Sn$ (Adr 05), ^{26}Ne (Gib 08), ^{68}Ni (Ros 13), etc., has been studied using the Coulomb break-up. In a pioneering experiment at GSI using 500 MeV/u ^{130}Sn and ^{132}Sn projectiles, the existence of pygmy resonance has been detected in both the tin isotopes at excitation energies of around 10 MeV, well above the one neutron separation energy (6.2 MeV for ^{132}Sn) but much lower than the GDR peak observed at an energy of ~16 MeV. The projectiles were Coulomb excited using a Pb target which subsequently decay by neutron emission. The excitation energy is determined by deducing the relative energy following the same procedure as described earlier and measuring the energies of the decay γ-rays by using the 4π Crystal Ball positioned around the target that consists of 160 NaI detectors that gave an energy resolution of about 15% for the γ peaks after Doppler correction. The Coulomb cross-section for the Pb target was deduced after subtracting the nuclear cross-section measured with a Ni target (after appropriate scaling), as explained above. The energy differential Coulomb break-up cross-section for both ^{130}Sn and ^{132}Sn showed two peaks: one around 10 MeV (FWHM ~ 2.5–3 MeV) which corresponds to the PDR and the other around 16 MeV (FWHM ~ 4.8 MeV) corresponding to GDR. The deduced $B(E1)$ values for PDR were 3.2 e^2fm^2 and 1.9 e^2fm^2 for ^{130}Sn and ^{132}Sn respectively. The large values gave strong indication for a collective oscillation of the core against the neutron skin.

7.11 Measurement of Cross-Sections for Nuclear Astrophysics

In Chapter 1, the important role played by the radiative capture reactions in stellar nucleo-synthesis was discussed. In this section, we briefly discuss some of the experimental techniques that are used for measuring these reaction rates involving β-unstable nuclei. It should be recalled that reactions involving exotic nuclei are important mainly in explosive stellar scenarios where the temperature of the stellar environment is around 10^9 K or even higher. In the rp process we need mainly to determine the proton capture reactions (p, γ) involving relatively light and medium heavy ($A \sim 20$ to less than 100) exotic p-rich nuclei. The astrophysically relevant energies (energies around the Gamow peak) at which the proton capture reactions need to be determined always lie well below the Coulomb barrier. So, the cross-sections are low. The cross-sections can often be estimated by extrapolating cross-sections (actually the astrophysical S factor) measured at much higher energy down to astrophysically relevant energies. But such estimates, although they are often reliable for non-resonant capture, become very unreliable if there are resonances at sub-barrier energies. In such cases the assumption of weak energy dependence of the S factor completely fails, and the cross-sections are dominated by resonance absorption rather than non-resonant direct capture. It is important to note that the nuclei in the rp process path are typically light p-rich nuclei almost on the proton drip line. The density of states in these light exotic nuclei is low and the capture is often dominated by isolated resonances. So, these cross-sections need to be determined experimentally at the astrophysically relevant energies in the region covered by the Gamow peak.

In the case of the r process nucleo-synthesis (especially for the cold r process), the neutron capture reactions (n, γ) involving light as well as heavy nuclei need to be determined. As discussed in Chapter 1, the neutron capture rates have hardly any influence in the hot r process nucleo-synthesis in which capture rates are much faster than the β decay rates but are important in cold r process nucleo-synthesis in which there is a competition between the neutron capture and the β decay. The compound nuclear reaction codes do not give a reliable estimate of these reaction cross-sections since the exotic nuclei have low density of states. In the r process path, the n-rich exotic nuclei have low neutron separation energies ($S_n \sim 2$–3 MeV) and therefore the density of states is rather low requiring analysis based on individual states; also at the shell closures the density of states decreases further. Thus, most of these cross-sections need to be determined experimentally. Since there is no Coulomb barrier for neutrons, the energy at which these reaction rates need to be determined is governed only by the temperature of the astrophysical site.

To determine the cross-sections experimentally, the most obvious problem is that targets made of short-lived isotopes cannot be prepared. So, the experiments are to be done in inverse kinematics using radioactive projectiles of these exotic nuclides produced in an ISOL or PFS facility. For proton capture reactions the radioactive projectiles can interact with a hydrogen target and the reaction in inverse kinematics can be studied, but for neutron capture reactions a neutron target cannot be made since the neutron itself is unstable. So, a source of neutrons (such as a spallation neutron source, a reactor, etc.) producing strong neutron flux is needed with which the n-rich RIB can interact. In either case, the RIBs can only be produced with intensities several orders of magnitude lower than the intensity of a primary beam of protons or flux of neutrons produced in a reactor or in a spallation neutron source. Thus, the inverse kinematics experiments suffer from much lower production rates compared to the usual kinematic situation of a proton or neutron beam hitting a stable target. For proton capture reactions hydrogen targets are used, but the target thickness can hardly exceed 10^{18}–10^{19}/cm². This is because nuclear astrophysics experiments are low energy experiments (Gamow peak typically around a few hundred keV), which is well below the Coulomb barrier and this limits the target thickness. The cross-sections for proton capture are also quite low, often less than or about 1 μb. The low intensity of the radioactive projectile, thin target and low cross-section together make the yield very small and make the experiments difficult.

In view of these difficulties in the direct method, a number of indirect techniques have been developed to measure the radiative capture cross-sections. These include the Surrogate reaction technique (Cram 70, Bau 86a, Esc 112), the Coulomb dissociation technique (Bau 86, Moto 91), and the Asymptotic Normalization Co-efficient (ANC) method (Xu 94, Banu 09). In many cases the indirect methods offer the only possibility to deduce the cross-section data experimentally. However, the indirect methods overcome the problem of low yield by using a different nuclear reaction (e.g., single nucleon transfer reaction in the ANC method) from which structure information relevant to the extraction of capture reaction cross-section of interest is obtained. This involves theoretical calculations and often some assumptions; and the accuracy of the cross-section data extracted depends on the validity of the calculations and the assumptions for the particular case. In this section, we concentrate mainly on the direct cross-section measurement techniques which, although more difficult experimentally, have been successfully employed for the determination of a number of important reaction rates of astrophysical importance. We also mention some advanced techniques that have been proposed involving storage rings for studying radiative capture reactions involving both light charged particles (p, α, etc.) and neutrons. Among the indirect techniques, only the Coulomb dissociation method would be discussed. There are a large number of excellent review articles on experimental techniques for the determination of reaction rates of astrophysical significance using RIBs. Readers may consult, among others, the articles by S. Kubono (Kub 01), R. Reifarth et al. (Rei 14) and A. C. Larsen et al. (Lars 19) and the references therein.

7.11.1 Measurement of Proton Capture Cross-Section, Direct Methods

The first successful direct measurement of proton capture cross-section was carried out at Louvain-la-Neuve for the ^{13}N (p, γ) ^{14}O reaction. This is a key reaction for the hot CNO cycle whose cross-section would decide the competition between the formation of ^{14}O and the β^+ decay of ^{13}N and therefore the suitable stellar environments (temperature) in which the normal CNO cycle would switch over to the hot CNO cycle. In the experiment a radioactive beam of ^{13}N was allowed to bombard a polyethylene $(CH_2)_n$ target of about 180 μg/cm² thickness (Dec 91, Del 93). The experimental arrangement is schematically shown in Figure 7.41 (left). The ^{13}N radioactive beam was produced in the ^{13}C (p, n) ^{13}N reaction by bombarding a graphite target with 30 MeV proton beam of intensity up to 200 μA delivered by the cyclotron Cyclone 30. The reaction products diffusing out of the target were ionized in an ECR ion source and after magnetic separation, the ^{13}N ion along with other contaminant ions were axially injected in the post-accelerator $K = 110$ Cyclotron (Cyclone 110). The cyclotron, being an excellent mass separator itself, produced a clean beam of ^{13}N ion beam (purity exceeding 99%) accelerated to 8.2 MeV. The achieved ion beam intensity was about 3×10^8 ions/s.

In the stellar temperature range, the ^{13}N (p, γ) ^{14}O reaction is expected to be dominated by resonant capture of s $(l_p = 0)$ wave protons to the 5.173 MeV 1⁻ excited state in ^{14}O, which is about 528 keV above the proton threshold in ^{14}O. The total width of the 5.173 MeV state, largely dominated by the proton width (Γ_p), was known from previous measurements to

Experimental Techniques 241

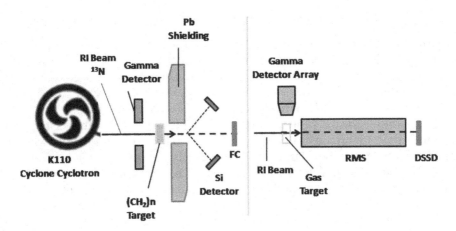

FIGURE 7.41
Schematic presentation of typical experimental set-ups for measurement of (p, γ) cross-sections: left, using γ spectroscopy only; right: using detection of reaction product in coincidence with γ-rays using a Recoil Mass Separator.

be about 38.1 (1.8) keV (Γ_p usually dominates for cases where the resonance state lies well above the particle threshold, as in this case). The target thickness brought down the energy of the incident beam from 8.2 to 5.8 MeV. This covered an excitation energy range in the c.m. system from 0.585 to 0.414 MeV scanning more than the resonance energy zone. A surface barrier silicon detector placed at an angle with respect to the incident beam direction was used to monitor the beam intensity and to measure the number of ^{13}N nuclides incident on the target. The surface barrier detector detected ^{13}N ions elastically scattered from ^{12}C in the target also the recoil protons and the ^{12}C recoils from the (CH$_2$)$_n$ target. A portion of the scattered beam not intercepted by the silicon detectors was deposited in the chamber walls. The rest of the ^{13}N beam was collected in the Faraday cup downstream. The γ-ray measurement was done using large volume HPGe detectors positioned in the backward hemisphere of the target (at an average angle of 129°). Appropriate Pb shielding was used to prevent the 511 keV γ-rays resulting from positron decay of ^{13}N (annihilation of positrons) reaching the γ detectors (Figure 7.41 (left)).

The cross-section was determined from the ratio of the peak area of the 5.173 γ-peak (after subtraction of the random events and normalizing for efficiencies, etc.) to the number of ^{13}N incident on the target, as calculated from the number of ^{13}N detected in the silicon detector and assuming pure Rutherford scattering since the center of mass energy of the ^{13}N + ^{12}C system was well below the Coulomb barrier. The average cross-section over the energy range of 5.8–8.2 MeV was determined to be 106 (30) μb. Assuming the resonant capture to be predominant over the direct capture in the vicinity of the resonance, the partial γ width (Γ_γ) of the state was deduced assuming Breit–Wigner resonance and it came out to be 3.8 (1.20) eV. The reaction rate for single, narrow resonance is proportional to the product of exp($-E_r / kT$) and ωγ (called resonance strength), where E_r is the resonance energy, $\omega = (2J_r + 1)/(2J_{proj.} + 1)(2J_{target} + 1)$ is the spin factor with J_r, J_{proj} and J_{target} representing the angular momentums of the resonant state, projectile and the target respectively, and $\gamma = (\Gamma_p \Gamma_\gamma / \Gamma_{total}) \sim \Gamma_\gamma$ for $\Gamma_p \gg \Gamma_\gamma$. Thus, the reaction rate is mainly decided by Γ_γ when the proton width Γ_p is much greater than the γ width. In cases in which the resonant state lies very close to the proton threshold, the γ width dominates and the reaction rate is decided mainly by the proton width.

7.11.1.1 *Study of Charged Particle Capture Reactions Using Recoil Mass Separators*

It is important to note that the reaction ^{13}N (p, γ) ^{14}O is a comparatively easy one to study because the partial γ width of the resonant state for this reaction is exceptionally large, several orders of magnitude higher than in most of the other cases of astrophysical interest (Γ_γ is often ~meV (=10^{-3} eV) or lower instead of ~eV in this case). Also, the high γ energy of 5.173 MeV in this case allowed relatively background-free measurement of the γ-spectrum. Most often, therefore, to study the reaction rates of interest to astrophysics more sensitive detection techniques with lower background and also higher beam intensity are needed. The sensitivity increases by a large factor if recoil is detected taking advantage of the inverse kinematics. Recoil determination is, as such, background-free and has almost 100% detection efficiency but the problem comes from the need to reject the incident beam (the radioactive projectile), which is about a trillion times higher in intensity. For example, if we assume a cross-section of about 1 μb and a target thickness of about 10^{18}/cm^2, a separation better than 1 in 10^{12} is needed to suppress the beam and detect only the reaction product, often termed as the recoil. This is indeed a challenging task made worse by the finite energy width of the incident beam and the energy straggling in the hydrogen target. Advanced Recoil Mass Separators, like the Detector of Recoils And Gammas Of Nuclear reactions (DRAGON) at TRIUMF have been designed and constructed for a high degree of beam suppression (D'Auria 02). The DRAGON uses a windowless hydrogen gas (recirculating) target and it is a two-stage separator with a total length of ~21 m, with each stage comprising of a pair of magnetic and electrostatic quadrupoles. A DSSD or an ionization chamber placed at the end of the separator is used to detect the recoil/reaction product. Coincidence between γ-rays detected using a BGO array around the gas target and recoil detection at DSSD is used to achieve a much better background reduction compared to beam suppression achieved by the optics. Additionally, a ToF measurement can also be made to further reduce the random events.

We briefly discuss the measurement of the cross-section for the ^{21}Na (p, γ) ^{22}Mg reaction carried out using the DRAGON at TRIUMF (Bis 03). In this reaction s wave resonant capture of a proton to the 2$^+$ state at 5.714 MeV in ^{22}Mg (212 keV above the proton threshold) is expected to give the dominant contribution to the capture cross-section. In the experiment ^{21}Na was produced by bombarding a SiC target with 500 MeV protons. After ionization, separation and acceleration in ISAC linear accelerators, the ^{21}Na radioactive beam of intensity ~5 xx 10^8 ions/s was allowed to incident upon a windowless gas target of hydrogen maintained at a pressure of 4.6 torr. Surface barrier silicon detectors placed at appropriate angles within the gas cell (not shown in Figure 7.41,(right)) were used to measure the elastic scattering rates of protons and the ^{21}Na. This allowed monitoring and measurement/normalization of the beam intensity. The array of BGO detectors surrounding the target covered a large solid angle (up to ~90% of 4π). For narrow resonances, it was possible to deduce the exact location of the reaction inside the gas target from the hit pattern of γ-rays in the different BGO crystals. The energy loss in the gas cell was measured using the first di-pole magnet in the DRAGON beam line. Combining the location of the reaction and the energy loss resonance energy can be accurately determined.

In the experiment, data recording was done both for singles and coincidence modes. In the coincident mode, the start and stop signals for ToF measurement were taken from the BGO and DSSD respectively. Although most of the unreacted beam ions of ^{21}Na could be suppressed/rejected by the DRAGON optical elements and the selection slits at the focal planes, a small fraction of it could find its way through the separator to reach the DSSD. These particles would be in random coincidence with the background γ-rays (resulting from the decay of radioactive beam ^{21}Na upstream of the target and the background due

Experimental Techniques

to cosmic rays and room radioactivity) detected in the BGO detectors and would have distributed ToF. However, the true proton capture events resulting in ^{22}Mg recoils would be strongly bunched in time. Putting a gate around the peak in the ToF spectrum that represented the true events, it was possible to extract or sift out the true capture events from events recorded in the DSSD. The energy spectrum of pure ^{22}Mg recoil was thus obtained and from the area the total number of true events was deduced, after correcting for various efficiencies. In a typical experiment at low beam energy, the maximum charge state fraction for a single charge state (the one to be selected by RMS) can be about 40%. The other major source of efficiency loss is in γ detection for which detection efficiency for a large array could be around 50%; other efficiencies such as transmission efficiency through RMS, efficiency for charge particle detection in DSSD, etc., are all close to 100%, resulting in an overall efficiency close 20% for recoil detection. In the ^{21}Na (p, γ) ^{22}Mg experiment, the resonant strength calculated from the yield per incident ^{21}Na came out to be 1.03 ± 0.16 (stat) ± 0.14 (sys) meV and the resonance energy was found to be 205.7 ± 0.5 keV instead of the expected value of 212 keV.

7.11.1.2 Study of Charged Particle Capture Reactions Using Low-Energy Ion Storage Rings

The study of (p, γ), (α, γ) and other kinds of direct reactions using storage rings has great potential. For determination of reaction rates for nuclear astrophysics, the beam energy should be around the Gamow peak energy, which depending upon the atomic number of the heavy ion, would lie in the range of ~100 keV/u to a few MeV/u. It is thus necessary to have a storage ring in which low energy RIBs can be stored for long enough. Indeed, there are already proposals to shift and combine two highly successful low-energy storage rings: the Test Storage Ring (TSR) at the Max Planck Institute, Heidelberg, with the HIE-ISOLDE facility at CERN and the CRYRING at Stockholm with the FAIR facility at GSI respectively. At ISOLDE, the low energy RIBs accelerated to suitable energies from a few tens of keV/u to a few MeV/u would be directly injected into the TSR (Gri 12). The CRYRING would be combined with the existing ESR at GSI to carry out experiments with cooled highly stripped ions down to a few tens of keV/u (Les 12, Lit 13, Hill 13).

The target would also contribute to the deterioration of the beam quality. Thus, only windowless thin gaseous internal targets can be used. As an example, for studying (p, γ) reactions a hydrogen target with a typical thickness of about 10^{14}/cm^2 would be used. This thickness is however typically about four orders of magnitude less than the target thickness (N_t) that can be used in non-ring experiments, such as the ones described earlier. However, this loss is more than compensated by many (~10^6) orders of magnitude gain in the luminosity, which is given by $f_R \times I \times N_t$, where f_R is the revolution frequency (usually ~1 MHz) and I is the ion beam intensity. The detection of the product ion, which would have the same velocity as the original ion circulating in the ring, could be done with SMS, as described in Section 7.6.1, if the frequency of the product ions is within the storage acceptance of the ring. Otherwise, the product ions will be deflected out of the central orbit in which the original ions are circulating, after passing through a di-pole magnet downstream of the reaction zone and can be detected by using suitably positioned in-ring charged particle detectors (Such as DSSD, Si(Li), etc.). In either case, the detection sensitivity is high: 100% for SMS and close to 100% for particle detection. The gain in the luminosity, coupled with high sensitivity of detection in storage rings, results in a figure of merit, not achievable by any other technique. The use of charged particle detectors within the ring aperture, however, requires special R&D efforts (Str 11) since the low energy storage rings require a very high level of vacuum for long enough survival of low-energy ions in the ring.

7.11.1.3 Direct Measurement of (n, γ) Cross-Sections Using Storage Rings

Direct measurement of radiative neutron capture reaction involving unstable n-rich nuclei is extremely difficult because of the low intensity of unstable nuclei and the absence of a neutron target. For a neutron target a strong source of neutrons is needed. A new approach for studying the radiative neutron capture reaction in inverse kinematics has recently been suggested (Rei 14a, Rei 17). The approach, schematically shown in Figure 7.42, involves the use of a storage ring in which rather low energy radioactive ions (typically in the range of 100–500 keV/u) circulate and the source of neutrons is provided by a reactor (Figure 7.42a) or a spallation neutron source (Figure 7.42b). The storage ring allows the most efficient use of the n-rich ions and a high neutron flux could be obtained by allowing the storage ring beam pipe to pass through the core of a research reactor; the space for the beam pipe can be created by removing some central fuel elements from the fuel assembly. Alternately, the ring can pass through the periphery of the core which is simpler from a practical point of view but at the cost of flux reduction by an order of magnitude. The reactor will provide a strong source of thermal neutrons which will act as an ideal target of neutrons (neutrons can easily penetrate the storage ring pipe) practically at rest with respect to the beam. The end product of the capture reaction would be with one extra neutron but at the same velocity as the beam ion. Thus, the revolution frequency for the product ion would be reduced in the ratio $A/(A + 1)$, where A is the mass number of the RIB injected into the storage ring from an ISOL-based RIB facility. The change in the revolution frequency can be detected

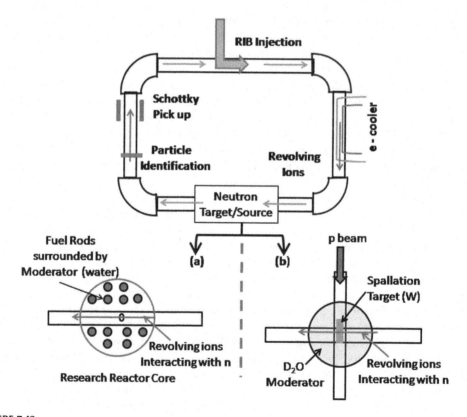

FIGURE 7.42
Schematic of typical experimental set-ups for measurement of (n, γ) cross-sections with storage ring (a) with reactor as neutron source (b) with spallation reaction as neutron source.

Experimental Techniques 245

in the Schottky detectors and the mass-to-charge ratio of all ions can be determined, provided they are cooled. The use of a spallation neutron source in place of the reactor is probably a better bet in terms of obtaining safety clearances. This also simplifies the coupling of a storage ring to the source to a considerable extent but a moderator is required, such as heavy water to shift the neutron spectrum to lower energy. Simulation showed that for a spallation-based source using a 100 µA, 800 MeV p beam, areal density of neutron would be only two orders of magnitude less than compared to a reactor-based neutron source (Rei 17).

7.11.2 Coulomb Dissociation Technique for Measuring (p, γ) and (n, γ) Reaction Rates

An indirect way to overcome the low yields in radiative proton capture reactions is to determine the cross-sections using the reverse reaction. It has been shown that the Coulomb dissociation cross-section could be related to the radiative capture cross-section in a rather simple way (Bau 86). Thus, for example, the cross-section of the reaction ^{13}N (p, γ) ^{14}O can be determined by a Coulomb dissociation experiment in which a beam of ^{14}O is coulomb excited to the 5.173 MeV resonant state in the Coulomb field of a, say, heavy Pb nucleus which subsequently decays into proton and ^{13}N (^{14}O* → ^{13}N + p). At the low relative energy of about 0.545 MeV, the E1 excitation by virtual photons would be dominant and the time-reversed Coulomb dissociation can be studied to extract the cross section for the capture reaction, as suggested originally by Baur, Bertulani and Rebel (Bau 86). The direct (radiative capture) and the time-reversed (Coulomb dissociation) reactions are connected by the detailed balance theorem (Bau 86):

$$\sigma(A + a \rightarrow B + \gamma) = \left[2(2J_B + 1)/(2J_A + 1)(2J_a + 1) \right] (k_\gamma^2 / k^2) \cdot \sigma(B + \gamma \rightarrow A + a);$$

where k and k_γ are the wave numbers in the $A + a$ and the $B + \gamma$ channels and $k \gg k_\gamma$ except for a few exceptional cases where the resonant state lies very close to the threshold. Thus k_γ^2/k^2 is very small and thus the phase space favors the photo-absorption/Coulomb dissociation and its cross-section is enhanced by large factors, often by two orders of magnitude compared to radiative capture. Apart from favorable phase space factors this technique, although an indirect one, allows the use of high energy radioactive projectiles, allowing the use of targets that are about 1000 times thicker. Overall, the increase in the yield is huge and can compensate for the low yield in the direct method. The first experiment using this technique was carried out by Motobayashi and collaborators at RIKEN in which they determined the cross-section of ^{13}N (p, γ) ^{14}O reaction (Moto 91). The experiment used a 91.2 MeV ^{14}O beam and a 350 mg/cm^2 Pb target for Coulomb excitation of ^{14}O beam. ^{14}O was produced by fragmentation of 135 MeV/u ^{16}O beam using a Be target and the ^{14}O fragments were selected using the RIPs at RIKEN. The experiment basically determined the invariant mass of the ^{13}N + p system resulting from the Coulomb dissociation. The ^{13}N nuclei and protons were detected in coincidence and their energy and positions were measured using an array $\Delta E - E$ telescopes comprising of position-sensitive ΔE silicon detectors followed by CsI (Tl) E detectors and two plastic scintillator hodoscopes placed in front of the telescopes. The relative energy spectrum showed a strong peak at $E_{rel} = 0.545$ MeV, as expected. The deduced γ width Γ_γ was 3.1 ± 0.6 eV, which agrees quite well with the width determined using the direct experiment at Louvain-la-Neuve (both the results were reported in August 1991 almost concurrently).

Following this pioneering experiment, the Coulomb dissociation (CD) technique has been used to determine the cross-sections of a number of radiative proton capture reactions (e.g., refer, among others to Moto 94, Gomi 04, Cag 01, Tog 11). The technique has been applied for measurement of radiative neutron capture reactions as well, for example for the ^{14}C (n, γ) ^{15}C reaction (Nak 09a). As mentioned in Section 7.10, the CD technique has been proved to be very effective in determining the γ-strength distributions of exotic n-rich nuclei (e.g., ^{68}Ni, ^{130,132}Sn, etc.) above the one neutron threshold and are therefore very useful to constrain the theories predicting the (n, γ) reaction rates. In fact, the CD technique makes possible measurements of radiative capture cross-sections otherwise impossible by the direct measurement techniques. However, the contributions from the nuclear interaction, mixing of other multi-polarities and coupling to other channels needs to be carefully considered while extracting reliable capture cross-section data.

7.12 EDM Experiments

The existence of Electric Di-pole Moment signifies CP violation. This has already been discussed in Chapter 1 (Section 1.4.1). The Standard Model (SM) predicts electric di-pole moment (EDM) to be infinitesimally small in magnitude, roughly 10^{-34} e cm for quarks and 10^{-38} e cm for electrons, while "beyond SM" theories such as SUSY predict much higher values. Thus, precise measurement of EDM would validate the basis of having "beyond Standard Model" theories apart from testing the SM and particularly the CP violation. Experimental validation of such a minuscule quantity requires innovative and precise measurement techniques. The very first measurement was carried out in 1957 on neutrons by Smith, Purcell and Ramsay at ORNL putting an experimental upper limit of 5×10^{-20} e cm (Smi 57) while SM prediction is $<10^{-32}$ e cm.

EDM experiments with atoms/nuclei hinge on accurately measuring the energy shift due to coupling of the di-pole moment with the applied electric field. The di-pole moment signal, being infinitesimally small, is transformed into frequency. For this, the atom is subjected to magnetic field (B) and electric field (E), with the net interaction Hamiltonian being:

$$H = -(\vec{\mu}.\vec{B} + \vec{d}.\vec{E}),$$

where μ and d are the magnetic and electric di-pole moments respectively. The magnetic moments are made to polarize in a particular direction with respect to the applied magnetic field. The di-pole moment coupled to E would modify the frequency of the Larmor precession of the magnetic moment around the magnetic field. In the case of a spin of ½ particle, the difference in precession frequency for non-zero EDM with E parallel to B and anti-parallel to B is given by $\Delta\omega = 4dE/h$, h being Planck's constant. The experimental goal is to measure the difference in precession frequency with the maximum achievable accuracy using different techniques. Usually, the effect of an electric field is much weaker (small EDM values) than that of the magnetic field, even for a magnetic field 10^{10} times smaller than Earth's field. So, effort and care is mostly concentrated on having a constant B field during the time of measurement. As an example, reducing the leakage current through the surface connecting the high voltage plates is of utmost importance. Reversal of the electric field would also induce reversal of leakage of current-induced magnetic field generating a systematic error in the measurement (For 03).

Experimental Techniques 247

Experimental apparatus for measuring EDM of ^{199}Hg consists of four mercury vapor cells inside a common magnetic field. The middle cells have oppositely directed electric fields while the outer cells are placed in a zero-electric field. The end cells with only the magnetic field are to correct for the gradient B field noise and spurious HV leakage-induced B field. A 254 nm (transition between $^1S_0$ to $^3P_1$ state of mercury atoms) circularly polarized laser light is used for pumping the atoms creating the population in the $m_F = +1/2$ sublevel. The polarized atom is now allowed to precess freely in the magnetic field. Linearly polarized light is used to probe the precession frequency in the cells with an oppositely directed electric field. The frequency difference is measured while the atoms are allowed to precess freely in the magnetic field separated by two probe periods (Gra 16). Using the experimental set-up, the upper limit for EDM of ^{199}Hg has been placed at 7.4×10^{-30} e cm with a 95% confidence limit.

TRIUMF has been working towards developing capabilities and apparatus for measuring EDM of ^{223}Rn. Neutron-rich ^{223}Rn isotope will be produced through spallation of actinide target at the ISAC facility using 500 MeV, 100 µA protons. The produced Rn will be collected in a foil and transferred to EDM measurement apparatus using a cryogenic transfer line. The transfer of the noble radon isotope to the pumping cell will be achieved using the nitrogen jet technique. The low-energy radon beam will be deposited on a foil for a time duration which is usually about two half-lives. Subsequently the foil will be heated to about 1000°C to efficiently release the deposited noble atom. During that time, a cold finger placed downstream of the transfer tube will be cooled to liquid nitrogen temperature. So, the released radon atoms from the foil will be gradually adsorbed onto the cold surface of the cold finger. Once all the atoms are adsorbed, the LN2 supply is stopped allowing it to warm up, releasing the radon atoms. Simultaneously opening the valve after the nitrogen gas manifold will allow a fresh burst of nitrogen gas pushing the released radon atom into the cell at the end of the transfer tube. With ^{120}Xe (in situ decay of 30 keV ^{120}Cs beam deposited on Zr foil) the apparatus results in transfer efficiency of around 80% (Tard 14). HPGe detectors placed near the foil, cold finger and cell are been used to monitor the activity of Xe.

Spin polarization of the radon gas in the measurement cell is achieved by collision with polarized rubidium atoms. The mechanism is called Spin Exchange Optical pumping which can polarize atoms of noble gas via collisional spin exchange with laser-polarized alkali atoms (Wal 97). Using the technique, at ISOLDE, CERN, it was possible to polarize lighter noble gases, especially xenon, with an efficiency of almost 70% (Kit 88).

The EDM measurement cell is placed in between current-carrying coils for generating a weak and uniform magnetic field and integrated electrodes for generating a strong electric field parallel or anti-parallel to the magnetic field direction. The radon nuclei polarized by the spin collision exchange process will start precessing about the magnetic field axis and the task will be to measure the anisotropy of the emitted γ-rays of the precessing radon using γ-ray detectors. The γ-ray anisotropy-based detection of Rn precession was chosen since measurement of β anisotropy would require placement of β-detectors inside the cell which is technically complicated. The angular distribution of emitted γ from polarized radioactive nuclei can be theoretically predicted in terms of spin states, multi-polarity of transition and other parameters (Har 54).

A simulation based on GEANT4 (Ago 03) was carried out to have an estimate of achievable precision or sensitivity of the planned EDM experiment (Ran 11). For the simulation, a simplified cell and oven design with selectable mu-metal shielding was chosen and eight large volume clover HPGe detectors (as in the GRIFFIN Spectrometer (Sve14)) was

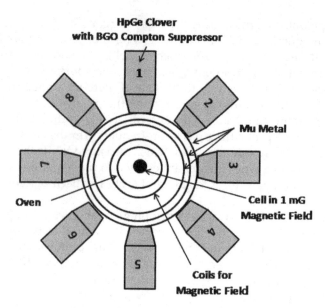

FIGURE 7.43
GRIFFIN detector array around the cell used for GEANT simulation.

considered to be placed in a ring configuration around the cell (Figure 7.43) in a plane which would provide the highest efficiency for the emitted γ-rays. Events were recorded by considering that detectors would be operating in the suppressed mode (rejecting γ events in crystal which are in coincidence with events in BGO).

The GEANT simulation predicts that for 100 days of running around 10^{12} photo-peak counts can be detected by the HPGe detector which can yield a sensitivity of 10^{-26} e cm (comparable with that of ^{199}Hg if the octupole enhancement factor of 600 is taken into account).

8
Overview of Major RIB Facilities Worldwide

8.1 Introduction

The study of exotic nuclei is well recognized as the emerging frontier in nuclear physics and radioactive ion beams are the means to achieve it. The wealth of new and unexpected properties of unstable nuclei has been revealed using first-generation rare isotope beam (RIB) facilities and has prepared a strong case for next-generation upgrades and massive investments in the construction of new "engines of discovery." In this chapter we give an overview of some major RIB facilities worldwide. Our aim is to highlight facilities that have played a pivotal role in key discoveries related to the structure of exotic nuclei and have extended our reach to unexplored regions of the nuclear landscape.

For a historical survey and discussion of world RIB facilities at various laboratories, readers are referred to published review articles (Blu 13, Mue 00, Kam 13, Sch 15). Table 8.1 lists major RIB facilities around the world.

8.2 Major ISOL-Type RIB Facilities

Various isotope separators built around particle accelerators and nuclear reactors in the 1950s were the precursors to dedicated isotope separator online (ISOL) type RIB facilities in later decades. The first post-accelerated RIB was produced at **Louvain-la-Neuve** in 1989 by coupling an isotope separator to two cyclotrons (Figure 8.1)—the Cyclone-30 driver and Cyclone-110 accelerator cyclotrons were operated simultaneously (Dar 90). As most of the RIBs were produced via low energy (30 MeV) proton-induced fusion evaporation reactions, persistent contaminants from target vapors demanded clean separation. Using a cyclotron as a post-accelerator ensured good isobaric separation along with acceleration, and a pure RIB could be delivered for experiments. Innovative target ion source techniques, e.g., transport of activity as gaseous molecules, led to quite a high intensity of post-accelerated beams.

In the first successful post-acceleration experiment, around 10^8 particles per second (pps) intensity was obtained for RIBs of ^{13}N ($T_{1/2} \sim 9.9$ min) that was accelerated to 8.2 MeV. This beam was used for the first ever direct measurement of a proton-capture cross-section for ^{13}N (p, γ) ^{14}O reaction of astrophysical interest (Dec 91). Thereafter, several RIBs such as ^{6}He, ^{19}Ne and ^{35}Ar were accelerated with intensities in the range of 10^4 to 10^9 pps and used for studies in nuclear physics and nuclear astrophysics. Later, under the Accelerated Radioactive Elements for Nuclear, Astrophysical and Solid-State Studies (ARENAS3)

TABLE 8.1
Representative List of Operational and Planned RIB Facilities in the World

Facility Name	Location	Type	Primary Accelerator; Particle(s); Energy; Beam Current	Post-Accelerator/Separator; Typical RI Beam Energy	Status	Ref.
ISOLDE	CERN, EU	ISOL	Synchrotron; protons; 1.4 GeV, 2 μA (average)	Linac; REX-ISOLDE; 3.5 MeV/u, HIE-ISOLDE; 5.5 MeV/u	Operating	Bor 17, Kad 17
CRC	KU Leuven, Belgium	ISOL	Cyclotron; protons; K30-Cyclone; 30 MeV, 300 μA	Cyclotron; K110-Cyclone; 8 MeV/u K44-Cyclone; 0.2–0.8 MeV/u	No longer in operation	Dar 90, Ryc 02
RARF	RIKEN, Japan	PFS	Cyclotron; heavy ions; K540-RRC; 7–135 MeV/u	Projectile Fragment Separator RIPS; near pri. beam energy	Operating	Kami 86, Kub 92
RIBF	RIKEN, Japan	PFS	Cyclotron; heavy ions; K2600-SRC; 345 MeV/u for U	Projectile Fragment Separator Big-RIPS; near pri. beam energy	Operating	Yan 07, Kub 07
NSCL	MSU, USA	PFS	Cyclotron; heavy ions; K500 + K1200 CCF; 90 MeV/u for U	Projectile Fragment Separator A1900; near pri. beam energy	Operating	Bol 04
FRIB	MSU, USA	ISOL/PFS	Linac; protons & heavy ions; 200 MeV/u for U, 0.6 GeV protons (ph-1) 400 MeV/u for U, 1 GeV protons (ph-2)	Linac; 21 MeV/u Cr; 12 MeV/u U	First segment of driver Linac commissioned	Gad 16, She 18
SPIRAL I	GANIL, France	ISOL/PFS	Cyclotron; heavy ions; C01/C02 + CCS1 + CCS2; 95 MeV/u	Cyclotron; CIME 25 MeV/u PF Separator LISE; 90 MeV/u	Operating	Mue 91
SPIRAL II	GANIL, France	ISOL/PFS	Linac; heavyions; 40 MeV, 5 mA deuterons; 14.5 MeV/u, 1 mA heavy ions	CIME Cyclotron for 1.7 to 25 MeV/A with A < 80 and 1.7 to 10 MeV/A for A ~ 100–150	Linac under construction; upgrades	Dol 19
GSI	Germany	PFS	Synchrotron; heavy ions; UNILAC+SIS18; 2 GeV/u for U	PF Separator FRS; near pri. beam energy	Operating	Bla 94, Gei 92
FAIR	Germany	PFS	Synchrotron; heavy ions; UNILAC + SIS18 + SIS100; 2 GeV/u for U	PF Separator Super-FRS; near pri. beam energy	Under construction	Fai 09
ISAC I & II	TRIUMF, Canada	ISOL	Cyclotron; protons; K500; 500 MeV, 100 μA	Linac; 1.5 MeV/u ISAC-I; 6 MeV/u ISAC-II	Operating	Lax 02

(Continued)

TABLE 8.1 (CONTINUED)
Representative List of Operational and Planned RIB Facilities in the World

Facility Name	Location	Type	Primary Accelerator; Particle(s); Energy; Beam Current	Post-Accelerator/Separator; Typical RI Beam Energy	Status	Ref.
ARIEL	TRIUMF, Canada	ISOL	Superconducting Electron Linac; 35/50 MeV, 10 mA; 500 MeV protons from Cyclotron	Linac; 1.5 MeV/u ISAC-I; 6 MeV/u ISAC-II	E-Linac commissioned	Mer 11
HRIBF	ORNL, USA	ISOL	Cyclotron; proton, deuteron, alpha; ORIC up to 80 MeV	25 MV Tandem; ~5 MeV/u for Sn	No longer in operation	Tat 97
CARIBU	ANL, USA	ISOL	Fission fragments from ^{252}Californium (1 Ci)	Linac; ATLAS SC linac; 20 MeV/u	Ph-1 Operating 70 mCi ^{252}Cf	Sav 08
IGISOL	JYFL, Finland	ISOL	Cyclotron; light and heavy ions; K130 heavy ions & K30 light ions;	Ion-guide Isotope Separator IGISOL; 30 keV	Operating	Ays 01
BEARS	LBNL, USA	ISOL	Cyclotron; Protons; K11 proton cyclotron	Cyclotron; Berkley 88-inch Cyclotron; 10 MeV/u	Operating	Pow 00
DRIBS	FLNR-JINR, Russia	ISOL/PFS	Cyclotron; light heavy ions; electrons U400M ~ 49 MeV/u; 25 MeV Microtron	Cyclotron; U400; 29 MeV/u PFS ACCULINA; 40 MeV/u	Operating	Ter 04Oga 02
SPES	LNL-INFN, Italy	ISOL	Cyclotron; protons; K70 with two exit ports; 40 MeV, 200 µA	Linac; ALPI; 10 MeV/u	Under construction	Com 20
HIRFL	IMP, China	PFS	Cyclotron; light and heavy ions; K450 SSC; 1 GeV proton, 100 MeV/u HI	PF Separator RIBLL-1, RIBLL-2; Storage ring CSRe & CSRm	Operating driver	Zho 16
VEC-RIB	VECC, India	ISOL	Cyclotron; protons and alpha particles; K130; 20–75 MeV; few µA	RFQ Linac; up to 1 MeV/u	Operating up to 0.4 MeV/u	Cha 98Nai 13
ANURIB	VECC, India	ISOL/PFS	30 MeV, 2 mA E-linac photofission driver; 50 MeV, 100 µA proton driver	Linac; 7 MeV/u; Cyclotron 100 MeV/u	Design study	Cha 13Ban 15
RAON	RISP, Korea	ISOL/PFS	Cyclotron; protons; 70 MeV, 500 µA protons (ph-1); 600 MeV protons (ph-2)	Linac; 200 MeV/u; Projectile Fragment Separator; near pri. Beam energy	Under construction	Kim 20
ALTO	Orsay, France	ISOL	Electron-linac photofission driver; 50 MeV, 10 µA	Isotope separator; 30 keV	Operating	Mha 08

(Continued)

TABLE 8.1 (CONTINUED)
Representative List of Operational and Planned RIB Facilities in the World

Facility Name	Location	Type	Primary Accelerator; Particle(s); Energy; Beam Current	Post-Accelerator/Separator; Typical RI Beam Energy	Status	Ref.
EURISOL	EU	ISOL	Linac; 1 GeV protons with up to 5 MW & multiple 100 kW targets	Linac; 150 MeV/u; Projectile Fragment Separator	Design study	Blu 09
ISOL@MYRRHA	SCK•CEN, Mol Belgium	ISOL	Linac; protons; 600 MeV, 200 µA	Isotope separator; 60 keV	Under construction	Pop 14
BRIF	CIAE, China	ISOL	Cyclotron; K100; 100 MeV, 200 µA	Linac; 2 MeV/u	Under construction	Zha 13
SOLEROO	ANU, Australia	IF	Tandem; 14 MV; 32 MeV for ^{7}Li;	6.5 T SC Solenoid; 27 MeV ^{8}Li	Operating	Raf 11

The RIB facility at Louvain-la-Neuve

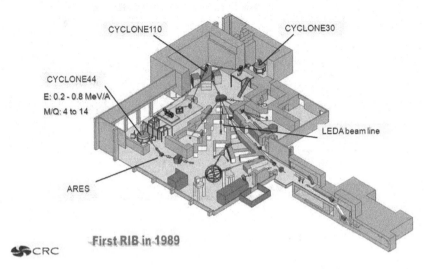

FIGURE 8.1
Schematic layout of RIB facility at Louvain-la-Neuve, KU Leuven, Belgium. The two cyclotrons CYCLONE 110 and CYCLONE 44 are post-accelerators for radioactive nuclei produced using a primary proton beam accelerated in CYCLONE 30. (Courtesy of Marc Loiselet/ Mark Huyse, Université catholique de Louvain (UCLouvain) © all rights reserved.)

project a high-acceleration efficiency K44 cyclotron was built for the post-acceleration of RIBs in the energy range of 0.2–0.8 MeV/u for nuclear astrophysics experiments (Ryc 02). Apart from basic science, RIBs accelerated in the facility have been used for preparation of radioactive targets (Gae 02) and applied science (Bor 08).

However, the key ISOL-type facility which has the maximum impact on exotic nuclei physics, has undoubtedly been the **Isotope Separator On-Line (ISOLDE)** facility. It was built in the 1960s around the 600 MeV proton synchro-cyclotron at CERN and has since been upgraded in various stages, starting from moving to the 1.4 GeV PS-booster in the early 1990s, to the more recent energy upgrades for RIBs. At the ISOLDE facility (Figure 8.2) a primary beam of high energy protons (1–1.4 GeV; average beam intensity up to 2 μA) bombards thick targets to produce exotic isotopes via target spallation, fission and fragmentation reactions. Products diffusing out of the heated target are transferred to an ion source via a short transfer tube. Depending on the chemical nature of the ion beam of interest, surface ionization, FEBIAD or laser ionization source is used. The RIB of interest is selected in a high-resolution separator (HRS) with a mass resolution of 7000. Downstream of the HRS is a radio frequency quadrupole (RFQ) cooler and buncher that reduces the emittance and energy spread of the low energy beam delivered for various precision measurements. The typical measured intensity for volatile elements after the mass separator is around 10^9 pps at energy up to a maximum of 60 keV. Several low energy experimental stations focusing on nuclear decay spectroscopy, solid-state and atomic physics and radiobiology studies are installed and cater to a global user base.

Post-acceleration has been done in stages at ISOLDE. The REX-ISOLDE post-accelerator delivers beams with energy of around 2.8 MeV/u and 3.5 MeV/u for $A/q < 4.5$ and 3.5 respectively, using a combination of RFQ linac, room temperature IH-Linac and super-conducting

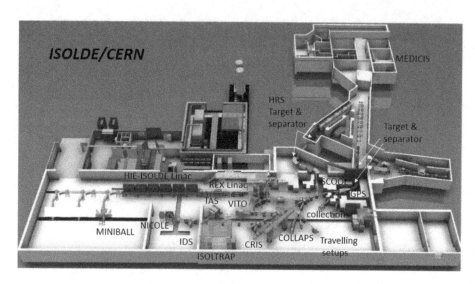

FIGURE 8.2
The ISOLDE facility at CERN. (Courtesy of Karl Johnston CERN © all rights reserved.)

linac module. The REX-EBIS charge breeder has been added to deliver highly charged ion beams to the post-accelerator. Ion beams of more than 100 different isotopes over a wide mass range from ^{6}He up to ^{224}Ra have been accelerated using REX-ISOLDE and used for nuclear reaction studies (Bor 17). A further upgrade in terms of higher beam energy and intensity is underway under the High Intensity & Energy ISOLDE (HIE-ISOLDE) project. Done in stages, this includes an increase of beam energy up to 5.5 MeV/u and later to 10 MeV/u by adding a series of superconducting resonators, and an increase of the primary beam intensity on the target to 6 µA. The first phase has been successfully completed with the acceleration of beams to energy of around 5.5 MeV/u (Kad 17).

ISOLDE has produced beams of around 1300 different isotopes of 75 chemical elements—the highest in the world at any ISOL facility. Discovery of shape co-existence in neutron-rich magnesium isotopes in the island of inversion, shape coexistence and deformation in neutron deficient mercury isotopes, precision mass measurement of exotic nuclei using ISOLTRAP and MR-ToF techniques for fundamental interaction studies and beta-decay spectroscopy of halo nuclei are some of the highlights of the physics program at ISOLDE (Bor 16, Bla 15). Recently using HIE-ISOLDE, RIBs of ^{206}Hg were accelerated to about 1.52 GeV (7.34 MeV/u) and used in an inverse kinematics reaction on a deuterium target to study a rapid neutron capture reaction relevant for element synthesis in the universe (Tan 20). Apart from the core program in nuclear physics a substantial fraction of RIB experiments at ISOLDE focuses on applied research in solid-state physics and radio-biology using techniques such as Mössbauer spectroscopy, perturbed angular correlation (PAC), emission channeling and tracer diffusion studies (Wah 11, Joh 17).

An interesting ongoing program with direct societal benefit is the CERN Medical Isotopes Collected from ISOLDE (CERN-MEDICIS) program which utilizes the "unspent" high-energy proton beam in ISOLDE experiments for isotope harvesting in a dedicated beam-dump target. This way, a large number of medical isotopes will be available for research and applications. Recently CERN-MEDICIS produced its first medical isotope—Terbium-155, an isotope used for a well-known nuclear imaging technique called

Overview of Major RIB Facilities Worldwide

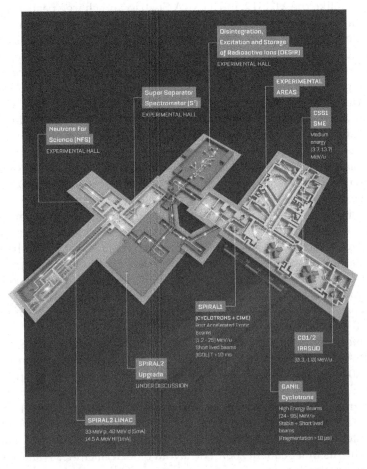

FIGURE 8.3
Artist's impression of the accelerator complex at GANIL showing SPIRAL I and SPIRAL II facilities and the various experiment beam lines. (Courtesy of Alahari Navin, GANIL © all rights reserved.)

single-photon emission computed tomography (SPECT), demonstrating the successful commissioning of the facility and its potential for large-scale isotope production.

The other major ISOL-type RIB facility in Europe is the **SPIRAL** at **GANIL**, France. SPIRAL I makes use of the GANIL coupled cyclotrons CSS1 and CSS2 as driver accelerators delivering heavy ion beams with energy up to around 100 MeV/u (Figure 8.3). Radioactive isotopes are produced through fragmentation reaction of an incident beam stopped in a thick target. Experiments can be done at low energy experiment stations after the separator or with high energy RIBs after the post-accelerator which in this case is again a cyclotron CIME which also acts as a separator. At up to 25 MeV/u SPIRAL I has the highest energy RIBs delivered to experiments conducted at the current ISOL-type facilities.

Research highlights using the SPIRAL I facility include the search of neutron clusters, study of halo nuclei, fusion and transfer reactions with very n-rich beams, discovery of a heavy hydrogen isotope ^{7}H, the proof of shell gap weakening around magic neutron numbers $N = 20$ and 28, study of evolution of spin-orbit splitting, shape coexistence in the neutron-deficient Kr isotopes and measurement of astrophysical reaction rates related with the physics of X-ray bursts or novae (Aza 08). These were done using several spectrometers and detectors like the MUST2 and TIARA Si-strip and charge particle arrays, INDRA detector for evaporation residues, the germanium detector gamma array EXOGAM, the MAYA target-detector chamber and the VAMOS, SPEG and LISE magnetic spectrometers.

The SPIRAL II facility upgrade, to deliver a higher intensity and variety of beams, is underway (Dol 19). Here, a new high current driver accelerator is being added, which is aimed at increasing the available RIB intensity by up to two orders. The new driver is a super-conducting linac delivering up to 5 mA (CW) deuterons of 40 MeV, protons up to 33 MeV and 1 mA of heavy ions up to 14.5 MeV/u. The charge-breeder has also been upgraded by adding a FEBIAD target ion source to produce a wider variety of beams including condensable radioactive species (Mau 20).

In the first phase of SPIRAL II, experiments will focus on studying exotic heavy ion fusion reaction products selected in the Super Separator Spectrometer (S3). The low energy (keV/nucleon) exotic nuclei produced at S3, and also at SPIRAL I, can be transported to a new experimental station called DESIR (decay, excitation and storage of radioactive nuclei) and will focus on the beta-decay and laser spectroscopy of exotic nuclei and experiments with trapped exotic nuclei. The Neutron For Science (NFS) facility will utilize the neutrons coming from the accelerated deuteron beam for research in nuclear and applied science. In the next phase RIBs will be produced using the ISOL method from fast neutron fission of the uranium target. Fission fragments so produced will be separated and transported to the existing experiment halls. n-Rich exotic beams over an energy range of a few keV/u up to 9 MeV/u will be produced (Nup 17).

The main North American ISOL-type RIB facility is the **Isotope Separator and Accelerator (ISAC)** at **TRIUMF** in Canada. Built around a 500 MeV proton cyclotron delivering up to a 100 μA beam, this is the highest power ISOL facility in operation. Targets such as SiC, TiC, ZrC, Nb, UCx, etc., have been used with a variety of ion sources such as surface ion source (SIS), the resonant ionization laser ion source (RILIS), ion guide laser ion source (IG-LIS) and the FEBIAD. The post-accelerators have been developed in stages. ISAC-I accelerated its first RIB in 1998 which was followed by the higher-energy upgrade ISAC-II delivering its first beam in 2007. The layout of ISAC facilities is shown in Figure 8.4. The ISAC-I post-accelerator consists of an 8 m-long RFQ linac that accelerates beams from 2 keV/u to 153 keV/u followed by drift tube linacs (DTL) accelerating beams to an energy up to 1.53 MeV/u (Lax 02). ISAC-II comprises a superconducting linac designed to accelerate RIBs with $A/q < 30$ to energy of about 6 MeV/u. The rich physics program at ISAC is focused on nuclear structure, nuclear astrophysics and materials studies.

Mass measurements, laser spectroscopy and studies on the test of fundamental symmetries have been conducted with short-lived RIBs using, TRIUMF's Neutral Atom Trap (TRINAT), as well as TRIUMF's Ion Trap for Atomic and Nuclear science (TITAN). The high resolution recoil separator Detector of Recoils And Gammas On Nuclear Reactions (DRAGON) is a key detector used for nuclear astrophysics experiments such as the ^{21}Na$(p,\gamma)^{22}$Mg and ^{33}S$(p,\gamma)^{34}$Cl reactions. The other detectors used for nuclear spectroscopy studies are the gamma-detector array TIGRESS, charged particles detector SHARC, an array of neutron detectors DESCANT and a recoil mass spectrometer EMMA (Dil 14).

TRIUMF aims to significantly expand its RIB program with the construction of the Advanced Rare Isotope Laboratory (ARIEL). This upgrade is to meet growing demand for beams by users for nuclear physics, nuclear astrophysics, medicine and materials science. ARIEL will be running three simultaneous RIB experiments by adding two new

Overview of Major RIB Facilities Worldwide

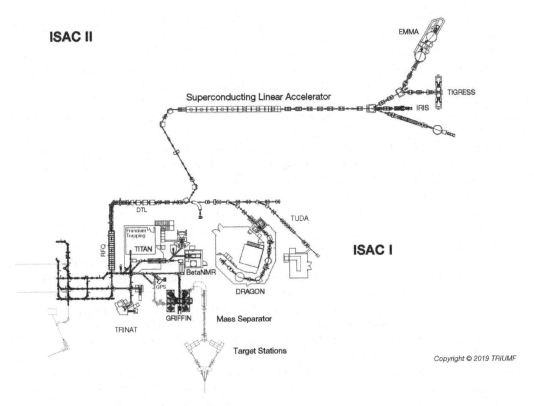

FIGURE 8.4
Layout of ISAC-I and ISAC-II and the low- as well as high-energy experiment stations. (Courtesy of Ann Y. W. Fong © 2020 TRIUMF. All rights reserved.)

target stations, a new superconducting electron linac photofission driver and a second proton beam line from the TRIUMF cyclotron. In addition, an EBIS charge breeder and two new mass separator beam lines to the ISAC-I and ISAC-II accelerator complex are being added with dedicated low-energy experimental stations (Mer 11). The TRIUMF facilities have been used extensively for producing radioisotopes used in medical diagnostics and research. New laboratories for novel isotope harvesting are being built under the recently funded Institute for Advanced Medical Isotopes (IAMI) to be constructed on campus under the next five-year plan (Bag 18).

8.3 Major Projectile Fragment Separator (PFS) Type RIB Facilities

The discovery of the neutron halo in the fragmentation of 1 GeV/u ^{20}Ne beams in the late 1980s using the Bevelac at Berkeley (Tan 85) opened a floodgate of opportunities in exotic nuclei research with energetic heavy ion beams. All the accelerator laboratories that had these machines, such as NSCL/MSU (USA), GSI (Germany), GANIL (France) and RIKEN (Japan) invested heavily in installing powerful in-flight fragment separators while simultaneously planning for the next generation, higher energy and intensity upgrades.

The pioneering Ligne des Ions SuperEpluchés (LISE, beam line for highly stripped ions) spectrometer built in the mid-1980s at GANIL was the first such machine and a precursor for the newer projectile separators at other laboratories such as NSCL/MSU, RIKEN, GSI, etc. The LISE spectrometer incorporated a novel double achromatic design with an intermediate energy degrader for the first time, which facilitated clear A/Z identification of fragments using their momentum-loss analysis (Mue 91, Duf 86).

The LISE facility has allowed the identification and study of many exotic nuclei for the first time, including the highly sought doubly magic nuclei ^{48}Ni (Bl 00) and ^{100}Sn (Lew 94). Over the years several upgrades have been incorporated. In the LISE3 beam line pure radioactive fragments can be selected using the $B\rho$-ΔE selections and a velocity selection by a Wien filter. The LISE2K line is dedicated to experiments requiring higher magnetic rigidity and larger angular acceptance. The radioactive beams are delivered to a large variety of experimental stations and detectors such as ACTAR active target detectors (TPC), gamma detection arrays (EXOGAM) and charged particle arrays (MUST2) to name a few.

The largest and the leading next-generation projectile fragment separator (PFS) type facility is the **RI Beam Factory (RIBF)** at RIKEN Nishina Centre in Japan (Yan 98, Yan 07). Commissioned in 2007, it built upon the existing capabilities at that time in the RIKEN Accelerator Research Facility (RARF). The RARF accelerator complex has a K540 MeV ring cyclotron (RRC) with two different injectors, a frequency-variable heavy ion linac (RILAC) and a K70 MeV AVF cyclotron (Kam 86). Coupled with the RIKEN projectile fragment separator (RIPS), the RARF facility has been used for several significant studies on weakly bound drip line nuclei, especially the nuclear structure of very n-rich nuclei (Sak 03). The RILAC, along with a gas-filled recoil separator (GARIS) was used for the historic discovery of Nihonium with 113 protons—the "first element discovered in Asia" (Mor 04, Mor 12).

A schematic layout of RIKEN RIBF is shown in Figure 8.5. In the RIBF, a cascade of an intermediate-stage ring cyclotron (IRC) and a superconducting ring cyclotron (SRC) has been added as post-accelerators for the RRC along with a new heavy ion injector (RILAC2) equipped with a 28 GHz superconducting ECR ion source (SC-ECRIS). The SRC (Figure 8.6)

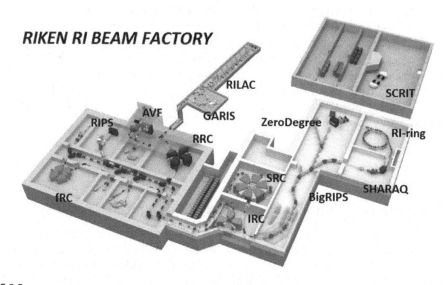

FIGURE 8.5
Schematic layout of RIKEN RI Beam Factory accelerator complex. (Courtesy of Osamu Kamigaito, RIKEN © all rights reserved.)

Overview of Major RIB Facilities Worldwide

FIGURE 8.6
The superconducting ring cyclotron (SRC) magnet weighs 8300 Tonnes. With a maximum magnetic field of 3.8 Tesla SRC is the heart of the largest cyclotron in the world. (Courtesy of Osamu Kamigaito, RIKEN © all rights reserved.)

has six superconducting sector magnets that generate a maximum magnetic field of 3.8 T, corresponding to a K value of 2600 MeV, which is the highest in the world and is capable of accelerating a uranium beam to 345 MeV/u (Kam 13). High-energy beams accelerated in an SRC impinge on a beryllium or carbon target and projectile fragments are separated using a new superconducting projectile fragment separator, named BigRIPS (Ku 07).

Continuous upgrades in terms of adding novel gas strippers and increasing the bending power of the fixed-frequency Ring Cyclotron (fRC), which is used for the intermediate stripping of heavy ions, has ensured acceleration of some of the highest intensity heavy ion beams currently in the world at the RI Beam Factory. For example, >300 pnA of ^{78}Kr, 100 pna of ^{124}Xe and 28 nA of ^{238}U beams have been routinely delivered for user experiments (Kam 16). The result is that gamut of new experiments with nuclei near the drip line are now accessible. The RIBF now boasts of discovering 73 previously unknown isotopes, which include 36 new isotopes ranging from rubidium to lanthanum observed recently in the EURICA campaign involving in-flight fission of 345 MeV/u ^{238}U (Shi 18).

Major experimental devices operational at the RIBF include SAMURAI—a large acceptance spectrometer, high-resolution spectrometer SHARAQ for a new type of missing mass spectroscopy using a RIB as a probe; a low-energy facility SLOWRI, an ion-guide facility delivering low-energy high-purity RIBs, a storage ring called a Rare RI-ring for precision mass measurement of short-lived exotic nuclei, SCRIT—a new electron storage ring for electron-RIB scattering experiments and high-efficiency gamma array DALI2 coupled with MINOS—a new device built with CEA Saclay for fast in-beam spectroscopy of exotic nuclei with a thick hydrogen target and a time projection chamber vertex tracker (Nak 17). Using these devices some of the recent major discoveries include direct spectroscopic evidence of the doubly magic nature of ^{78}Ni, a possible waiting point nucleus in r-process nucleo-synthesis (Tan 19) and corroboration of 34 as a neutron magic number in ^{54}Ca studied through neutron knock-out reactions with these exotic nuclei (Che 19).

In North America the largest PFS-type RIB facility is the **National Superconducting Cyclotron Laboratory (NSCL)** at Michigan State University (MSU), USA. RIBs are produced using high-energy heavy ion beams accelerated in the Coupled Cyclotrons Facility (CCF) followed by a projectile fragment separator A1900 (previously A1200). In the CCF ion beams are first accelerated to an energy of about 10 MeV/u in a K500 superconducting cyclotron and thereafter in the K1200 superconducting cyclotron to around 200 MeV/u for light heavy ions (90 MeV/u for uranium) (Bol 04). Downstream from the A1900 the RIB is transported to several measurement stations where experiments can be performed with fast beams (fragments retaining the initial momentum), stopped and trapped ions or reaccelerated rare isotope beams—the world's first facility where energetic fragments are stopped in a gas cell, purified and re-accelerated to near Coulomb barrier energies. The NSCL facility has been used for several pioneering studies on exotic nuclei, particularly the measurement of structural and decay properties of nuclei near the drip lines, nuclear structure data for nuclei of astrophysical interest, precision atomic mass measurements and reaction studies with energetic rare isotope beams. More than 1000 RIBs have been produced and delivered to multinational user base.

The major detectors and experimental facilities for fast beam experiments include a superconducting high-resolution magnetic spectrograph, S800, that can be coupled with High-Resolution charged-particle Array (HiRA), gamma-ray arrays such as GRETINA, Segmented Germanium Array (SeGA) or arrays for neutron time of flight measurements, e.g., the Modular Neutron Array (MoNA) and Large multi-Institutional Scintillator Array (LISA) for fast neutrons. The stopped beam station is equipped with a facility for collinear laser and atomic spectroscopy with cooled ions, called BECOLA (BEam-COoler and LAser spectroscopy), a Penning trap for precision mass measurements of exotic nuclei and a total absorption gamma-ray spectrometer, SuN.

A unique, newly added facility is the re-accelerator ReA3. Here, fast beams from the A1900 are stopped in a gas cell. From the gas cell surviving 1+ ions are guided to a cooler buncher and mass separator and thereafter injected in an Electron Beam Ion Trap (EBIT) charge state booster. After another q/A separation stage a selected RIB is post-accelerated in a room temperature RFQ linac followed by superconducting linac modules. The ReA3 is designed to deliver beams in the energy range from 0.3 to 6 MeV/u, with maximum energy of 3 MeV/u for uranium, and up to 6 MeV/u for lighter ions with $A < 50$ (Kes 09). The ReA3 post-accelerators were commissioned in 2013 with acceleration of the first radioactive beam of ^{37}K and the first experiment was run with the full ReA3 in 2015 demonstrating the capability of the facility. The ReA3 secondary target stations are followed by experiment stations such as the Active-Target Time-Projection Chamber (AT-TPC), JENSA gas-jet target and a general-purpose beam line.

MSU has embarked on the construction of the next generation RIB facility, namely the Facility for Rare Isotope Beams (FRIB) which is adjacent to the current NSCL laboratory (Figure 8.7). It will be a major new national facility with a linear accelerator driver designed to accelerate all beams from protons to uranium to beam power of 400 kW—potentially the highest beam power available so far. Beam energy for heaviest ions, i.e., uranium, will be 200 MeV/u. The FRIB is expected to produce nearly 5000 different isotopes and will substantially enhance the multi-user capabilities at MSU with fast, stopped and re-accelerated beams of rare isotopes at intensities exceeding NSCL capabilities by three orders of magnitude (Gad 16).

The FRIB superconducting linac is the first of its kind and uses a large quantity (340) of low-beta quarter-wave resonator (QWR) niobium cavities operated at superfluid helium (2 K). Recently the first segment of the linac has been successfully tested and four ion beam

Overview of Major RIB Facilities Worldwide

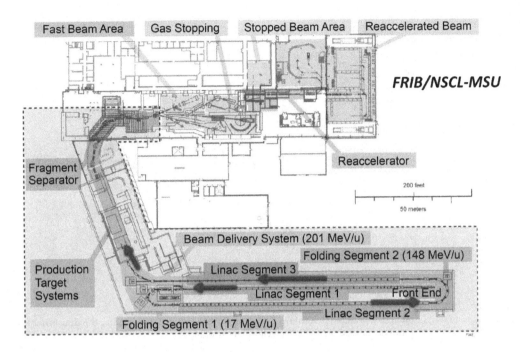

FIGURE 8.7
Layout of the upcoming Facility for Rare Isotope Beams (FRIB) at NSCL, MSU, USA. (Courtesy of Erin O'Donnell, MSU-NSCL © all rights reserved.)

species (Ne, Ar, Kr and Xe) were accelerated up to about 20 MeV/u (Ost 19). Construction of the FRIB is expected to be complete by 2022. Thereafter the NSCL primary accelerator will cease operations and existing as well as future experimental facilities will be served by RIBs from FRIB (Gad 16).

The SI18 synchrotron at **GSI (Helmholtzzentrum für Schwerionenforschung GmbH), Darmstadt, Germany** sits in the midst of a large accelerator complex (Bla 94) and is the driver accelerator for the largest PFS-type RIB facility in Europe (Figure 8.8). The UNILAC, the injector linac for the synchrotron, accelerates ion species from protons to uranium with maximum energy of up to around 11.4 MeV/u. The final energy of SI18 is 2 GeV/u and uranium ions have been accelerated with intensity higher than 10^9 pps (for Au) (Nup 17). The beam from the UNILAC may either be injected into the SI18 via a transfer line or delivered to an experimental hall. The main facility in the low energy experimental area is the large acceptance velocity filter, Separator for Heavy Ion reaction Products (SHIP), which has been the workhorse for the pioneering experiments on super heavy elements (SHE) at GSI. An ion Penning trap, namely SHIPTRAP, has been set up downstream from the velocity filter for precision experiments on very heavy ions produced at the SHIP facility. Apart from this a gas-filled recoil separator, TASCA, and ancillary experimental set-ups, e.g., for nuclear decay spectroscopy, laser spectroscopy and chemistry of heavy and super-heavy elements have been installed in the low energy experimental area.

The 2 GeV/u heavy-ion beam from SI18 is delivered to a high energy experimental area comprising the projectile fragment separator (FRS) (Gei 92) and an experimental storage ring (ESR) and ancillary reaction studies. In the FRS the primary heavy ion beam impinges on a target and projectile-like fragments emerging in the forward direction are separated in-flight, identified and either stopped in detectors for measurement or injected into the

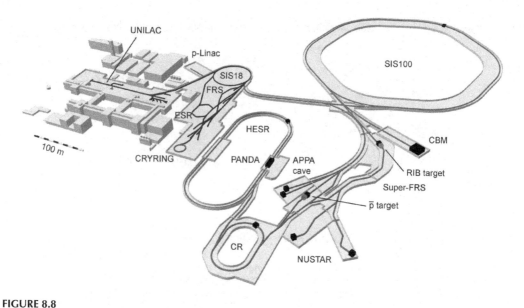

FIGURE 8.8
The accelerator complex at GSI, Darmstadt, Germany, showing the existing RIB facility and the upcoming FAIR facility. (Courtesy of Carola Pomplun, GSI/FAIR © all rights reserved.)

ESR ring. The injected radioactive ions in the ESR are cooled, stored and used for high precision experiments, or directed to the set-up for reaction studies downstream. The comparatively smaller average intensity of a primary beam is more than compensated by the advantage offered by higher energy, and the production yield for exotic nuclei in FRS is comparable to other RIB facilities built around cyclotrons or linacs. This is because the higher velocity of projectile fragments enables better capture of the particles in the forward momentum window of the separator, while at the same time allowing the use of a thicker production target.

The UNILAC-SHIP and FRS has been at the forefront of discoveries—the six heaviest elements with proton numbers Z = 107–112 have been discovered using the SHIP, whereas to date 442 new isotopes have been discovered at Darmstadt (Tho 20). Bound-state beta-decay of fully stripped ions stored and cooled in the ESR ring was observed for the first time at GSI. For example, a drastic reduction of beta-decay half-life from 42 Gy for neutral ^{187}Re atoms to 33 years for fully stripped ^{187}Re ions was observed and as a consequence, modifications were necessary in employing ^{187}Re/^{187}Os pair as cosmo-chronometers (Arn 84). The other pioneering research has been the direct mass measurement of a large number of short-lived nuclei, many with half-lives down to 100 nanoseconds, with an unprecedented precision of better than 10^{-6}. Along with this, nuclear structure studies on exotic nuclei and single-atom chemistry of trans-actinides and super-heavy elements are the other highlights of research using the GSI accelerator facility (Mun 13).

With the aim of expanding research avenues many-fold using much higher intensity heavy ions and anti-proton beams, GSI has undertaken a mega science project namely the Facility for Antiproton and Ion Research (FAIR). The FAIR is being constructed as a multinational facility with a staged approach and an agreed Modularized Start Version. This comprises construction of a new synchrotron called the SIS100 with the upgraded high intensity UNILAC and the SI18 ring as the injector, a superconducting fragment separator (Super FRS), a complex of storage-cooler rings and experiment stations and an antiproton beam line. The baseline research program of the FAIR will cover experiments on the four topics namely—NuSTAR focusing on structure and reactions of exotic nuclei, nuclear astrophysics and radioactive ion beams; atomic and plasma physics (APPA), and applied sciences in the bio, medical and materials sciences; CBM—physics of hadrons and quarks in compressed baryonic matter and PANDA, covering hadron structure and spectroscopy, strange and charm physics and hyper-nuclear physics with antiproton beams (Nup 17).

In preparation for the FAIR the existing accelerator complex at GSI is undergoing upgrades in terms of new instrumentation and controls, and the addition of advanced ion sources and components to accelerate the high brilliance beam of uranium from the UNILAC injector (Bar 17). A low energy storage ring, CRYRING, has been relocated from Sweden as an in-kind contribution to the FAIR, and has been recently commissioned with a beam from the ESR. Under the modularized start approach (Fai 09) FAIR Phase-0 has recently commenced with the goal to restart science experiments after the shutdown for upgrades while the developments related to the FAIR accelerators and detectors will be continuing in parallel (Bai 18).

The in-flight separator RIB facility at the **Heavy Ion Research Facility in Lanzhou (HIRFL)**, China, has been operating since the late 1990s. The Radioactive Ion Beam Line in Lanzhou (RIBLL-1) delivers light-mass RIBs with energies below 100 MeV/u. The driver is a $K450$ Separated Sector Cyclotron (SSC) with a $K70$ sector-focused cyclotron (SFC) as the injector. The SSC accelerates protons to around 2.8 GeV and heavy ion beams from carbon to uranium in the energy range of 1 GeV/u to 100 MeV/u respectively. The RIBLL-1 has produced RIBs of more than 100 species which have been extensively used via the RIBLL collaboration comprising over ten domestic institutions as well as international collaborators. Downstream from the RIBLL-1 is the cooler storage rings (CSR) complex comprising two rings: the main storage ring CSRm and experimental storage ring CSRe constructed for precise nuclear physics and atomic physics study with higher energy pulsed beam (Yua 18).

A second RIB, namely the RIBLL-2, has been installed between the CSRm and CSRe. The projectile fragment separator in RIBLL-2 selects fragments of incoming 300 MeV/u beams from the CSRm and delivers short-lived exotic nuclei beams to the CSRe for mass measurements. About 80 nuclei have been produced and identified unambiguously using the RIBLL-2. With external targets placed at an intermediate focal plane of the asymmetric achromatic separator on the RIBLL-2, rare isotopes with the proton number $Z < 30$ could be uniquely identified. Using the CSR facility masses of 16 short-lived exotic nuclei have been measured for the first time, and for over 27 nuclides these were measured with improved precision of around 10^{-6} to 10^{-7} (Yuh 15).

Owing to the growing demand for beam time for users, a second linac-based injector for the SSC and another independent linac injector for the CSR complex are under construction. With the new injectors it is expected that the maximum numbers of ions stored in the CSRm could be increased by a factor of five through 100 compared to the SFC injector with the number for uranium reaching 4.8×10^9 ions (Zho 16).

8.4 Specialized Facilities

In this section we discuss a few specialized RIB facilities, which although somewhat limited in scope, have been utilized for high impact research on exotic nuclei. In this section we also include two new RIB facilities that are coming up in India and South Korea.

The low-energy RIB facility at the **University of Jyvaskyla**, Finland, is based on an ion-guide technique discovered in the 1980s by Juha Aysto's group at the Department of Physics (Ay 01 and references therein). Nuclear recoils from thin targets bombarded by energetic light or heavy ion beams are thermalized in a gas cell maintained at optimum gas pressure so that a reasonable fraction of recoils retain a 1+ charge state. These ions are pulled out of the gas cell through a small hole by an electrostatic puller electrode and low-energy RIB of typically a few tens of keV is transported to various experimental stations. The advantage of this so-called Ion Guide Isotope Separator On-Line (IGISOL) technique is that no ion source is needed and element-independent thermalization of 1+ ions is obtained within milliseconds. IGISOL is therefore ideally suited for the study of short-lived fission products, especially refractory elements that are difficult to produce at conventional ISOL facilities. The Jyvaskyla facility has led to the discovery and detailed nuclear decay spectroscopy of more than 40 new n-rich fission products.

The current IGISOL separator is connected to two primary accelerators—the heavy ion cyclotron K130, and the high-intensity light ion (protons and deuterons) cyclotron MCC30. Cooled and bunched radioactive ions with energy up to 30 keV are delivered to a Penning ion trap and a Multi-Reflection Time-of-Flight Mass Spectrometer. Using the JYFLTRAP double Penning trap mass spectrometer connected to the IGISOL facility, mass excess measurement for 90 ground state and eight isomeric states of neutron-deficient nuclides have been determined (Kan 11). The other studies involve laser ionization, high resolution collinear laser spectroscopy and beta-delayed gamma and particle spectroscopy of short-lived neutron-deficient and n-rich nuclei, as well as isomers.

The Lawrence Berkeley National Laboratory (LBNL) Nuclear Science Division, at UC Berkeley, USA, has been at the forefront of exotic nuclei research using light and heavy ion beams from the 88-inch cyclotron. The pioneering studies on new modes of radioactive decay of p-rich nuclei, two-proton radioactivity, mass measurements of very unstable nuclei and the nuclear structure of exotic nuclei are some of the research highlights.

For production of radioactive ion beams two innovative methods, namely the **Berkeley Experiments with Accelerated Radioactive Species (BEARS)** and the **Re-cyclotron** method have been developed. The BEARS facility is used for acceleration of short-lived light-mass isotopes of gaseous species and involves production of nuclei in a gas target bombarded by 11 MeV protons accelerated in a cyclotron also used for medical isotope production. The nuclear reaction products are transported as a gas through a 300 meter-long capillary to a cryogenic trap connected to the AECR-U ECR ion source. Low energy radioactive ions extracted from the ion source are axially injected in the main 88-inch cyclotron that finally accelerates the RIB to an energy of around 10 MeV/u (Pow 00). The BEARS facility has delivered RIBs such as ^{11}C, ^{14,15}O, ^{18}F, etc., typically with intensity of 10^8 ions/s on target for nuclear experiments. Several studies in nuclear astrophysics and nuclear reactions have been conducted using the BEARS facility (Joo 00, Guo 05, Lee 07).

For longer-lived radioactive isotopes, e.g., ^{76}Kr (half-life 14.8 hr), a RIB is produced as well as accelerated using the 88-inch cyclotron by operating the accelerator in so-called re-cyclotron mode (Koo 04). Radioactive krypton is produced by bombarding a selenium target with alpha particles accelerated in the 88-inch cyclotron. Gaseous krypton released from the molten target is collected in the cryogenic trap, ionized in the AECR-U and injected into the 88-inch cyclotron for accelerating ^{76}Kr RIB to around 230 MeV. About 10^8 particles/s of ^{76}Kr RIB accelerated using the Re-cyclotron method was used for the first ever g-factor measurement using a RIB (Kum 04). The unique advantage of the method allowed for acceleration of a stable isotope beam of ^{78}Kr in the same set-up under almost identical kinematic conditions, thus enabling comparison of measurements between stable isotopes and much less intense radioactive isotope beams.

A unique RIB facility delivering beams of n-rich nuclei unreachable to date is the **Californium Rare Ion Breeder Upgrade (CARIBU)** at ATLAS Laboratory, USA (Sav 08). n-Rich fission-fragments emitting from an approximately 1 Curie (Ci) ^{252}Cf are thermalized in a gas catcher followed by an RFQ cooler. Up to 50% of the fission fragments can be extracted as a beam of 1+ (or 2+) ions with typical energy of 50 keV. After purification in a magnetic isobar separator, the pure low-energy beam is delivered to "stopped beam" experiment stations or sent to a charge breeder for multiplying the charge state to suit post-acceleration in the Argonne Tandem Linac Accelerator System (ATLAS) linac. The process of fission fragment capture to beam formation takes around 30 milliseconds and is element-independent, opening up the unique advantage to study short-lived nuclei of refractory elements that are difficult to produce at conventional ISOL facilities.

The CARIBU facility has been commissioned with an effectively 70 mCi ^{252}Cf source and its low energy as well as post-accelerated beams have been used in several experiments. Efforts are ongoing to add a thin 1 Ci source that will enhance the intensity of the available beams manyfold. The low energy program focuses on mass measurement and decay spectroscopy of nuclei of interest to r-process nucleo-synthesis. For example, a successful campaign of high accuracy mass measurement of over 150 of the close to 500 n-rich isotopes available at the facility has been completed using the Canadian Penning Trap mass spectrometer in one of the low energy experiment stations (Par 16). The other low energy detectors comprise an X-array spectrometer for lifetime and gamma-decay studies and a Beta Paul Trap (BPT) spectrometer for beta-delayed neutron measurements on short-lived n-rich exotic nuclei.

An important component upstream of the ATLAS post-acceleration section is the charge breeder delivering ion beams with the required mass to charge ratio of seven or smaller. Although initially commissioned with an ECR ion source as the breeder, the facility has been recently upgraded by replacing the ECR ion source with an EBIS to overcome the issue of background contamination emanating from the ECR plasma chamber walls. The new EBIS-based charge breeder has demonstrated breeding efficiencies of >18% into a single charge state for radioactive beams with $A/q < 6$, with significantly reduced background contamination and enhanced radioactive ion contribution of greater than 76% against 3% in the ECR case (Von 18).

The ATLAS is the world's first superconducting accelerator for projectiles heavier than the electron. It comprises three sections: the positive ion injector (PII) with a high-power CW RFQ, the "booster" with low-beta super-conducting half-wave resonators followed by the 20 MV "ATLAS" section. Heavy ion beams from hydrogen to uranium are accelerated to energies up to 20 MeV/u. With the ATLAS, post-acceleration efficiency to around 80% has been measured for the CARIBU with a total post-acceleration efficiency, including charge breeding, of roughly 10% (Sav 15). The experiments with re-accelerated beams include gamma-spectroscopy and Coulomb excitation measurements with a GAMMASPHERE array and the GRETINA/CHICO2 system. For transfer reaction studies with RIBs in inverse kinematic, a large bore superconducting solenoid-based spectrometer, namely the Helical Orbit Spectrometer (HELIOS) has been recently commissioned and will be used with the CARIBU beams.

The **Holifield Radioactive Ion Beam Facility (HRIBF)** facility at Oak Ridge National Laboratory, USA, now decommissioned and designated as an APS Historic Physics Site, has hosted pioneering experiments using light p-rich as well as heavy n-rich RIBs for studying nuclear structure and reactions governing astrophysical processes. In the HRIBF facility a RIB is produced using the ISOL method with the Oak Ridge Isochronous Cyclotron (ORIC) as the primary accelerator (Tat 97). Light ion beams from the ORIC, such as protons, deuterons and alpha particles with energy of up to around 80 MeV bombard suitable targets to produce low charge state ions in an integrated ion source. After undergoing two stages of mass analysis, the low energy radioactive ions pass through a charge exchange cell to produce negatively charged radioactive ions that are finally accelerated in a 25 MV DC-tandem accelerator.

More than 200 radioactive ion beams have been accelerated at the HRIBF which include p-rich nuclei such as ^{7}Be and ^{17,18}F and n-rich beams such as ^{82}Ge and ^{132}Sn. Several studies on nuclear structure and the nuclear reactions of astrophysical interest as well as Coulomb excitation experiments in inverse kinematics on n-rich tin and tellurium isotopes have been performed with these exotic beams (Bar 06, Rad 04, Ryk 14, Rad 02). Typical beam energy is up to 10 MeV/u for light RIBs and around 5 MeV/u for beams near ^{132}Sn. These studies were mainly done at three experiment stations: a Recoil Mass Spectrometer that is mainly used for nuclear structure studies, the Daresbury Recoil Separator used for nuclear astrophysics investigations and an Enge split-pole spectrograph for reaction studies. Detectors include the Clover Array for Radioactive Decay Studies (CARDS), plastic scintillator-based Versatile Array for Neutrons at Low Energy (VANDLE), the Modular Total Absorption Spectrometer (MTAS), the CsI-based HyBall, the germanium-based CLARION and the BaF2-based ORNL-MSU-TAMU gamma-ray detector arrays.

The HRIBF was the world's first facility to post-accelerate beams of heavy n-rich nuclei and with this achievement experimental confirmation of the doubly magic nature of ^{132}Sn was made. Using the facility, the first neutron-transfer measurement on the n-rich nucleus of ^{82}Ge was performed and new or improved beta-decay measurements on more than 70 uranium fission products were studied. These include the first measurement of beta decay half-lives of exotic n-rich nuclei such as ^{82,83}Zn, ^{85,86}Ga and ^{86}Ge. Although no longer in operation, together with the ORIC, the HRIBF facility in its half-century of operation demonstrated immense discovery potential and enriched the field of exotic nuclei physics in a unique way.

The **Flerov Laboratory of Nuclear Reactions** (FLNR) in the Joint Institute for Nuclear Research (JINR), Dubna, Russia has been at the forefront of super-heavy element and exotic nuclei studies. Pioneering experiments on the synthesis of super heavy elements (SHE) led to the discovery of five new elements with proton numbers 114–118 at Dubna. Recently the next generation "SHE factory" has been commissioned with the aim of synthesizing the heavier elements 119 and 120 and to study in detail the properties of the elements already known. The SHE factory is built around a high intensity universal DC-280 cyclotron accelerating ions with mass number $A \leq 238$ and final energy of ≤ 10 MeV/u with beam intensity OF up to 20 particle µA. It is expected that the rate of super-heavy nuclei production will be enhanced by up to two orders of magnitude compared with the present facility (Oga 15).

Along with the SHE program, studies on short-lived exotic nuclei are conducted in the **Dubna Radioactive Ion Beams** (DRIBs-1) complex. This comprises **inflight separator ACCULINNA** built around the U-400M cyclotron and an **ISOL-type RIB facility** using coupled cyclotrons, U-400M and U-400, that have been in operation for the last two decades (Ter 04). Subsequent upgrades in terms of construction of a further powerful ACCULINA-2 separator and DRIBs-2 and DRIBs-3 projects have considerably diversified and strengthened the frontline research program, addressed using radioactive beams accelerated in the facility.

In the ACCULINA fragment separator light heavy ion beams such as ^{7}Li, ^{11}B, ^{13}C, ^{15}N and ^{18}O accelerated in the U-400M cyclotron with energy between 32 and 49 MeV/u and intensity of typically a few particle µA bombard a beryllium target to produce light n-rich RIB. Beams such as ^{6}He, $^{8,\,9,11}$Li, ^{11,14}Be have been produced with typical intensity in the range of a few thousand to 10^5 pps. Several significant studies on transfer reactions with n-rich hydrogen and helium beams were conducted using the ACCULINNA set-up. New experimental evidence about the structure of halo nuclei and the properties of resonance states for heavy hydrogen and helium nuclei were obtained (Ter 04). The higher momentum acceptance in the upgraded ACCULINA-2 facility aims to produce n-rich RIBs with higher intensity by a factor of 10–15 for enhanced purity beams (Kru 10).

From the DRIBs ISOL-type facility a typically 10^6 pps intensity ^{6}He RIB was accelerated at the U-400 post-accelerator cyclotron. This is produced by bombarding a thick beryllium target with about one particle µA of ^{11}B and ionization of reaction products in the DECRIS ion source. The DRIB-2 project added a second option of a 25 MeV electron microtron MT-25 as driver accelerator and acceleration of n-rich RIBs from photofission of uranium in the U-400 cyclotron. DRIBs-2 aims to produce n-rich beams such as ^{132}Sn and ^{142}Xe with an intensity of up to 10^7 pps in the energy range from 5 to 18 MeV/u whereas the DRIBs-3 upgrade is focusing on the construction of a high current cyclotron DC-280 and the SHE factory (Oga 02).

One of the major accelerator laboratories for nuclear physics and ion beam-based applied research in Italy is the **Laboratori Nazionali di Legnaro (LNL) at INFN (Istituto Nazionale di Fisica Nucleare)**, Legnaro. Its Tandem–ALPI–PIAVE accelerator complex houses a 15 MV Van de Graaff XTU Tandem accelerator installed in the late 1980s, which can be used alone or coupled to the superconducting linear accelerator Acceleratore Lineare Per Ioni (ALPI). The ALPI can also be operated in standalone mode with the positive ion injector PIAVE comprising an ECR ion source and superconducting an RFQ linac pre-accelerator. The ALPI presently houses 74 Nb/Cu-NB QWR cavities that accelerate ions from ^{12}C to ^{197}Au at an equivalent high voltage of about 45–52 MV (Ur 13). A total of 12 beam lines with experimental set-ups for basic nuclear physics, applied and interdisciplinary research are available.

A major experimental set-up for exotic nuclei research is the in-flight facility, named EXOTIC, for the production of secondary radioactive beams via inverse kinematics reactions induced by accelerated light stable ion beams on gas targets. Secondary beams such as ^{17}F, ^{8}B and ^{7}Be have been produced with intensity in the range from 10^3 to 10^5 pps. The major detectors for nuclear physics experiments with heavy ion beams include the gamma-ray detector array GALILEO, PRISMA, a large acceptance magnetic separator and a 4-π charged particle array GARFIELD.

Selective Production of Exotic Species (SPES), an ISOL-type RIB facility with a high current proton cyclotron driver accelerator, is presently under construction at LNL-INFN. The SPES is expected to add new dimensions for research on exotic nuclei with post-accelerated RIBs along with currently available stable isotope beams accelerated in the Tandem–ALPI–PIAVE accelerator complex. In SPES, n-rich radioactive isotopes will be produced in proton-induced fission in ^{238}U. The proton driver is a commercial 70 MeV, 750 µA cyclotron with two exit ports from Best Cyclotron Systems. For RIB production a 40 MeV, 200 µA beam delivered by the cyclotron a rate of 10^{13} fission/s is expected. After selection in a high-resolution mass separator low-energy RIB will be accelerated in the existing ALPI superconducting accelerator to an energy up to 10 MeV/u for $A/q = 7$ produced in a charge breeder that has been constructed in collaboration with the SPIRAL II team. A new CW normal-conducting RFQ is also under development. The RFQ is designed to accelerate beams with $A/q \leq 7$ to an energy of 727 keV/u. Between the RFQ and existing ALPI linac, a new beam line including two normal-conducting quarter-wave resonators (QWR) for longitudinal matching with the ALPI linac is under construction.

In the SC linac ALPI itself, many upgrades have been incorporated with the aim of better performance and reliability. The baseline design of SPES is to accelerate reference RIBs of ^{132}Sn to a final energy of 10 MeV/u with maximum possible intensity (Com 20).

The **ALTO RIB facility** at **IPN Orsay** is the first operational photofission-based ISOL-type RIB facility. The primary accelerator is a 50 MeV, 10 μA electron accelerator and radioactive isotopes are produced via gamma-induced fission of a uranium target. Low charge state ions are produced in an integrated target ion source—surface, plasma or laser ionization is employed depending on species. Selected species are separated and delivered to various experiment stations for nuclear decay spectroscopy, mass measurement and other studies. The online mass separator, PARRNe, with a mass resolution of about 1500 provides low-energy (30 keV) n-rich radioactive beams (Mha 08). A dedicated beam line, BEDO, has been set up for beta-gamma decay spectroscopy of short-lived nuclei. A new beam line for collinear laser spectroscopy is being set up for measurement of ground-state properties of exotic nuclei. This facility, laser-induced nuclear orientation (LINO), for measuring nuclear moment, charge radii, etc., also has the provision for polarized beta-decay studies, e.g., beta–NMR and/or NQR. A Penning trap for high-precision mass measurements and trap-assisted decay spectroscopy has been also constructed (Fra 15).

The ALTO facility is being used as a testbed for R&D on RIB detectors and instrumentation for facilities such as SPIRAL II and is an associated member of the EURISOL-DF (distributed facility) consortium along with the other core and associated members, such as JYFL, GANIL-SPIRAL II, CERN-ISOLDE and INFN-SPES (Nup 17).

At the **Variable Energy Cyclotron Centre (VECC)**, India, the planning for a low energy ISOL-type RIB facility started in the late 1990s prompted by the exciting physics of exotic nuclei and enormous activity worldwide in the field (Cha 98). The **VECC-RIB** facility is built around the K130 variable energy cyclotron as the driver accelerator. The K130 cyclotron is the first working cyclotron in India and was built indigenously in the late 1970s following the design of the 88-inch cyclotron at Lawrence Berkeley Laboratory.

Studies on exotic nuclei started with the development of a gas-jet transport system (Cha 88) and later a gas-jet coupled ISOL facility (Cha 92). The RIB facility with post-accelerators has been set up in one of the available experiment caves where an online ECR ion source and a new mass separator have been constructed. Low energy ion beams with $A/q \leq 14$ from the separator are accelerated in a four-rod-type RFQ linac to about 100 keV/u. This is followed by three modules of IH-Linacs accelerating the beam further to around 415 keV/u. Much of the initial physics design of the accelerator components has been done in collaboration with RIKEN RIBF group. Another two linac modules aimed at boosting the energy to about 1 MeV/u have been added. A gas-jet coupled ECR ion source has been used for online production of RIBs and radioactive ion beams of ^{14}O (half-life 71 s), ^{42,43}K (12.4 / 22 hr) and ^{41}Ar (1.8 hr) have been produced with typical intensity of a few times 10^3 pps using this method (Naik 13). Beams of longer-lived isotopes such as ^{111}In (half-life 2.8 days) have also been developed using the offline plasma sputtering method.

A major upgrade is planned in terms of increasing the intensity and variety of RIBs with the construction of an Advanced National Facility for Unstable and Rare Isotope Beams (ANURIB), to be built at one of VECC's upcoming campuses (Cha 07, Cha 13, Ban 15). The aim is to accelerate both p-rich and n-rich short-lived nuclei that will be produced using a 50 MeV proton injector and a 50 MeV, 100 kW super-conducting electron linac (e-Linac) photofission drivers. To be built in stages, initially a low-energy experimental facility for research with stopped beams is planned with new high-resolution separators and an RFQ cooler buncher. In the next phase RFQ-Linac post-accelerators and a ring cyclotron (along the lines of the RIKEN-RRC) will be added for accelerating ion beams up to ^{40}Ar to energy of about 7 MeV/u and 100 MeV/u respectively. The ANURIB project has received seed funding to complete the physics and engineering design of the facility, R&D on the high-power target module and construction of an injector for a superconducting e-linac. The latter two activities are being done in collaboration with the TRIUMF laboratory and a 10 MeV Injector Cryomodule for the e-linac has been recently tested at TRIUMF. The funding also covers the preparation of a Technical Design Report (TDR) detailing the design study and results of R&D activities with a completion timeline by the middle of 2022 when the funding of the construction of the ANURIB will be decided.

One major upcoming RIB facility in Asia is the **Rare Isotope Science Project (RISP)** in South Korea. It was established in 2011 for the construction of a Rare isotope Accelerator complex for ON-line experiments (RAON) comprising both an in-flight fragment separator (IF) and ISOL-type RIB beam lines. The RAON will deliver both stable and rare isotope beams with energy in the range from a few keV to a few hundred MeV per nucleon and research facilities for basic as well as applied sciences. The first beam from the RAON is expected in 2021.

For the ISOL facility the primary accelerator planned for the initial phase is a commercial proton cyclotron delivering a 70 MeV, 500 μA proton beam that will produce n-rich radioisotopes from fission of a uranium target. At a later stage, 600 MeV protons accelerated in a superconducting linac will be used to produce a wider variety and higher intensity RIBs. The low energy RIBs at the exit of the ISOL separator will be delivered to a low energy experimental station or sent to the post-accelerator section. The post-accelerator comprises an RFQ that will accelerate isotopes with $A/q \leq 7$ to around 500 keV/u followed by a superconducting linac accelerating the ion beam to 18.5 MeV/u. Beyond this stage a tripper will be added to produce highly charged ions with $A/q \leq 3$ which will be further accelerated in another superconducting linac to an energy of 200 MeV/u (Jeo 16).

The in-flight beam line of the RAON will be built around a 400 kW superconducting linear accelerator designed to accelerate heavy ion beams up to uranium with maximum energy up to 200 MeV/u. The rare isotope beams generated by heavy ion projectile fragmentation and fission will be separated in an in-flight fragment separator consisting of pre and main separators and delivered to a high energy experimental hall for the experiments. The experimental facility under construction at the RAON comprises a High Precision Mass Measurement System (HPMMS) with Multi-Reflection Time-of-Flight (MR-ToF) and Collinear Laser Spectroscopy (CLS), the KOrea Broad acceptance Recoil spectrometer and Apparatus (KOBRA), a Large acceptance Multi-Purpose Spectrometer (LAMPS) at the end of the fragment separator, a Nuclear Data Production System (NDPS), a Muon Spin Resonance (μSR) set-up for material science and a Beam Irradiation System (BIS) for biomedical science (Kim 20).

References

(Abb 17): B. P. Abbott, et al., (LIGO Scientific collaboration and VIRGO collaboration): *Phys. Rev. Lett.* 119 (2017) 161101.
(ABLA 07): A. Kelic, V. Ricciardi and K.-H. Schmidt, Joint ICTP-IAEA Advanced Workshop on Model Codes for Spallation Reactions, ICTP Trieste, Italy, 4–8 February 2008, https://arxiv.org/pdf/0906.4193.
(Abo 95): Y. Aboussir, et al., *At. Data Nucl. Data Tables* 61 (1995) 127.
(Adr 05): P. Adrich, et al., *Phys. Rev. Lett.* 95 (2005) 132501.
(Ago 03): S. Agostinelli, et al., *Nucl. Instrum. Methods A* 506 (2003) 250.
(Ahn 19): D. S. Ahn, et al., *Phys. Rev. Lett.* 123 (2019) 212501.
(Alk 83): G. D. Alkhazov, et al., *JETP Lett.* 37 (1983) 231.
(Alt 02): G. D. Alton, et al., *Indian J. Phys.* 76S(1) (2002) 9.
(Alt 98): G. D. Alton and H. H. Moeller, *Physics Division Progress Report*, ORNL-6957, (1998).
(Alv 55): L. V. Alvarez, et al., *Rev. Sci. Instrum.* 26 (1955) 111.
(Ame 14): F. Ames, et al., *ISAC and ARIEL: TRIUMF Radioactive Beam Facilities and the Scientific Program*, Springer (2014) 63.
(Ann 93): R. Anne, et al., *Phys. Lett. B* 304 (1993) 55.
(Ann 94): R. Anne, et al., *Nucl. Phys. A* 540 (1992) 341.
(Ar 04): P. Armbrusteret, et al., *Phys. Rev. Lett.* 93 (2004) 212701.
(Ar 70): A. G. Artukh, et al., *Phys. Lett.* 32B (1970) 43.
(Ar 71): A. G. Artukh, et al., *Nucl. Phys. A* 176 (1971) 284.
(Ar 87): E. Arnold, et al., *Phys. Lett. B* 197 (1987) 311.
(Arc 07): A. Arcones, et al., *Astronom. Astrophys.* 467 (2007) 1227.
(Arje 81): J. Arje and K. Valli, *Nucl. Instrum. Meth. A.* 179 (1981) 533.
(Arn 84): M.Arnould, et al., *Astronom. Astrophys.* 137 (1984) 51.
(Arn 85): W. Arnett, et al., *Astrophys. J.* 295 (1985) 589.
(Arn 87): E. Arnold, et al., *Phys. Lett. B* 197 (1987) 311.
(Arn 92): E. Arnold, et al., *Phys. Lett. B* 281 (1992) 16.
(As 92): A. Astier, et al., *Nucl. Instrum. Methods B* 70 (1992) 233.
(Asa 90): K. Asahi, et al., *Phys. Lett. B* 251 (1990) 488.
(Att 02): F. Attallah, et al., *Nucl. Phys. A* 701 (2002) 561c.
(Aub 03): S. Aubin, et al., *Rev. Sci. Instrum.* 74 (2003) 4342.
(Aub 12): S. Aubin, et al., IL Nuovo Cimento (2012), doi 10.1393/ncc/i2012-11264-y, Colloquia PAVI 2011.
(Augu 14): R. M. S. Augusto, et al., *Appl. Sci.* 4 (2014) 265.
(Augu 16): R. M. S. Augusto, et al., *Nucl. Intrum. Meth. B* 376 (2016) 374.
(Aun 00): B. Aune, et al., *Phys. Rev. ST-AB* 3 (2000) 092001.
(Avil 16): M. Avilov, et al., *Nucl. Instrum. Methods B* 376 (2016) 24.
(Ay 88): J. Aysto, et al., *Phys. Lett. B* 201 (1988) 201.
(Ay 01): J. Aysto, *Nucl. Phys. A* 693 (2001) 477.
(Aza 08): F. Azaiez, et al., "Research Report," GANIL (2008) 1–225. ffin2p3-00336915f.
(Ba 00): V. S. Barashenkov, *Comp. Phys. Commun.* 126 (2000) 28.
(Ba 71): S. Baba, et al., *Nucl. Phys. A* 175 (1971) 177.
(Bag 18): J. Bagger, et al., *International Particle Accelerator Conference*, Vancouver, BC, Canada (2018) 6.
(Bai 18): M. Bai, et al., *Proc. of IPAC2018*, Vancouver, BC, Canada (2018).
(Bal 01): G. C. Ball, et al., *Phys. Rev. Lett.* 86 (2001) 1454.
(Bal 16): G. C. Ball, et al., *Phys. Scr.* 91(9) (2016) 093002.
(Ban 00): V. Banerjee, et al., *Nucl. Instrum. Methods A* 447 (2000) 345.
(Ban 15): A. Bandyopadhyay, et al., *PRAMANA* 85 (2015) 505.

(Ban 20): D. Banerjee, et al., *Eur. Phys. J. A* 56 (2020) 201.
(Ban 93): P. Banerjee, et al., *Phys. Lett. B* 318 (1993) 268.
(Banu 09): A. Banu, et al., *Phys. Rev. C* 79 (2009) 025805.
(Bar 06): D. W. Bardayan, *Eur. Phys. J. A* 27 (2006) 97.
(Bar 13): J. Barnes and D. Kasen, *Astrophys. J.* 775 (2013) 18.
(Bar 17): W. Barth, et al., *Phys. Rev. Accel. Beams* 20 (2017) 050101.
(Bas 07): D. Bastin, et al., *Phys. Rev. Lett.* 99 (2007) 022503.
(Bau 07): T. Baumann, et al., *Nature (London)* 449 (2007) 1022.
(Bau 86): G. Baur, et al., *Nucl. Phys. A* 458 (1986) 188.
(Bau 86a): G. Baur, et al., *Phys. Lett. B* 178 (1986) 135.
(Bau 92): G. Baur, et al., *Nucl. Phys. A* 550 (1992) 527.
(Baz 03): D. Bazin, et al., *Nucl. Instrum. Methods B* 204 (2003) 629.
(Baz 08): D. Bazin, et al., *Phys. Rev. Lett.* 101 (2008) 252501.
(Be 70): M. J. Berger and S. M. Seltzer, *Phys. Rev. C* 2 (1970) 621.
(Be 94): M. Bernas, *Phys. Lett. B* 331 (1994) 19.
(Be 97): M. Bernas, et al., *Phys. Lett. B* 415 (1997) 111.
(Beg 71): K. Beg and N. T. Porile, *Phys. Rev. C* 3 (1971) 1631.
(Ben 99): J. Benlliure, et al., *Nucl. Phys. A* 660 (1999) 87.
(Ber 02): U. C. Bergmann, et al., *Nucl. Phys. A* 701 (2002) 363c.
(Ber 88): C. Bertulani, et al., *Phys. Rep.* 163 (1988) 299.
(Berg 02): I. Bergstrom, et al., *Nucl. Instrum. Methods A* 487 (2002) 618.
(Berg 03): U. C. Bergmann, et al., *Nucl. Phys. A* 714 (2003) 21.
(Bert 07): C. Bertulani, et al., *Phys. Lett. B* 650 (2007) 233.
(Bert 09): C. Bertulani, *arXiv: 0908.4307v [nucl-th]* August 29, 2009.
(Bert 63): Hugo W. Bertini, *Physical. Rev.* 131 (1963) 1801.
(Bert 81): G. Bertsch, *Phys. Rev. Lett.* 46 (1981) 472.
(Bethe 39): H. A. Bethe, *Physiol. Rev.* 55 (1939) 534.
(Bethe 90): H. A. Bethe, *Rev. Mod. Phys.* 62 (1990) 801.
(Bh 15): D. Bhowmick, et al., *Phys. Rev. C* 91 (2015) 044611.
(Bh 20): D. Bhowmick, et al., arXiv.2012.07102 (2020).
(Bi 79): K. Van Bibber, et al., *Phys. Rev. Lett.* 43 (1979) 840.
(Bis 03): S. Bishop, et al., *Phys. Rev. Lett.* 90 (2003) 162501.
(Bjor 87): T. Bjornstad, et al., *Nucl. Instrum. Methods B* 26 (1987) 174.
(Bl 00): B. Blank, et al., *Phys. Rev. Lett.* 84 (2000) 1116.
(Bl 03): K. Blaum, et al., *Nucl. Instrum. Methods B* 204 (2003) 331.
(Bl 52): J. Blatt and V. Weisskopf, *Theoretical Nuclear Physics*, Wiley, New York, NY (1952).
(Bla 02): K. Blaum, et al., *Eur. Phys. J. A* 15 (2002) 245.
(Bla 06): K. Blaum, et al., *Phys. Rep.* 425 (2006) 1.
(Bla 10): K. Blaum, *Contemporary Physics* 51 (2010) 149.
(Bla 13): K. Blaum, et al., *Phys. Scr. T* 152 (2013) 014017.
(Bla 15): K. Blaum, et al., *60 Years of CERN Experiments and Discoveries*, World Scientific Publishing Co. Ltd., Singapore.
(Bla 94): K. Blasche and B. Franzke, *Proc. 4th EPAC*, London (1994) 133.
(Bla 13): K. Blaum, et al., *Phys. Scr. T* 152 (2013) 014017.
(Blan 96): B. Blank, et al., *Phys. Rev. Lett.* 77 (1996) 2893.
(Blu 09): Y. Blumenfeld, et al., *Int. J. Mod. Phys. E* 18 (2009) 1960.
(Blu 13): Y. Blumenfeld, et al., *Phys. Scr. T* 152 (2013) 014023.
(Bo 16): M. J. Borge, *Nucl. Instrum. Methods B* 376 (2016) 408.
(Bo 36): N. Bohr, *Nature* 137 (1936) 344.
(Bo 39): N. Bohr and J. A. Wheeler, *Physiol. Rev.* 56 (1939) 426.
(Bol 04): G. Bollen, *Proc. of International Conference on Cyclotrons and Their Applications 2004*, Tokyo (Japan), October 18–22, 2004.
(Bol 06): G. Bollen, et al., *Phys. Rev. Lett.* 96 (2006) 152501.
(Bol 87): G. Bollen, et al., *Hyper. Inter.* 38 (1987) 793.

(Bol 96): G. Bollen, et al., *Nucl. Instrum. Methods A* 368 (1996) 675.
(Bor 08): V. Borcea, *Thesis: Radioactive Ion Implantation of Thermoplastic Elastomers*, 2008.
(Bor 16): M. J. G. Borge, *Nucl. Instrum. Methods B* 376 (2016) 408.
(Bor 17): M. J. G. Borge and B. Jonson, *J. Phys. G Nucl. Part. Phys.* 44 (2017) 044011.
(Bor 16): M. J. G. Borge, *Nucl. Instrum. Methods B* 376 (2016) 408.
(Borr 02): D. Borremans, et al., *Phys. Lett. B* 537 (2002) 45.
(Bos 96): F. Bosch, et al., *Phys. Rev. Lett.* 77 (1996) 5190.
(Bos 13): F. Bosch, et al., *Int. J. Mass Spectrom.* 349 (2013) 151.
(Bou 74): M. A. Bouchiat and C. Bouchiat, *J. Physiol. (Paris)* 35 (1974) 899.
(Bou 82): M. A. Bouchiat, et al., *Phys. Lett. B* 117 (1982) 358.
(Bow 73): J. D. Bowman, et al., *LBL Report LBL-2908* (1973).
(Boyd 08): R. N. Boyd, *An Introduction to Nuclear Astrophysics*; University of Chicago Press, Chicago (2008).
(Br 80): H. Breuer, et al., *Phys. Rev. C* 22 (1980) 2454.
(Bre 16): M. Breitenfeldt, et al., *Nucl. Instrum. Methods B* 376 (2016) 116.
(Bri 14): P. G. Bricault, et al., *ISAC and ARIEL*, edited by J. Dilling, R. Krucken and L. Merminga; Springer Science + Business Media Dordrecht, Springer (2014) 25.
(Bri 16): P. G. Bricault, *Nucl. Instrum. Methods B* 376 (2016) 3.
(Bro 82): L. S. Brown, et al., *Phys. Rev. A* 25 (1982) 2423.
(Bro 86): L. S. Brown and G. Gabrielse, *Rev. Mod. Phys.* 58 (1986) 233.
(Bru 12): T. Brunner, et al., *Nucl. Instrum. Methods A* 676 (2012) 32.
(Bud 75): G. I. Budker, et al., *IEEE Transactions on Nuclear Science* 22 (1975) 2093.
(Burb 57): E. M. Burbidge, et al., *Rev. Mod. Phys.* 29 (1957) 547.
(Burg 14): G. Burgunder, et al., *Phys. Rev. Lett.* 112 (2014) 042502.
(Burr 07): A. Burrows, et al., *Phys. Rep.* 442 (2007) 23.
(But 16): P. A. Butler, *J. Phys. G Nucl. Part. Phys.* 43 (2016) 073002.
(But 19): P. A. Butler, et al., *Nat. Commun.* 10 (2019) 2473.
(But 96): P. A. Butler and W. Nazarewicz, *Rev. Mod. Phys.* 68 (1996) 349.
(Cabb 63): N. Cabibbo, *Phys. Rev. Lett.* 10 (1963) 531.
(Cag 01): J. A. Caggiano, et al., *Phys. Rev. C* 64 (2001) 025802.
(Cam 16): P. Campbell, et al., *Prog. Part. Nucl. Phys.* 86 (2016) 127.
(Cam 57): A. G. W. Cameron, "Chalk River Report," CRL 41 (1957); *Publ. Astron. Soc. Pac.* 69 (1957) 201.
(Cam 81): J. Camplan, *Nucl. Instrum. Meth.* 186 (1981) 445.
(Car 78): G. Carron and L. Thorndahl, *Stochastic Cooling of Momentum Spread by Filter Techniques*, CERN-ISR-RF/78–12 (1978).
(Cath 17): R. Catherall, et al., *J. Phys. G: Nucl. Part. Phys.* 44 (2017) 094002.
(Ced 07): J. Cedercall, et al., *Phys. Rev. Lett.* 98 (2007) 172501.
(Cha 93): A. Chakrabarti, et al., *Z. Phys. A* 345 (1993) 401.
(Cha 07): A. Chakrabarti, et al., *Nucl. Instrum. Methods B* 261 (2007) 1018.
(Cha 13): A. Chakrabarti, et al., *Nucl. Instrum. Methods B* 317 (2013) 253.
(Cha 88): A. Chakrabarti, et al., *Nucl. Instrum. Methods A* 263 (1988) 421.
(Cha 92): A. Chakrabarti, et al., *Nucl. Instrum. Methods B* 70 (1992) 254.
(Cha 98): A. Chakrabarti, *Radiation Physics and Chemistry* 51 (1998) 497.
(Chac 19): A. Chacon, et al., *Nature (Scientific Reports)* (2019) 9:6537.
(Chat 17): A. Chatterjee, et al., www.cmeri.res.in; *Designing of Beam Stoppers for Super FRS in FAIR Project* (2017).
(Chau 07): G. Chaudhuri, et al., *Phys. Rev. C* 76 (2007) 067601.
(Chau 13): G. Chaudhuri, et al., *J. Phys. Conf.* 420 (2013) 012098.
(Che 07): B. Cheal, et al., *Phys. Lett. B* 645 (2007) 133.
(Che 10): B. Cheal, et al., *Phys. Rev. Lett.* 104 (2010) 252502.
(Che 19): S. Chen, et al., *Phys. Rev. Lett.* 123 (2019) 142501.
(Che 68): K. Chen, *Phys. Rev.* 166 (1968) 949.
(Chr 64): J. H. Christensen, et al., *Phys. Rev. Lett.* 13 (1964) 138.
(Cla 03): J. Clark, et al., *Nucl. Instrum. Methods A* 204 (2003) 487.

(Co 74): S. Cohen, F. Plasil and W. J. Swiatecki, *Ann. Phys. (N.Y.)* 82 (1974) 577.
(Com 20): M. Comunian, et al., *J. Phys. Conf. Ser.* 1401 (2020) 012002.
(Cor 04): J. C. Cornell, SPIRAL Collaboration, *17th International Conference on Cyclotrons and Their Applications*, 2004 (arXiv:nucl-ex/0501030).
(Cow 97): J. J. Cowan, et al., *Ap. J.* 480 (1997) 246.
(Cram 70): J. D. Cramer and H. C. Britt, *Nucl. Sci. Eng.* 41 (1970) 177.
(Cran 83): K. R. Crandall; "Los Alamos Memorandum AT-1," 83–3 (1983).
(Cu 87): J. Cugnon, *Nucl. Phys. A* 462 (1987) 751.
(Cun 78): J. B. Cunning, et al., *Phys. Rev. C* 17 (1978) 1632.
(Cun 90): J. B. Cunning, et al., *Phys. Rev. C* 42 (1990) 2530.
(D'Auria 02): J. M. D'Auria, et al., *Nucl. Phys. A* 701 (2002) 625.
(Dan 93): H. Danared, et al., *Nucl. Instrum. Methods* A335 (1993) 397.
(Dar 90): D. Darquennes, et al., *Phys. Rev. C* 42 (1990) R804.
(Dav 01): C. N. Davids, et al., *Hyperfine Interact.* 132 (2001) 133.
(Dav 08): C. N. Davids, et al., *Nucl. Instrum. Methods B* 266 (2008) 4449.
(Deb 78): S. Debrenev, *Fizyk Plazmy* 4 (1978) 492.
(Dec 13): S. Dechoudhury, et al., *Phys. Rev. ST-AB* 16 (2013) 052001.
(Dec 14): S. Dechoudhury, et al., *Phys. Rev. ST-AB* 17 (2014) 074201.
(Dec 20): S. Dechoudhury, et al., *JINST* 15 (2020) P11022.
(Dec 91): P. Decrock, et al., *Phys. Rev. Lett.* 67 (1991) 808.
(Dei 02): M. Deicher, *Indian J. Phys.* 76 S (2002) 173.
(Del 92): J. R. Delayen, et al., *Proceedings of Linear Accelerator Conference*, ON, Canada (1992) 692.
(Del 93): T. Delbar, et al., *Phys. Rev. C* 48 (1993) 3088.
(Det 79): C. Detraz, et al., *Phys. Rev. C* 19 (1979) 164.
(Di 14): J. Dilling, R. Krucken and L. Merminga: *ISAC and ARIEL: TRIUMF Radioactive Beam Facilities and the Scientific Program Hyperfine Interactions*, Springer (2014) 253.
(Di 99): W. T. Diamond, *Nucl. Instrum. Methods A* 432 (1999) 471.
(Dick 13): T. Dickel, et al., *Nucl. Instrum. Methods B* 317 (2013) 779.
(Dil 03): I. Dillman, et al., *Phys. Rev. Lett.* 91 (2003) 162503.
(Dil 06): J. Dilling, et al., *Int. J. Mass Spectrom.* 251 (2006) 198.
(Dil 14): J. Dilling, et al., *ISAC and ARIEL: TRIUMF Radioactive Beam Facilities and the Scientific Programme*, Springer Science (2014).
(Dol 19): P. Dolegieviez, et al., *Proc. 10th International Particle Accelerator Conference IPAC2019*, Melbourne, Australia (2019).
(Dom 03): M. Dombsky, *Nucl. Instrum. Methods B* 204 (2003) 191.
(Domb 98): M. Dombsky, et al., *Rev. Sci. Instrum.* 69 (1998) 1170.
(Don 69): E. D. Donets, et al., *Proc. First Int. Conf. on Ion Sources*, Saclay, France, (1969) 635.
(Don 81): E. D. Donets, et al., *Sov. Phys. JETP* 53 (1981) 466.
(Doo 14): P. Doornenbal, et al., *Phys. Rev. C* 90 (2014) 061304 (R).
(Duf 86): J. P. Dufor, et al., *Nucl. Instrum. Methods A* 248 (1986) 267.
(Dufl 95): J. Duflo, et al., *Phys. Rev. C* 52 (1995) R 23.
(Dun 16a): M. R. Dunlop, et al., *Phys. Rev. Lett.* 116 (2016) 172501.
(Dun 16b): M. R. Dunlop, et al., *Phys. Rev. C* 93 (2016) 062801.
(Dut 03): G. Dutto, et al., *Proceedings of Particle Accelerator Conference*, Portland, OR, USA (2003) 1584.
(Dzu 12): V. A. Dzuba, et al., *Phys. Rev. Lett.* 109 (2012) 203003.
(Dzu 95): V. A. Dzuba, et al., *Phys. Rev. A* 51 (1995) 3454.
(Eg 83): C. Eglehaff, et al., *Nucl. Phys. A* 405 (1983) 397.
(Eic 89): D. Eichler, et al., *Nature* 340 (1989) 126.
(Eli 11): S. Eliseev, et al., *Phys. Rev. Lett.* 107 (2011) 152501.
(Eli 13): S. Eliseev, et al., *Phys. Rev. Lett.* 110 (2013) 082501.
(Eph 79): M. Epherre, et al., *Phys. Rev. C* 19 (1979) 1504.
(Erl 12): J. Erler, et al., *Nature* 486 (2012) 509.
(Esc 12): J. E. Escher, et al., *Rev. Mod. Phys.* 84 (2012) 353.

References

(Ett 11): S. Ettenauer, et al., *Phys. Rev. Lett.* 107 (2011) 272501.
(Fac 11): A. Facco, et al., *Physical Review ST – Accelerators and Beams* 14 (2011) 070101.
(Fai 09): FAIR, "The Modularized Start Version—A Stepwise Approach to the Realization of FAIR Green Paper." (2009). www.yumpu.com/user/fair.center.eu.
(Fe 05): *FLUctuating KAscade Simulation Program; Ferrari: CERN 2005–10* (2005).
(Fe 12): V. N. Fedosseev, et al., *Phys. Scr.* 85 (2012) 058104.
(Fed 14): A. V. Fedetov, et al., *ICFA Beam Dynamics News Letter* 65 (2014) 22.
(Fer 97): R. Ferdinand, et al., *Proceedings of Particle Accelerator Conference*, Vancouver, Canada (1997) 2723.
(Feyn 58): R. P. Feynman and M. Gell-Mann, *Phys. Rev.* 109 (1958) 193.
(Fin 11): P. Finlay, et al., *Phys. Rev. Lett.* 106 (2011) 032501.
(Fla 09): K. T. Flanagan, et al., *Phys. Rev. Lett.* 103 (2009) 142501.
(Fla 13): K. T. Flanagan, et al., *Phys. Rev. Lett.* 111 (2013) 212501.
(Flam 03): V. V. Flambaum and V. G. Zelevinsky, *Phys. Rev. C* 68 (2003) 035502.
(Flam 84): V. V. Flambaum, et al., *Phys. Lett. B* 146 (1984) 367.
(For 03): N. Fortson, et al., *Physics Today* 56(6) (2003) 33.
(Fr 47): S. Frankel and N. Metropolis, *Phys. Rev.* 72 (1947) 914.
(Fr 81): G. Friedlander, J W. Kennedy, E S. Macias and J. M. Miller, *Nuclear and Radiochemistry*, 3rd edition, John Wiley & Sons, New York (1981).
(Fra 15): Serge Franchoo, *Proc. Conf. Advances in Radioactive Isotope Science (ARIS2014) JPS Conf. Proc* (2015) 020041.
(Fra 87): B. Franzke, et al., *Nucl. Instrum. Methods B* 24/25 (1987) 18.
(Fra 08): B. Franzke, et al., *Mass Spectrom. Rev.* 27 (2008) 428.
(Fre 99): C. Freiburghaus, et al., *Astrophys. J. Lett.* 525 (1999) L121.
(Frid 05): J. Fridmann, et al., *Nature* 435 (2005) 922.
(GA 007-A): *SPIRAL II Project (Electron Option) GANIL/SPI2/007-A* (2002).
(Ga 91): J. J. Gaimard and K. H. Schmidt, *Nucl. Phys.* A531 (1991) 709.
(Gab 89): G. Gabrielse, et al., *Int. J. Mass Spectrom. Ion Process.* 88 (1989) 319.
(Gad 16): A. Gade and B. M. Sherrill, *Phys. Scr.* 91 (2016) 053003.
(Gae 02): M. Gaelens, *Proceedings of PROROB 2001, Ind. J. Phys.* 76S (2002) 113.
(Gae 02): M. Gaelens, et al., *Rev. Sci. Instrum.* 73 (2002) 714.
(Gav 80): A. Gavron, *Phys. Rev. C* 21 (1980) 230.
(Gei 01): H. Geissel, et al., *Nucl. Phys. A* 693 (2001) 19.
(Gei 92): H. Geissel, et al., *Nucl. Instrum. Methods B* 70 (1992) 286.
(Gei 03): H. Geissel, et al., *Nucl. Instrum. Methods B* 204 (2003) 71.
(Gell 76): R. Geller, *IEEE Trans. Nucl. Sci.* NS-23 (1976) 904.
(Gell 96): R. Geller, *Electron Cyclotron Resonance Ion Sources and ECR Plasmas*, IOP, Bristol (1996).
(Ger 09): F. Gerigk, et al., *Proceedings of Particle Accelerator Conference*, Vancouver, BC, Canada (2009) 4881.
(Ghos 02): P. Ghosh, et al., *Indian J. Phys.* 76S (2002) 199.
(Gib 08): J. Gibelin, et al., *Phys. Rev. Lett.* 101 (2008) 212503.
(Gla 97): T. Glasmacher, et al., *Phys. Lett. B* 395 (1997) 163.
(Gol 48): A. S. Goldberger, *Phys. Rev.* 74 (1948) 1269.
(Gol 74): A. S. Goldhaber, *Phys. Lett.* 53B (1974) 306.
(Gol 78): A. S. Goldhaber, et al., *Annu. Rev. Nucl. Part. Sci.* 28 (1978) 161.
(Gol 86): K. S. Golovanivsky, *Instrum. Exp. Tech.* 28, No. 5, part 1 (1986) 989.
(Gom 06): E. Gomez, et al., *Rep. Prog. Phys.* 69 (2006) 79.
(Gome 06): M. B. Gomez Horillos, et al., *AIP Conf. Proc.* 819 (2006) 159.
(Gomi 04): T. Gomi, et al., *Nucl. Phys. A* 734 (2004) E 77.
(Gor 10): S. Goriely, et al., *Phys. Rev. C* 82 (2010) 035804.
(Gos 77): J. Gosset, et al., *Phys. Rev. C* 16 (1977) 629.
(Gott 16a): A. Gottberg, *ARIEL e-to γ Converter/Target Concept Study: TRI-DN-00-00* (2016).
(Gott 16b): A. Gottberg, *Nucl. Instrum. Methods B* 376 (2016) 8.

(Gott 17): A. Gottberg, *Design Note TRI-DN-00-00* (2017).
(Gr 10): P. Barros Graiciany, et al., *Brazil. J. Phys.* 40 (2010) 414.
(Gra 16): B. Graner, et al., *Phys. Rev. Lett.* 116 (2016) 161601.
(Gri 08): G. F. Grinyer, et al., *Phys. Rev. C* 77 (2008) 015501.
(Gri 12): M. Grieser, et al., *Eur. Phys. J. Spec. Top.* 207 (2012) 1.
(Gri 75): D. E. Griener, et al., *Phys. Rev. Lett.* 35 (1975) 152.
(Gro 15): R. P. de Groote, et al., *Phys. Rev. Lett.* 115 (2015) 132501.
(Gro 17): R. P. de Groote, et al., *Hyperfine Interact.* 238 (2017) 5.
(Gu 80): D. Guerreau, et al., *Z. Phys. A* 295 (1980) 105.
(Gu 83): D. Guerreau, et al., *Phys. Lett. B* 131 (1983) 293.
(Gua 13): G. Guastalla, et al., *Phys. Rev. Lett.* 110 (2013) 172501.
(Guo 05): F. Q. Guo, et al., *Phys. Rev. C* 72 (2005) 034312.
(Ha 00): B. Harss, et al., *Rev. Sci. Instrum.* 71 (2000) 380.
(Haa 96): H. Hass, *Nucl. Instrum. Methods B* 107 (1996) 349.
(Hag 06): U. Hager, et al., *Phys. Rev. Lett.* 96 (2006) 042504.
(Hag 12): G. Hagen, et al., *Phys. Rev. Lett.* 108 (2012) 242501.
(Hak 12): J. Hakala, et al., *Phys. Rev. Lett.* 109 (2012) 03250.
(Han 87): P. G. Hansen and B. Johnson, *Europhys. Lett.* 4 (1987) 409.
(Han 95): J. S. Hangst, *Phys. Rev. Lett.* 74 (1995) 4432.
(Har 54): C. D. Hartogh, et al., *Physica* 20 (1954) 1310.
(Hard 15): J. C. Hardy and I. S. Towner, *Phys. Rev. C* 91(2015) 025501.
(He 71): H. H. Heckman, et al., *Science* 174 (1971) 1130.
(He 72): H. H. Heckmann, et al., *Phys. Rev. Lett.* 8 (1972) 926.
(Heg 05): A. Heger, et al., *Phys. Lett. B* 606 (2005) 258.
(Hel 03): K. Helariutta, et al., *Eur. Phys. J. A* 17 (2003) 181.
(Her 01): F. Herfurth, et al., *Nucl. Instrum. Methods A* 469 (2001) 254.
(Her 10): F. Herfurth, et al., *Acta Physiol. Pol. B* 41 (2010) 457.
(Hi 00): D. Hitz, et al., *Rev. Sci. Instrum.* 71 (2000) 839.
(Hi 17): K. Hirose, et al., *PRL* 119 (2017) 222501.
(Hi 84): A. S. Hirsch, et al., *Phys. Rev. C* 29 (1984) 508.
(Higu 12): Y. Higurashi, et al., *Proceedings 20th International Workshop on Electron Cyclotron Resonance Ion Sources 2012*, Sydney, Australia (2012) 159.
(Hill 13): P. M. Hillenbrand, et al., *Phys. Scr. T* 156 (2013) 014087.
(Hof 01): S. Hofmann, et al., *Eur. Phys. J. A* 10 (2001) 5.
(Hoj 05): S. Hojo, et al., *Nucl. Instrum. Methods B* 240 (2005) 75.
(Hol 14): B. J. Holzer, *arXiv* 1404:0927 (2014).
(Hor 01): C. J. Horowitz, et al., *Phys. Rev. Lett.* 86 (2001) 5647.
(Horn 18): C. Hornung, et al., *Tenth International Conference on Charged Particle Optics*, Florida, USA (2018).
(Hoto 18): K. Hotokezaka, et al., *arXiv: 1801.01141v1* [astro-ph. HE] January 3, 2018.
(Hu 05): C. Huet-Equilbec, et al., *Nucl. Instrum. Methods B* 240 (2005) 752.
(Hu 85): A. Huck, et al., *Phys. Rev. C* 31 (1985) 2226.
(Hur 07): A. Hurst, et al., *Phys. Rev. Lett.* 98 (2007) 072501.
(Huy 11): M. Huyse and R. Raabe, *J. Phys. G* 38 (2) (2011) 24001.
(Hy 64): E. K. Hyde, *The Nuclear Properties of Heavy Elements III. Fission Phenomena*, Prentice-Hall, New Jersey (NJ) (1964).
(Ilg 03): T. Ilg, et al., *Proceedings of Particle Accelerator Conference*, Portland, USA (2003) 2841.
(Irn 95): H. Irnich, et al., *Phys. Rev. Lett.* 75 (1995) 4182.
(Is 08): S. Isaev, et al., *Nucl. Phys.* A809 (2008) 1.
(Iwa 01): H. Iwasaki, et al., *Phys. Lett. B* 522 (2001) 227.
(Iwa 05): H. Iwasaki, et al., *Phys. Lett. B* 620 (2005) 118.
(Jackson 98): J. D. Jackson, *Classical Electrodynamics*, 3rd edition, Published by John Wiley & Sons (asia) pte ltd, Singapore, reprint by Wiley (India) Ltd, New Delhi (1998).

References

(Janc 17): A. Jancso, et al., *J. Phys. G: Nucl. Part. Phys.* 44 (2017) 064003.
(Jar 12): P. Jardin, et al., *Rev. Sci. Instrum.* 83 (2012) 02A911.
(Je 16): S. Jeong, IPAC 2016, Busan, Korea (2016).
(Jeo 16): S. Jeong, *Proceedings of IPAC2016*, Busan, Korea (2016).
(Joh 17): Karl Johnston, et al., *J. Phys. G: Nucl. Part. Phys.* 44 (2017) 104001.
(Jok 06): A. Jokinen, et al., *Int. J. Mass Spectrom.* 251 (2006) 204.
(Jon 10): K. L. Jones, et al., *Nature* 465 (2010) 454.
(Joo 00): R. Joosten, et al., *Phys. Rev. Lett.* 84 (2000) 5066.
(Jur 07): B. Jurado, et al., *Phys. Lett. B* 649 (2007) 43.
(Kad 17): Y Kadi, et al., *J. Phys. G: Nucl. Part. Phys.* 44 (2017) 084003.
(Kal 00): R. Kalpakchieva, et al., *Eur. Phys. J. A* 7 (2000) 451.
(Kam 13): O. Kamigaito, *Proc. IPAC2013*, Shanghai, China, 4000.
(Kam 16): O. Kamigaito, *Proc. IPAC2016*, Busan, Korea.
(Kam 19): O. Kamigaito, et al., *Proc. of Cyclotrons 2019*, Cape Town, South Africa.
(Kami 86): H. Kamitsubo, *Proc. of 11th Cyclotrons and Their Applications*, Tokyo (1986) 17.
(Kam 99): O. Kamigaito, et al., *Rev. Scient. Instr.* 70 (1999) 4523.
(Kan 03): R. Kanungo, et al., *Phys. Lett. B* 571 (2003) 21.
(Kan 05): A. Kankainen, et al., *Eur. Phys. J. A* 25 (2005) 129.
(Kan 09): R. Kanungo, et al., *Phys. Rev. Lett.* 102 (2009) 152501.
(Kan 11): Anu Kankainen and the JYFLTRAP Collaboration, *AIP Conference Proceedings* 1409 (2011) 9.
(Kapp 11): F. Kappeler, et al., *Rev. Mod. Phys.* 83 (2011) 157.
(Kap 70): I. M. Kapchinskii, et al., *Prib. Tekh. Eskp.* 4 (1970) 17.
(Kas 13): D. Kasen, et al., *Astrophys. J.* 775 (2013) 25.
(Ke 71): O. Kester, et al., *Nucl. Phys. A* 701 (2002) 71.
(Kel 12): M. Kelly, *Rev. Accel. Sci. Technol.* 5 (2012) 185.
(Kell 01): A. Kellerbauer, et al., *Nucl. Instrum. Methods A* 469 (2001) 276.
(Kes 09): O. Kester, et al., *Proc. SRF2009*, (2009) Berlin, Germany.
(Kh 16): F. A. Khan, et al., *Phys. Rev. C* 94 (2016) 054605.
(Ki 76): R. Kirchner and E. Roecki, *Nucl. Instrum. Methods* 133 (1976) 187.
(Kli 11): H. Klingbell, *Proceedings of the CAS - CERN Accelerator School 2010: RF for Accelerators*, CERN 2011-007, 299.
(Kim 08): S.-H. Kim, *Proceedings of Linear Accelerator Conference*, Canada (2008) 11.
(Kim 20): Y. J. Kim, *Nucl. Instrum. Methods B* 463 (2020) 408.
(Kim 97): T. Kim, *Buffer Gas Cooling of Ions in Radio Frequency Quadrupole Ion Guide*, PhD Thesis, McGill University, Canada (1997).
(King 84): W. H. King, *Physics of Atoms and Molecules*, Springer, Boston, MA (1984) 63.
(Kir 81): R. Kirchner, *Nucl. Instrum. Methods* 186 (1981) 275.
(Kit 88): M. Kitano, et al., *Phys. Rev. Lett.* 60 (1988) 2133.
(Kla 94): I. Klaft, et al., *Phys. Rev. Lett.* 73 (1994) 391.
(Klu 13): H. Jurgen Kluge, *Int. J. Mass Spectrom.* 349 (2013) 26.
(Kö 97): U. Köster, et al., *Nucl. Instrum. Methods B* 126 (1997) 253.
(Kö 98): U. Köster and PIAFE Collaboration, *AIP Conference Proceedings* 447 (1998) 119. https://doi.org/10.1063/1.56733.
(Kob 88): T. Kobayashi, et al., *Phys. Rev. Lett.* 60 (1988) 2599.
(Kob 89): T. Kobayashi, et al., *Phys. Lett. B* 232 (1989) 51.
(Koba 73): M. Kobayashi and T. Maskawa, *Prog. Theor. Phys.* 49 (1973) 652.
(Koo 04): J. R. Kooper, et al., *Nucl. Instrum. Methods A* 533 (2004) 287.
(Kor 09): S. R. Koscielniak, *Proceedings of SRF 2009*, Dresden, Germany (2009) 907.
(Kor 94): A. A. Korsheninnikov, et al., *Phys. Lett. B* 326 (1994) 31.
(Kr 93): K. L. Kratz, et al., *Astrophys. J.* 402 (1993) 216.
(Kru 10): S. A. Krupko, et al., *AIP Conference Proceedings* 1224 (2010) 516.
(Ku 03): T. Kubo, *Nucl. Instrum. Methods B* 204 (2003) 97.
(Ku 13): T. Kubo, *Nucl. Instrum. Methods B* 317 (2013) 373.

(Ku 92): T. Kubo, et al., *Nucl. Instrum. Methods B* 70 (1992) 309.
(Ku 07): T. Kubo, et al., *IEEE Trans. Appl. Supercond.*, 17 (2007) 1069.
(Ku 16): T. Kubo, *Nucl. Instrum. Methods B* 376 (2016) 102.
(Ku 12): T. Kubo, et al., *Prog. Theo. Exp. Physiol.* 03 (2012) C003.
(Kub 01): S. Kubono, *Nucl. Phys. A* 693 (2001) 221.
(Kum 04): G. Kumbartzki, et al., *Phys. Lett. B* 591 (2004) 213.
(Lab 08): F. Labrecque, et al., *Nucl. Instrum. Methods B* 266 (2008) 4407.
(Lam 02): T. Lamy, et al., *Proceedings of EPAC2002*, Paris (2002) 1724.
(Lam 98): T. Lamy, et al., *Rev. Sci. Instrum.* 69 (1998) 1322.
(Lam 06): T. Lamy, et al., *Rev. Sci. Instrum.* 77 (2006) 03B101.
(Lars 19): A. C. Larsen, et al., *Prog. Part. Nucl. Phys.* 107 (2019) 69.
(Latt 74): J. M. Lattimer and D. N. Schramm, *Atrophys. J. Lett.* 192 (1974) L145.
(Latt 76): J. M. Lattimer and D. N. Schramm, *Atrophys. J.* 210 (1976) 549.
(Lax 02): R. E. Laxdal, *Proceedings of PROROB 2001, Ind. J. Phys.* 76S (2002) 153.
(Lax 02): R. E. Laxdal, *Proceedings of LINAC2002*, Gyeongju, Korea.
(Lax 14): R. E. Laxdal, et al., *Hyperfine Interactions* 225 (2014) 79.
(Le 08): N. Lecesne, et al., *Rev. Sci. Instrum.* 79 02A907 (2008) 1.
(Le 13): Y. L. Ye, *Nucl. Instrum. Methods B* 317 (2013) 201–203.
(Le 78): M. Lefort and C. Ngo, *Ann. Phys.* 3 (1978) 5.
(Led 17): X. Ledoux, et al., *EPJ Web of Conferences* 146 (2017) 03003.
(Lee 07): D. W. Lee, et al., *Phys. Rev. C* 76 (2007) 024314.
(Lei 03): M. Leino, *Nucl. Instrum. Methods B* 204 (2003) 129.
(Lek 93): K. Leki, et al., *Phys. Rev. Lett.* 70 (1993) 730.
(Ler 04): R. Leroy, et al., *Proceedings of the 17th International Conference on Cyclotrons and Their Applications*; Tokyo, Japan (2004) 261.
(Les 12): M. Lestinsky, et al., "CRYRING@ESR: A Study Group Report," *GSI and FAIR Report* (2012). www.gsi.de/fileadmin/SPARC/documents/Cryring/ReportCryring_40ESR.PDF.
(Lev 14): C. D. P. Levy, et al., *Hyperfine Interactions.* doi:10.1007/s10751-013-0896-4.
(Lev 81): M. A. Levine, et al., *Phys. Scr. T* 22 (1981) 157.
(Lew 94): M. Lewitowicz, et al., *Phys. Lett. B* 332 (1994) 20.
(Li 11): W. P. Liu, et al., *Sci. China Phys. Mech. Astron.* 54 suppl 1 (2011) 14.
(Li 98): L.–X. Li, et al., *Astrophys. J. Lett.* 507 (1998) 159.
(Lic 17): R. Lica, et al., *J. Phys. G* 44 (2017) 054002.
(Lid 16): S. N. Liddick, et al., *Phys. Rev. Lett.* 116 (2016) 242502.
(Lie 18): E. Liestenschnieder, et al., *Phys. Rev. Lett.* 120 (2018) 062503.
(Lin 04): M. Lindroos, *Proceedings European Particle Accelerator Conference*, Edinburgh, Scotland (2004) 45.
(Lit 10): Y. Litvinov, et al., *Acta Physic. Polo. B* 41 (2010) 511.
(Lit 11): Y. A. Litvinov, et al., *Rep. Prog. Phys.* 74 (2011) 016301.
(Lit 13): Y. A. Litvinov, et al., *Nucl. Instrum. Methods B* 317 (2013) 603.
(Lit 15): Y. A. Litvinov, *Joliot-Curie School (EJC) 2015 Conference*, France (2015).
(Liu 19): H. N. Liu, et al., *Phys. Rev. Lett.* 122 (2019) 072502.
(Lom 11): I. Lombardo, et al., *Nucl. Instrum. Methods B* 215 (2011) 272.
(Lop 08): A. M. L. Lopes, et al., *Phys. Rev. Lett.* 100 (2008) 155702.
(Lor 15): G. Lorusso, et al., *Phys. Rev. Lett.* 114 (2015) 192501.
(Lun 17): D. Lunney, *J. Phys. G* 44 (2017) 064008.
(Lyn 14): K. M. Lynch, et al., *Phys. Rev. X* 4 (2014) 011055.
(Lyn 16): K. M. Lynch, et al., *Phys. Rev. C* 93 (2016) 014319.
(Macf 69): R. D. Macfarlane, et al., *Nucl. Instrum. Methods* 73 (1969) 285.
(Mal 15): J. A. Maloney, et al., *Nucl. Instrum. Methods B* 376 (2016) 135.
(Mane 11): E. Mane, et al., *Phys. Rev. Lett.* 107 (2011) 212502.
(Mar 97): R. E. March, *J. Mass Spectr.* 32 (1997) 351.
(Mar-Pin 12): G. Martnez-Pinedo, et al., *Phys. Rev. Lett.* 109 (2012) 251104.

(Marx 01): G. Marx, et al., *Hyp. Int.* 132 (2001) 459.
(Mau 20): L. Maunoury, et al., *Rev. Sci. Instrum.* 91 (2020) 023315.
(Me 39): L. Meitner and O. R. Frisch, *Nature* 143 (1939) 239.
(Me 74): W. D. Meyers and W. J. Swiatecki, *Ann. Phys.* 84 (1974) 186.
(Me 79): W. D. Meyer, et al., *Phys. Rev. C* 20 (1979) 1716.
(Me 83): W. D. Meyers and K. H. Schmidt, *Nucl. Phys. A* 410 (1983) 61.
(Mei 96): M. J. Meigs, et al., *Nucl. Instrum. Methods A* 382 (1996) 51.
(Mend 14): T. M. Mendonca, et al., *CERN-ACC-NOTE-2014-0028(2014)* (2014).
(Mer 11): Lia Merminga, et al., *Proc. of IPAC2011*, San Sebastián, Spain (2011).
(Met 10): B. D. Metzer, et al., *Mon. Not. R. Astron. Soc.* 406 (2010) 2650.
(Mha 08): M. Cheikh Mhamed, et al., *Nucl. Instrum. Methods B* 266 (2008) 4092.
(Mo 03): D. J. Morrissey, et al., *Nucl. Instrum. Methods B* 204 (2003) 90.
(Mol 12): P. Moller, et al., *Phys. Rev. Lett.*108 (2012) 052501.
(Mo 89): D. J. Morrissey, *Phys. Rev. C* 39 (1989) 460.
(Mo 95): P. Moller, et al., *At. Data Nucl. Data Tables* 59 (1995) 185.
(Moc 07): M. Mocko, et al., *Europhys. Lett.* 79 (2007) 12001.
(Moh 83): D. Mohl, CAS-CERN Accelerator School 84–15 (1983) 97.
(Moh 91): M. F. Mohar, et al., *Phys. Rev. Lett.* 66 (1991) 1571.
(Molt 80): D. M. Moltz, et al., *Nucl. Instrum. Methods* 172 (1980) 507.
(Mon 71): E. J. Monitz, et al., *Phys. Rev. Lett.* 26 (1971) 445.
(Mor 04): K. Morita, et al., *J. Phys. Soc. Japan* 73 (2004) 2593.
(Mor 12): K. Morita, et al., *J. Phys. Soc. Japan* 81 (2012) 103201.
(Mor 14): G. D. Morris, et al., *Hyper. Inter.* 225 (1–3) (2014) 173. doi: 10.1007/s10751-013-0894-6.
(Moto 91): T. Motobayashi, et al., *Phys. Lett. B* 264 (1991) 259.
(Moto 94): T. Motobayashi, et al., *Phys. Rev. Lett.* 73 (1994) 2680.
(Mu 85): D. G. Mueller, et al., *Z. Phys. A* 322 (1985) 415.
(Mue 00): A. C. Mueller, *Proceedings of EPAC 2000*, Vienna, Austria (2000) 73.
(Mue 91): A. C. Mueller and R. Anne, *Nucl. Inst. Meth.* B 56/57 (1991) 559.
(Mun 13): G. Munzenberg, et al., *J. Phys.: Conference Series* 413 (2013) 012006.
(MYRRHA 18): myrrha@sckcen.be; *World Nuclear News*, October 08 (2018).
(Na 04): P. Napolitani, et al., *Phys. Rev. C* 70 (2004) 054607.
(Naik 13): V. Naik, et al., *Rev. Sci. Instrum.* 84 (2013) 033301.
(Naik 01): V. Naik, et al., *Phys. Rev. C* 63 (2001) 024307.
(Naik 10): V. Naik, et al., *Proceedings of Linear Accelerator Conference* (2010) 727.
(Naik 13): V. Naik, et al., *Rev. Sci. Intrum.* 84 (2013) 033301.
(Nak 06): T. Nakamura, et al., *Phys. Rev. Lett.* 96 (2006) 252502.
(Nak 09): T. Nakamura, et al., *Phys. Rev. Lett.* 103 (2009) 262501.
(Nak 09a): T. Nakamura, et al., *Phys. Rev. C* 79 (2009) 035805.
(Nak 17): T. Nakamura, et al., *Prog. Part. Nucl. Phys.* 97 (2017) 53.
(Nak 94): T. Nakamura, et al., *Phys. Lett. B* 331 (1994) 296.
(Nak 99): T. Nakamura, et al., *Phys. Rev. Lett.* 83 (1999) 1112.
(Naka 91): T. Nakagawa, et al., *Jpn. J. Appl. Phys.* 30 (1991) 1588.
(Naka 96): T. Nakagawa, et al., *Jpn. J. Appl. Phys.* 35 (1996) 4077.
(Nay 99): R. C. Nayak, et al., *At. Data Nucl. Data Tables* 73 (1999) 213.
(Neu 17): R. Neugart, et al., *J. Phys. G* 44 (2017) 064002.
(Ney 05): G. Neyens, et al., *Phys. Rev. Lett.* 94 (2005) 022501.
(Nie 02): A. Nieminen, et al., *Phys. Rev. Lett.* 88 (2002) 094801.
(Nie 05): O. Niedermaier, et al., *Phys. Rev. Lett.* 94 (2005) 172501.
(Nik 14): E. N. Nikolaev, et al., *Mass Spectr. Rev.* (2014).
(Nis 11): S. Nishimura, et al., *Phys. Rev. Lett.* 106 (2011) 052502.
(No 07): M. Notani, et al., *Phys. Rev. C* 76 (2007) 044605.
(No 93): J. A. Nolen, *Proc. of the 3rd Int. Conf. on Radioactive Nuclear Beams*, East Lancing, MI, Edition Frontiers (1993).

(No 96): M. Notani, et al., *RIKEN Accel. Prog.* Rep. 29 (1996) 50.
(No 97): M. Notani, et al., *RIKEN Accel. Prog.* Rep. 30 (1997) 48.
(Nol 03): J. A. Nolen, et al., *Nucl. Instrum. Methods B* 204 (2003) 293.
(Nol 11): F. Nolden, et al., *Nucl. Instrum. Methods* A659 (2011) 69.
(Nup 17): NuPECC *Long Range Plan 2017 Perspectives in Nuclear Physics*, Europhysics News 48(4) (2017) 21–24.
(Oga 01): Y. T. Oganessian; *Nucl. Phys. A* 685 (2001) 17c.
(Oga 02): Y. T. Oganessian, et al., *Nucl. Phys. A* 701 (2002) 87c–95c.
(Oga 15): Y. T. Oganessian, *Proc. Conf. Advances in Radioactive Isotope Science (ARIS2014) JPS Conf. Proc.* (2015) 010024.
(Oh 10): T. Ohnishi, et al., *J. Phys. Soc. Japan* 79 (2010) 073201.
(Ohn 08): T. Ohnishi, et al., *J. Phys. Soc. Japan* 77 (2008) 083201.
(Oku 12): H. Okuno et al., *Prog. Theor. Exp. Phys.* 1 (2012) 03C002.
(Oku 14): H. Okuno, *Proceedings of High Brightness Hadron Beams* (2014) 340.
(Ost 19): P. N. Ostroumov, et al., *Phys. Rev. Accel. Beams* 22 (2019) 040101.
(Ota 16): M. Otani, et al., *Physical Rev. Accel. Beams* 19 (2016) 040101.
(Ots 05): T. Otsuka, et al., *Phys. Rev. Lett.* 95 (2005) 232502.
(Ots 10): T. Otsuka, et al., *Phys. Rev. Lett.* 105 (2010) 032501.
(Ots 13): T. Otsuka, *Phys. Scr. T* 152 (2013) 014007.
(Pa 18): A. Papageorgiou, et al., *J. Phys. G: Nucl. Part. Phys.* 45 (2018) 095105.
(Pad 14): H. Padamsee, "CAS-CERN Accelerator School: Superconductivity for Accelerators, Italy" (2013) (CERN–2014–005).
(Pan 16): I. V. Panov, *Phys Atom. Nuclei* 79 (2016) 159.
(Par 16): R. C. Pardo, G. Savard and R. V. F. Janssens (2016) *ATLAS with CARIBU: A Laboratory Portrait, Nuclear Physics News* 26(1) (2016) 5–11.
(Pas 96): S. Pastuzka, et al., *Nucl. Instrum. Methods A* 369 (1996) 11.
(Pau 55): W. Paul, et al., *Z. Phys.* 140 (1955) 262.
(Pau 58): W. Paul, et al., *Z. Phys.* 152 (1958) 143.
(Pau 90): W. Paul, *Rev. Mod. Phys.* 62 (1990) 531.
(Pe 05): D. B. Pelowitz: *LA-CP-05-0369* (2005).
(Pe 88): H. Penttila, et al., *Phys. Rev. C* 38 (1988) 931.
(Pelle 16): F. Pellemoine, *NAPAC 2016*; October 9–14, Chicago (2016).
(Pen 01): Y. U. Penionzhkevich, *Hyp. Int.* 132 (2001) 265.
(Per 09): J. Preira, et al., *Phys. Rev. C* 79 (2009) 035806.
(Per 11): D. Perez-Loureiro, et al., *Phys. Lett. B* 703 (2011) 552.
(Pf 95): R. Pfaff, et al., *Phys. Rev. C* 51 (1995) 1348.
(Pfe 97): B. Pfeiffer, et al., *Z. Phys. A* 357 (1997) 235.
(Pie 12): J. Piekarewicz, et al., *arXiv:1201.3807 v1 [nucl-th]* (2012).
(Pie 49): J. R. Pierce, *Theory and Design of Electron Beams*, D. van Nostrand Co., New York (NY) (1949) 41.
(Pod 13): H. Podlech, *CAS–CERN Accelerator School: Course on High Power Hadron Machines*, Spain (2011) 151 (CERN-2013-001).
(Pop 14): L. Popescu, *EPJ Web of Conferences* 66 (2014) 10 011.
(Pow 00): J. Powell, et al., *Nucl. Instrum. Methods A* 455 (2000) 452.
(Pr 89): *Prael: LA-UR-89-3014* (1989).
(Pur 50): E. M. Purcell and N. F. Ramsey, *Physiol. Rev.* 78 (1950) 807.
(Ra 79): H. L. Ravn, *Phys. Rep.* 54 (1979) 203.
(Ra 89): H. L. Ravn and B. W. Allardyce, *Treatise on Heavy-Ion Science*, edited by D. Allan Bromley, Plenum Press, New York, NY (1989) vol. 8; 363–439.
(Rad 00): T. Radon, et al., *Nucl. Phys. A* 677 (2000) 75.
(Rad 02): D. C. Radford, C. Baktash, J. R. Beene, et al., *Phys. Rev. Lett.* 88 (2002) 222501.
(Rad 04): D. C. Radford, et al., *Nucl. Phys. A* 746 (2004) 83.
(Rad 97): T. Radon, et al., *Phys. Rev. Lett.* 78 (1997) 4701.
(Raf 11): R. Rafiei, et al., *Nucl. Instrum. Methods A* 631 (2011) 12.

References

(Rai 91): C. Raiteri, et al., *Ap. J.* 367 (1991) 228.
(Rai 97): H. Raimbault-Hartmann, et al., *Nucl. Instrum. Methods B* 126 (1997) 378.
(Ram 16): J. P. Ramos, et al., *Nucl. Instrum. Methods B* 376 (2016) 81.
(Ran 11): E. T. Rand, et al., *J. Phys. conf. S.* 312 (2011) 102013.
(Ran 81): J. Randrup and S. E. Koonin, *Nucl. Phys.* A 356 (1981) 223.
(Rat 05): U. Ratzinger, *CAS–CERN Accelerator School: Radio Frequency Engineering*, Germany, May 8–16, (2000) 351 (CERN-2005-003).
(Rau 13): T. Rauscher, et al., *arXiv: 1303. 26666v3 [astro-ph.SR]* April 23, 2013.
(Rei 14): R. Reifarth, et al., arXiv:1403.5670v1 *[astro-ph.IM]* March 22, 2014.
(Rei 14a): R. Reifarth, et al., *Phys. Rev. Accel. Beams* 17 (2014) 014701.
(Rei 17): R. Reifarth, et al., *Phys. Rev. Accel. Beams* 20 (2017) 044701.
(Reif 14): R. Reifarth, et al., *J. Phys. G: Nucl. Part. Phys.* 41 (2014) 053101.
(Ri 03): M. V. Ricciardi, et al., *Phys. Rev. Lett.* 90 (2003) 212302.
(Ri 06): M. V. Ricciardi, et al., *Phys. Rev. C* 73 (2006) 014607.
(Roc 11): X. Roca-Maza, et al., *Phys. Rev. Lett.* 106 (2011) 252501.
(Rod 05): E. Padilla-Rodal, et al., *Phys. Rev. Lett.* 94 (2005) 122501.
(Rolf 88): C. E. Rolfs and W. S. Rodney, *Cauldrons in the Cosmos*, Chicago Press, Chicago, IL (1988).
(Ros 13): D. M. Rossi, et al., *Phys. Rev. Lett.* 111 (2013) 242503.
(Roth 13): S. Rothe, et al., *Nat. Commun.* (2013). doi: 10.1038/ncomms2819.
(Rub 94): C. Rubbia, *International Conference on Accelerator Driven Transmutation, Technology & Applications AIP Conference Proceedings* 346 (1994), Las Vegas, US.
(Rui 06): C. Ruiz, et al., *Phys. Rev. Lett.* 96 (2006) 252501.
(Rui 16): R. F. Garcia Ruiz, et al., *Nat. Phys.* 12 (2016) 594.
(Rus 18): P. Rusotto, et al., *J. Phys. Conf. S.* 1014 (2018) 012106.
(Ryc 02): G. Ryckewaert, et al., *Nucl. Phys. A* 701 (2002) 323c.
(Ryk 14): K. P. Rykaczewski , *Nucl. Data Sheets* 120 (2014) 16.
(Sa 03): G.. Savard, *Nucl. Instrum. Methods B* 204 (2003) 582.
(Sak 03): H. Sakurai, *JAERI-Conf 2003-017* (2003).
(Sak 67): A. D.. Sakharov, *JETP Lett.* 5 (1967) 24.
(Sam 17): S. Saminathan, et al., *Proceedings of International Particle Accelerator Conference*, Denmark (2017) 648.
(Sat 83): L. Satpathy, et al., *Phys. Rev. Lett.* 51 (1983) 1243.
(Sav 03): G. Savard, et al., *Nucl. Instrum. Methods B* 204 (2003) 582.
(Sav 08): G. Savard, et al., *Nucl. Instrum. Methods B* 266 (2008) 4086.
(Sav 15): G. Savard, et al., *Proc. Conf. Advances in Radioactive Isotope Science (ARIS2014), JPS Conf. Proc.* (2015) 010008.
(Sav 91): G. Savard, et al., *Phys. Lett. A* 158 (1991) 247.
(Sav 96): A. Savalle, et al., *European Particle Accelerator Conference*, Spain (1996) 2403.
(Sc 02): K. H. Schmidt, et al., *Nucl. Phys. A* 701 (2002) 115.
(Sc 81): D. Schardt, et al., *Nucl. Phys. A* 368 (1981) 153.
(Sc 92): K. H. Schmidt, et al., *Nucl. Phys. A* 542 (1992) 699.
(Sc 93): K. H. Schmidt, et al., *Phys. Lett. B* 300 (1993) 313.
(Sch 06): P. A. Schmelzbach, et al., *Proceedings of High Brightness HB2006* Tsukuba, Japan (2006) 274.
(Sch 15): C. Scheidenberger, *Proc. Conf. Advances in Radioactive Isotope Science (ARIS2014) JPS Conf. Proc.* (2015) 010027.
(Sch 78): B. Schinzler, et al., *Phys. Lett. B* 79 (1978) 209.
(Sch 92): A. Schempp, CERN Accelerator School, CERN 92-03, vol II (1992) 522.
(Sch 96): H. Scheit, et al., *Phys. Rev. Lett.* 77 (1996) 3967.
(Scha 01): H. Schatz, et al., *Phys. Rev. Lett.* 86 (2001) 3471.
(Schar 10): D. Schardt, et al., *Rev. Mod. Phys.* 82 (2010) 383.
(Schi 01): G. Schiwietz, et al., *Nucl. Instrum. Methods B* 177 (2001) 125.
(Schl-Th 97): B. Schlitt, PhD Thesis (1997).
(Schm 87): K.-H. Schmidt, et al., *Nucl. Instrum. Methods A* 260 (1987) 287.

(Schu 07): P. Schury, et al., *Phys. Rev. C* 75 (2007) 055801.
(Se 47): R. Serber, *Phys. Rev.* 72 (1947) 1114.
(Seth 87): K. K. Seth, et al., *Phys. Rev. Lett.* 58 (1987) 1930.
(Sh 91): B. M. Sherrill, et al., *Nucl. Instrum. Methods B* 57 (1991) 1106.
(Sha 89): K. S. Sharma, et al., *Nucl. Instrum. Methods A* 275 (1989) 123.
(Shi 12): K. Shimada, et al., *Phys. Lett. B* 714 (2012) 246.
(Shi 18): Y. Shimizu, et al., *J. Phys. Soc. Japan* 87 (2018) 014203.
(Shim 92): K. Shima, et al., *At. Nucl. Data Tables* 51 (1992) 173.
(Shu 12): D. Shubina, PhD Thesis, University of Heidelberg, Germany (2012).
(Shua 16): P. Shuai, *Nucl. Instrum. Methods* B376 (2016) 311.
(Shy 92): R. Shyam, et al., *Nucl. Phys. A* 540 (1992) 341.
(Si 65): G. Sidenius, *Nucl. Instrum. Methods* 38 (1965) 19.
(Sim 12): M. C. Simon, et al., *Rev. Sci. Instrum.* 83 (2012) 02A912.
(Slo 31): D. H. Sloan, et al., *Phys. Rev.* 38 (1931) 2021.
(Smi 08): M. Smith, et al., *Phys. Rev. Lett.* 101 (2008) 200501.
(Smi 57): J. H. Smith, et al., *Phys. Rev.* 108(1) (1957) 120.
(Smi 16): G. J. Farook-Smith, et al., *Phys. Rev. C* 94 (2016) 054305.
(Sor 02): O. Sorlin, et al., *Phys. Rev. Lett.* 88 (2002) 092501.
(Spe 97): V. Spevak, et al., *Phys. Rev. C* 56 (1997) 1357.
(Spi 08): P. Spiller, *Proceedings of Hadron Beam*, Tennessee, USA (2008) 393.
(Spi 14): P. Spiller, *Proceedings of High Intensity, High Brightness and High Power Hadron Beams*, East Lansing, USA (2014) MOZLR08.
(St 67): V. M. Strutinsky, *Nucl. Phys. A* 95 (1967) 420.
(Sta 90): J. W. Staples, "RFQ's - An Introduction," *LBL-29472* (1990) 29.
(Ste 09): M. Steck, et al., *Int. J. Mod. Phys. E* 18 (2009) 411.
(Ste 11): M. Steck, CERN Accelerator School, Greece (2011).
(Step 13): D. Steppenbeck, et al., *Nature* 502 (2013) 207.
(Sto 13): T. Stora, *Nucl. Instrum. Methods B* 317 (2013) 402.
(Str 11): B. Streichner, et al., *Nucl. Intrum. Meth.* 654 (2011) 604.
(Su 00): K. Summerer and B. Blank, *Phys. Rev. C* 61 (2000) 034607.
(Su 12): K. Summerer, *Phys. Rev. C* 86 (2012) 014601.
(Sun 07): B. Sun, et al., *Eur. Phys. J. A* 31(3) (2007) 393.
(Sun 10): B. Sun, et al., *Nucl. Phys. A* 834 (2010) 476.
(Suz 95): T. Suzuki, et al., *Phys. Rev. Lett.* 75 (1995) 3241.
(Sve 14): C. E. Svenson, et al., *Hyp. Int.* 225 (2014) 127.
(Sy 79): T. J. Symons, et al., *Phys. Rev. Lett.* 42 (1979) 40.
(Sym 82): E. Symbalisty and D. N. Schramm, *Astrophys. Lett.* 22 (1982) 143.
(Ta 85): I. Tanihata, et al., *Phys. Rev. Lett.* 55 (1985) 2676.
(Ta 85a): I. Tanihata, et al., *Phys. Lett. B* 160 (1985) 380.
(Ta 88): I. Tanihata, et al., *Phys. Lett. B* 206 (1988) 592.
(Ta 89): I. Tanihata, *In Treatise on Heavy-Ion Science*, (1989) edited by D. Allan Bromley, vol. 8.
(Ta 92): I. Tanihata, et al., *Phys. Lett. B* 289 (1992) 261.
(Tan 19): R. Taniuchi, et al., *Nature* 569 (2019) 53.
(Tan 20): T. L. Tang, et al., *Phys. Rev. Lett.* 124 (2020) 062502.
(Tana 10): K. Tanaka, et al., *Phys. Rev. Lett.* 104 (2010) 062701.
(Taniu 19): R. Taniuchi, et al., *Nature* 569 (2019) 53.
(Tar 03): O. B. Tarasov, et al., *Nucl. Instrum. Methods B* 204 (2003) 293.
(Tar 09): O. B. Tarasov, et al., *Phys. Rev. Lett.* 102 (2009) 142501.
(Tar 13): O. B. Tarasov, et al., *Phys. Rev. C* 87 (2013) 054612.
(Tard 14): E. Tardiff, et al., *ISAC and ARIEL: The TRIUMF Radioactive Beam Facilities and the Scientific Program*, edited by J. Delling, R. Krucken and L. Merminga, (2014) 197 (*also Hyperfine Interactions*, vol. 225, 2014).
(Tat 97): B. A. Tatum, et al., *Proc. Part. Accel. Conf*, Vancouver, BC, Canada, *Am. Phys. Soc.*, May (1997).

(Ter 04): G. M. Ter-Akopiana, et al., *Nucl. Phys. A* 734 (2004) 295.
(Th 04): M. Thoennessen, *Rep. Prog. Phys.* 67 (2004) 1187.
(Th 16): R. Thies, et al., *Phys. Rev. C* 93 (2016) 0546010.
(Thi 75): C. Thibault, et al., *Phys. Rev. C* 12 (1975) 644.
(Tho 20): M. Thoennessen, "Discovery of Nuclides Project," https://people.nscl.msu.edu/~thoennes/isotopes/ (2020).
(Tho 99): M. Thoennessen, et al., *Phys. Rev. C* 59 (1999) 111.
(Thom 17): J. C. Thomas, *GANIL Colloquium* October 16–20, 2017.
(To 96): K S. Toth, et al., *Z. Phys. A* 355 (1996) 225.
(Tog 11): Y. Togano, et al., *Phys. Rev. C* 84 (2011) 035808.
(Tog 16): Y. Togano, et al., *Phys. Lett. B* 761 (2012) 412.
(Tow 10): I. S. Towner and J. C.. Hardy, *Rep. Prog. Phys.* 73 (2010) 046301.
(Tro 92): J. Trotscher, *Nucl. Instrum. Meth.* B70 (1992) 455.
(Tu 18): X. L. Tu, *Phys. Rev. C* 97 (2018) 014321.
(Uen 05): H. Ueno, et al., *Phys. Lett. B* 615 (2005) 186.
(Ur 13): C. A. Ur, *AIP Conference Proceedings*, 1530 (2013) 35.
(Vi 79): Y. P. Viyogi, et al., *Phys. Rev. Lett.* 42 (1979) 33.
(Vill 01): A. C. Villari, et al., *Nuclear Physics A* 693 (2001) 465 and references therein.
(Vo 78): V. V. Volkov, *Physics Reports* 44 (1978) 93.
(Vo 89): V. V. Volkov Vadim, *Treatise on Heavy Ion Science*, vol 8, edited by D. Allan Bromley, Plenum Press, New York (1989) 112.
(Von 16): R. Vondrasek, *Nucl. Instrum. Methods B* 376 (2016) 16.
(Von 18): R. C. Vondrasek, et al., *Rev. Sci. Instrum.* 89 (2018) 052402.
(Wa 04): M. Wada, et al., *Nucl Instrum Methods A* 532 (2004) 40–47.
(Wa 62): A. C. Wall, et al., *Phys. Rev.* 126 (1962) 1112.
(Wah 11): U. Wahl, *Nucl. Instrum. Methods B* 269 (2011) 83014.
(Wal 07): J. Van de Walle, et al., *Phys. Rev. Lett.* 99 (2007) 142501.
(Wal 97): T. G. Walker, et al., *Rev. Mod. Phys.* 69 (1997) 629.
(Wall 81): R. K. Wallace and S. E. Wooseley, *Astrophys. J. Suppl. Ser.* 45 (1981) 389.
(Wan 10): N. Wang, et al., *Phys. Rev. C* 81 (2010) 044322.
(Wan 18): H. B. Wang, et al., *Nucl. Instrum. Methods A* 908 (2018) 244.
(Wang 08): Thomas Wangler, *RF Linear Accelerator*, 2nd edition, Wiley-VCH Verlag Gmbh & Co. KGaA, Weinheim (2008).
(We 76): G. D. Westfall, et al., *Phys. Rev. Lett.* 37 (1976) 1202.
(We 79): G. D. Westfall, et al., *Phys. Rev. Lett.* 43 (1979) 1859.
(Web 94): M. Weber, et al., *Nucl. Phys. A* 578 (1994) 659.
(Wei 12): J. Wei, et al., *Proceedings of Linac Conference*, Tel Aviv (2012) 417.
(Wei 35): C. F. von Weizsäcker, *Z. Phys.* 96 (1935) 431.
(Wen 08): F. Wenander, *Nucl. Instrum. Methods B* 266 (2008) 4346.
(Wen 10): F. Wenander, *JINST* 5 (2010) C10004.
(Wi 76): B. D. Wilkins, et al., *Phys. Rev. C* 14 (1976) 1832.
(Wie 15): F. Wienholtz, et al., *Phys. Scr. T* 166 (2015) 014068.
(Win 15): D. Winters, et al., *Phys. Scr. T* 166 (2015) 010408.
(Wolf 16): R. N. Wolf, et al., *Nucl. Instrum. Methods B* 376 (2016) 275.
(Wolf 12): R. N. Wolf, et al., *Int. J. Mass Spectrom.* 313 (2012) 8.
(Wolf 13): R. N. Wolf, et al., *Phys. Rev. Lett.* 110 (2013) 041101.
(Wolf 13a): R. N. Wolf, et al., *Int. J. Mass Spectrom.* 349 (2013) 123.
(Woll 87): H. Wollnik, *Optics of Charged Particles*, Academic Press (1987).
(Woll 90): H. Wollnik, et al., *Int. J. Mass Spectrom. Ion Process.* 96 (1990) 267.
(Woll 95): H. Wollnik, *Nucl. Instrum. Methods A* 363 (1995) 393.
(Wood 97): C. S. Wood, et al., *Science* 275 (1997) 1759.
(Woos 02): S. E. Woosley, et al., *Rev. Mod. Phys.* 74 (2002) 1015.
(Woos 78): S. E. Woosley and W. M. Howard, *Astrophys. J. Suppl. Ser.* 36 (1978) 285.

(Woos 90): S. E. Woosley and W. M. Howard, *Astrophys. J. Lett.* 354 (1990) 21.
(Wu 17): J. Wu, et al., *Phys. Rev. Lett.* 118 (2017) 072701.
(Xie 97): Z. Q. Xie, *Particle Accelerator Conference 1997*, Vancouver, Canada (1997) 2662.
(Xu 94): H. M. Xu, et al., *Phys. Rev. Lett.* 73 (1994) 2027.
(Ya 79): Y. Yariv and Z. Fraenkel, *Phys. Rev. C* 20 (1979) 2227.
(Yan 07): Y. Yano, *Nucl. Instrum. Methods Phys. Res. B* 261 (2007) 1009.
(Yan 98): Y. Yano, *Proc. 15th Int. Conf. Cyclotrons and Their Applications*, Caen, France (1998).
(Yan 13): J. C. Yang, et al., *Nucl. Instrum. Methods B* 317 (2013) 263.
(Yan 16): X. L. Yan, et al., *J. Phys. Conf.* 665 (2016) 012503.
(Yav 15): M. Yan, "Ion Optics of Spatially Dispersing Magnetic Mass Analyser" in *Euroschool on Exotic Beam*, Dubrovnik, Croatia (2015).
(Yor 09): R. C. York, et al., *Proceedings of SRF 2009*, Dresden, Germany (2009) 888.
(Yua 18): Y. J. Yuan, et al., *Proc. of 14th Int. Conf. on Heavy Ion Accel. Tech. (HIAT2018)*, Lanzou, China.
(Yuh 15): Y. Zhang, et al., *Proc. of Conf. Advances in Radioactive Isotope Science (ARIS2014)*, Japan, *JPS Conf. Proc.* 6 (2015) 010019.
(Zha 13): T. Zhang, *Proceedings of IPAC2013*, Shanghai, China (2013) 345.
(Zha 17): P. Zhang, et al., *Phys. Lett. B* 767 (2017) 20.
(Zho 16): X. Zhou, "The Heavy Ion Research Facility in Lanzhou," *Nucl. Phys. News* 26(2) (2016) 4.
(Zs 14): G. Zschornack, et al., *arXiv: 1410.8014v1 [physics.acc-ph]* October 29 2014.

Index

Astrophysical cross section measurement, 239–246
    coulomb dissociation, 245–246
    proton capture, 240–241
    recoil mass separator, 242
    storage ring, 243–245

Beam cooling, 212
Beam dumps, 117–118
Beta delayed γ spectroscopy, 191

Charge-breeder, 120, 128–132
    EBIS as a charge-breeder, 130–132
    ECRIS as a charge-breeder, 128–130
Charge stripping, 160–162
    equilibrium charge state, 161
    gas stripper, 161
    low Z stripper, 161
Collinear laser spectroscopy, 220–223
    charge exchange cell, 222
    doppler broadening, 221
Collinear resonant laser excitation, 223–225
Coloumb excitation experiment, 233–236
    in PFS facility, 234–236
Core collapse supernovae, 24
Coulomb breakup studies, 236–238
Cut angle dipole, 171
Cyclotron, 138–144
    Azimuthally Varying Field, 140
    flutter factor, 141
    focusing limit, 141
    isochronism condition, 139
    isochronous field, 139
    limitations, 144
    mass resolution, 164
    separated sector, 141
    superconducting AVF cyclotron, 143
    Thomas and Spiral tunes, 141

DC/Tandem accelerator, 138, 164
Decay measurement, 191
Deep inelastic transfer reaction, 104–106
Dispersion, 160, 170–173
    addition, 172
    substractive (achromatic), 172
Drip-line, 4
    neutron/proton, 5
Dual charge acceleration, 162

$E(2^+)$, $B(E2)$ & shell closure, 15–17
ECR ion source (ECRIS), 125–130
EDM experiments, 246–248
Electron Beam Ion Source (EBIS), 130–132
Emittance, 119–120
    definition, 119
    normalized emittance, 120
Evolution of shell structure, 14–18
    island of inversion, 15–16
    new magic numbers, 8, 14, 17

Features of ISOL & PFS type facilities, 57
Fission, 85–97
    charge distribution ^{235}U (thermal n,f), 88
    converter thickness for photo-fission, 96
    cross-section of ^{238}U (γ,f), 94
    induced by energetic neutrons, 91–93
    induced by gamma rays, 93–96
    induced by protons/light ions, 89–91
    liquid drop model, 85–86
    mass distribution in ^{232}Th (α, f), 90
    mass distribution ^{235}U (thermal n,f), 88
    shell effects, 86, 89
    symmetric, asymmetric splitting, 86
    ^{238}U (γ,f) yield with electron energy, 95

Gas-jet recoil transport, 132–134

Heavy-ion fusion evaporation reaction, 97–104
    production of proton dripline nuclei between (50<Z<82), 102
    super heavy elements (SHE) production, 104
Hyperfine splitting, 219

In-flight fission of ^{238}U, 77–79
    production cross-section of exotic n-rich nuclei at GSI and RIKEN, 74
    production of exotic n-rich isotopes around r-process path, 78
Intensity of RIB, formula, 50
Interaction cross section measurement, 227–228
    nuclear radii, 228
Intermediate Mass Fragments, 64, 68
Ionization of radioactive atoms, 120
Isochronous Mass Spectrometry, 216–217
    ToF, 217
ISOL facility, Separation, 167–178
    cross contamination, 173

high resolution separator: ARIEL, 177
identification of products, 177–178
separation, 167, 172–173
ISOL type RIB facilities, 249–257
    ISAC and ARIEL at TRIUMF, 256–257
    ISOLDE at CERN, 253–254
    Louvain la Neuve, 249, 253
    SPIRAL at GANIL, 255–256
ISOL *vs.* PFS targets, 107
Isotopic shift, 218

Kilpatric limit, 153

Laser assisted beta NMR, 225–226
    asymmetry parameter, 225
LIGO/VIRGO collaboration, 31
Limits of nuclear stability, 3–9
Linear accelerator, 148–152, 156–157
    Alvarez, 149–150
    CH linac, 157
    coupled cavity linac, 157
    drift tube linac, 156
    H linac, 157
    IH linac, 157
    Wideroe, 148
Liouville's theorem, 119
Livingston plot, 135–136

Mass formulae, 7–9
Mass measurement, 189–197, 204–210
    cyclotron, 196
    direct method, 196–197
    indirect method, 190–192
    invariant mass, 194–196
    ISOLTRAP, 207–208
    missing mass, 192–194
    MR-TOF, 210
    Paul trap, 200
    Paul trap: FT-ICR, 206
    Penning trap: TOF-ICR, 204–205
    projectile fragments, 197
Measurement of half life, 229–232
    PFS method, 232
    RFQ Cooler and MR-TOF, 231
Medical radio-isotopes, 45–46
    ^{60}Co, 45
    ^{99}Mo, 45
    ^{99m}Tc, 45
    ^{131}I, 45
    alpha-emitters for in-vivo therapy, 46
Methods for producing RIB, 51–60
    combined approaches, 58–60
    ISOL-post accelerator approach, 52

PFS approach, 53
MR-TOF, 208–210
    mass separation, 209
Multiple charge acceleration, 162–163
    Alternate Phase Focusing, 163
Multi-walled carbon nanotubes, 113

Neutron skin, 11–12
Nier–Johnson configuration, 177
Nuclear halo, 9–14
    halo in light neutron-rich nuclei, 10
    neutron halo in ^{11}Li, 9–11
    proton halo, 11
    pygmy dipole resonance, 13
    soft dipole resonance, 10, 13
Nuclear multi-fragmentation, 64, 67
Nucleo-synthesis, 18–36
    big bang nucleo-synthesis, 19–21
    CNO cycle, 27, 28
    cold and hot r-process, 29
    gamma process, 33–35
    Gamow peak, 22
    hydrostatic burning stages in massive star, 23
    neutrino process, 36
    p-process, 32, 34–35
    rp process, 32–36
    r-process, 25, 28–32
    s-process, 25–28
    waiting points, 29

Paul trap, 198–200
    mass spectroscopy, 199
Penning trap, 200–204
    buffer gas cooling, 203
    invariance theorem, 202
    ion motion, 201
PFS facility, separation, 169, 178–189
    achromatic transport, 180
    degrader, 180–181, 183
    fragment energy at target, 181
    identification, 181, 185–189
    separation, 180–185
    two stage, 185
    wedge shaped degrader, 184–185
PFS type RIB facilities, 257–263
    FAIR at GSI, 262–263
    FRIB at MSU, 260–261
    GSI-FRS, 261–262
    HIRFL at Lanzhou, 263
    LISE at GANIL, 258
    NSCL-MSU, 260
    RI Beam Factory at RIKEN, 258–260

Index

Projectile fragmentation, 70–77
   historical overview, 70
   limiting fragmentation & factorization, 72
   momentum width, 72–74
   optimum choice of projectile energy, target and its thickness, 78–81
   production of exotic nuclei in Cold Fragmentation, 76–77
   production of n-deficient nuclei, 74–75
   production of n-rich nuclei, 75–77
   reaction process, 70–72
   schematic representation, 71
   secondary beams, of, 81–83
Projectile fragmentation cross section codes, 83–85
   ABRABLA, 83–84
   COFRA, 84
   EPAX, 85

Radio Frequency Quadrupole Linac, 152–155
   rod type, 154–155
   sections, 153–154
   vane type, 154–155
Resolving power, 169, 171, 172
Resonating cavities, 150–151, 155–157
   effective shunt impedance, 156
   modes (TE, TM, TEM), 151, 157
   Q factor, 156
   shunt resistance, 156
   standing & travelling wave, 150
Reticulated Vitreous Carbon Fibres, 112
RF power loss, 155
RFQ cooler, 173–176
   buffer gas collision, 173, 175
   confinement, 173–174
   ion motion, 175, 176
   ISOLDE, 175
   pseudo-potential, 175
   resolving power, 176
   RF heating, 176
RF transverse defocusing, 151
RIB for cancer therapy, 46–47
RIB in condensed matter physics, 41–45
   beta NMR spectroscopy, 44–45
   mossbauer spectroscopy, 42–43
   perturbed angular correlation studies, 43–44

Schottky mass spectrometry, 212–216
   electron cooling, 213–214
   Schottky noise, 214
   stochastic cooling, 212–213
Separatrix, 154
Sources for 1^+ charge state ions, 121–126

2.45 GHz ECR ion-source, 126
   FEBIAD ion-source, 124–125
   laser ion source, 122–124
   surface ion source, 121–122
Sources for high charge state ions, 127–132
   EBIS, 131–132
   high frequency ECRIS, 127–128
Space charge, 135
Spallation, 61–69
   followed by evaporation or fission, 64
   highly asymmetric fission vs multi-fragmentation, 67–68
   intra-nuclear cascade, 62–63
   mass distribution, 64
   measured yields at CERN-ISOLDE, 68–69
   production of n-deficient nuclei, 64–66
   production of n-rich nuclei, 66–67
   reaction codes, 69
   reaction process, 62–64
   schematic representation, 63
Specialized facilities, 263–269
   ACCULINNA at Flerov Lab, 266–267
   ALTO at IPN Orsay, 268
   ATLAS, 265
   BEARS at Berkeley, 264–265
   CARIBU at Argonne, 265
   HRIBF (former facility at ORNL), 265–266
   IGISOL at Jyvaskyla, 264
   RAON at RISP, 269
   SPES at INFN, 267–268
   VECC-RIB, 268–269
Spectroscopic quadrupole moment, 219
Storage ring, 211–212
Superconducting cavities, 158–160
   critical field, 158
   peak field, 158
   structures, 158–159
   transition energy from NC to SC, 160
Synchrotron, 144–147
   alternating gradient synchrotron, 144
   field index, 144
   heavy ion, 146
   longitudinal stability, 145
   momentum compaction, 145
   transition energy, 145

Target release properties, 108–109
   diffusion coefficient, 108
   effusion, 108
   radiation enhanced diffusion, 109
   sintering, 109
   thermal conductivity, 109
Targets for ISOL facilities, 108–114

Targets for PFS facilities, 114–118
Target type, 109–113
 carbides, 110, 111
 foils, 110–112
 molten, 109–111, 113
 oxides, 110–112
 powder, 108, 110–112
Test of fundamental symmetries, 36–41
 atomic parity violation, 38–40
 CP violation, 36–38
 CVC hypothesis, 40–41

Transfer tube, selectivity, 109
Transit time factor, 151

Uranium carbide target, 111–112

Valley of beta stability, 3
 neutron-rich & neutron-deficient nuclei, 5

Windowless liquid target, 116
Wish list of a versatile RIB facility, 50